The Anthropocene in Global Media

This book offers the first systematic study of how the 'Anthropocene' is reported in mass media globally, drawing parallels between the use (or misuse) of the term and the media's attitude towards the associated issues of climate change and global warming.

Identifying the potential dangers of the Anthropocene provides a useful path into a variety of issues that are often ignored, misrepresented, or sidelined by the media. These dangers are widely discussed in the social sciences, environmental humanities, and creative arts, and this book includes chapters on how the contributions of these disciplines are reported by the media. Our results suggest that the natural science and mass media establishments, and the business and political interests which underpin them, tend to lean towards optimistic reassurance (the 'good' Anthropocene), rather than pessimistic alarmist stories, in reporting the Anthropocene. In this volume, contributors explore how dangerous this 'neutralizing' of the Anthropocene is in undermining serious global action in the face of the potential existential risks confronting humanity. The book presents results from media in more than 100 countries in all major languages across the globe. It covers the reporting of key environmental issues, such as the impact of climate change and global warming on oceans, forests, soil, biodiversity, and the biosphere. We offer explanations for differences and similarities in how the media report the Anthropocene in different regions of the world. In doing so, the book argues that, though it is still controversial, the idea of the Anthropocene helps to concentrate minds and behaviour in confronting ongoing ecological (and Coronavirus) crises.

The Anthropocene in Global Media will be of interest to students and scholars of environmental studies, media and communication studies, and the environmental humanities, and all those who are concerned about the survival of humans on planet Earth.

Leslie Sklair is emeritus professor of Sociology at the London School of Economics and Political Science. His work has been translated into more than ten languages. He is the President of the Global Studies Association (UK), and, in 2016, the Czech Academy awarded him the František Palacký Medal for his contribution to Historical Sciences.

Routledge Studies in Environmental Communication and Media

Climate Change and Post-Political Communication
Media, emotion and environmental advocacy
Philip Hammond

The Discourses of Environmental Collapse
Imagining the end
Edited by Alison E. Vogelaar, Brack W. Hale, and Alexandra Peat

Environmental Management of the Media
Policy, industry, practice
Pietari Kääpä

Participatory Networks and the Environment
The BGreen Project in the US and Bangladesh
Fadia Hasan

Participatory Media in Environmental Communication
Engaging communities in the periphery
Usha Harris

Journalism, Politics, and the Dakota Access Pipeline
Standing Rock and the framing of injustice
Ellen Moore

Environmental Literacy and New Digital Audiences
Patrick Brereton

Reporting Climate Change in the Global North and South
Journalism in Australia and Bangladesh
Jahnnabi Das

Theory and Best Practices in Science Communication Training
Edited by Todd P. Newman

The Anthropocene in Global Media
Neutralizing the risk
Edited by Leslie Sklair

For more information about this series, please visit: https://www.routledge.com/Routledge-Studies-in-Environmental-Communication-and-Media/book-series/RSECM

The Anthropocene in Global Media

Neutralizing the Risk

Edited by Leslie Sklair

Routledge
Taylor & Francis Group
LONDON AND NEW YORK

earthscan
from Routledge

First published 2021
by Routledge
2 Park Square, Milton Park, Abingdon, Oxon OX14 4RN

and by Routledge
52 Vanderbilt Avenue, New York, NY 10017

Routledge is an imprint of the Taylor & Francis Group, an informa business

British Library Cataloguing-in-Publication Data
A catalogue record for this book is available from the British Library

Library of Congress Cataloging-in-Publication Data
A catalog record has been requested for this book

ISBN: 9780367375973 (hbk)
ISBN: 9780429355202 (ebk)

Typeset in Times New Roman
by Deanta Global Publishing Services, Chennai, India

Contents

List of figures vii
List of tables ix
List of contributors x
Acknowledgements xii
Abbreviations xiv

PART I
The Anthropocene and media 1

1 **Editor's introduction** 3
LESLIE SKLAIR

2 **Anthropocene in the mass media: The big picture** 22
LESLIE SKLAIR

PART II
Media coverage of the Anthropocene: A global survey 35

3 **Africa's Anthropocene in the media: A kaleidoscope of contradictions** 37
MERYL MCQUEEN AND LESLIE SKLAIR

4 **The Anthropocene in North American media: The pursuit of the 'good' Anthropocene** 55
LESLIE SKLAIR, CHAD STEACY, JONATHAN DEVORE, AND RON WAGLER

5 **Challenges and ideas of representations of the Anthropocene in Latin American and Caribbean media** 83
VIVIANE RIEGEL, SOFIA ÁVILA, AND JERICO FIESTAS-FLORES

6 **The Anthropocene in the media of North Asia** 97
LESLIE SKLAIR, KA HO MOK, AND YUYANG KANG

7 **South Asia: The 'provincializing' dilemma** 118
LESLIE SKLAIR, JAHNNABI DAS, AND SUNITHA KUPPUSWAMY

8 **Latecomers to capitalism, latecomers to the risks of the Anthropocene** 141
VLADIMIR VULETIĆ AND ENI BULJUBAŠIĆ

9 **Western Europe: Planetary Eurocentrism** 159
BORIS HOLZER AND LESLIE SKLAIR

10 **The Anthropocene in Middle Eastern media: Invisible oil?** 187
BARAN ALP UNCU AND RAMZI DAROUICHE

11 **Oceania: Big islands, small islands, and the Anthropocene** 202
LESLIE SKLAIR

PART III
From the Anthropocene to the Anthropo-scene 215

12 **Media coverage of the Anthropocene in the social sciences and environmental humanities** 217
VIVIANE RIEGEL

13 **Media coverage of Anthropocene-related creative arts** 232
LESLIE SKLAIR

14 **Conclusion: We need to talk about the Anthropocene** 252
LESLIE SKLAIR

Appendix 1: Countries in Regions 265
Index 266

Figures

1.1	Ichthyosaurs attending a lecture on fossilized human remains. Lithograph by Sir Henry de la Bèche, 1830, after his drawing. Creative Commons Attribution (CC BY 4.0). © Wellcome collection	8
2.1	Thinking like a planet, thinking like humans (Andrew Zolnai)	25
2.2	Emissions by country (© Union of Concerned Scientists)	30
3.1	'What do you think of HAT's word of the year?', Antroposeen in Afrikaans (2016)	41
4.1	Fabulous Anthropocene (2013) © Robyn Woolston www .robynwoolston.com	63
4.2	Infographic for 143 Countries (Mark Jacobson). www. thesolutionsproject.org	67
4.3	Word cloud USA, Boykoff et al. (2018)	79
5.1	Campaign poster '1.5 to stay alive' © Jonathan Gladding	90
5.2	Anthropocene display (2019) © Museum of Tomorrow, Rio	94
6.1	Chinese novel *Anthropocene* (2016) © Zhao Defa	103
6.2	Museums in the Anthropocene (Age of Man), conference poster (2016) © National Museum of Science Tokyo	109
7.1	Loss of green cover after the Gaja cyclone, Thanjavur district of Tamil Nadu © Ms. S. Mahalakshmi (2019)	122
8.1	Special issue on the Anthropocene in the Czech magazine *A2* (2016) © *A2*	145
8.2	Big polluters bankrolling COP 24, Poland (2018) © Corporate Accountability and Corporate Europe Observatory	152
9.1	*Le Monde* man (2016) © Nini La Caille	161
9.2	First public international conference on the Anthropocene (2011) © Geological Society of London	165
10.1	Palestinian, Jordanian, and Israeli teenagers campaign together to save the Jordan © Eco Peace Middle East	190
11.1	Anthropocene special issue, *Balairung* (Indonesia), (2018) © *Jurnal Balairung*	209
12.1	*L'Atlas de l'Anthropocene'* (2019) © Sciences Po Presse	219
13.1	Special exhibition on the Anthropocene at the Deutsches Museum Munich (2014) © Deutsches Museum	234

13.2 Destroying nature is destroying life (2018) © Robin Wood/ Illusion CGI Studio 240
13.3 Stain, Spain, Santander (2011) © Pejac @pejac art 241
13.4 Capitalist Ruins, acrylic, one of four high-density polystyrene panels, epoxy, hand-carved text (2020) © Justin Brice Guariglia 242
14.1 'Can I get mine with long duree?' (2015) © Felice Wynham, https://kuchka.org/ecomyopia 256
14.2 'Degrowth' (2015) © Barbara Castro (Barbaracastrourio.com) 258

Tables

2.1 Summary results by years and regions 23
2.2 Publications with 36+ Anthropocene items (beginning 2000 to end 2017) 24

Contributors

Sofia Avila is a PhD candidate at the Institute of Environmental Science and Technology and Research Assistant at the Environmental Justice Project, both at the Autonomous University of Barcelona.

Eni Buljubašić is a postdoctoral researcher and teaching assistant at the Faculty of Humanities and Social Sciences, University of Split, Croatia.

Ramzi Darouiche is a research assistant and student at the City University of Applied Sciences Bremen, Germany.

Jahnnabi Das is a Research Associate at the Climate Justice Research Centre, University of Technology, Sydney, Australia.

Jonathan DeVore is Visiting Assistant Professor of Anthropology and Latin American, Latino/a & Caribbean Studies at Miami University of Ohio, USA.

Jerico Fiestas Flores is a PhD student in Agricultural and Resource Economics and Research Assistant in Future Energy Systems. Both at the University of Alberta.

Boris Holzer is Professor of Sociology at the University of Konstanz, Germany.

Yuyang Kang is a PhD candidate in the Department of Sociology and Social Policy, Lingnan University, Hong Kong.

Sunitha Kuppuswamy is an Assistant Professor in the Department of media sciences of Anna University, India.

Astrid Kusumowidagdo is Dean of the School of Creative Industry and teaches in the Department of Interior Architecture at the Universitas Ciputra, Surabaya, Indonesia.

Meryl McQueen is a sociolinguist and independent researcher based in Seattle, USA.

Joshua Ka-ho Mok is Vice-President and Professor of Comparative Policy at Lingnan University, Hong Kong.

Viviane Riegel heads the Research Laboratory at Escola Superior de Propaganda e Marketing in São Paulo, Brazil.

Leslie Sklair is Emeritus Professor of Sociology at London School of Economics, London, UK.

Chad Steacy is completing his PhD in the Department of Geography at the University of Georgia, USA.

Baran Alp Uncu has a PhD in sociology from the LSE and is an independent researcher in Istanbul with a focus on social movements and climate politics.

Vladimir Vuletic is Professor of Sociology at the University of Belgrade, Serbia.

Ron Wagler is Associate Professor of Science Education at the University of Texas at El Paso, USA.

Acknowledgements

My first acknowledgement must be to the extraordinary team of volunteer researchers who, like me, believe that the Anthropocene Media Project, on which the book is based, might contribute something to raising awareness about the present and on-going ecological and existential crises we face. At the time of writing the Coronavirus global health and economic crises seem likely to displace the ecological crisis as a focus of media attention.

All those who searched newspapers, magazines, and online news sites for the project and helped to write the book are named in the chapters to which they contributed. Literally, this book could not have been written without them, especially, of course, the authors and co-authors of the chapters. The Anthropocene and what has been labelled as the 'Anthropo-scene' are both uniquely inter-disciplinary research topics and I have been fortunate to have had advice, help in finding volunteer researchers, and access to publications from a large number of people, most of whom I have never met personally. Some of them probably do not share the views expressed in my own chapters in the book, so their collegial generosity is much appreciated, as was the constructively critical feedback from four anonymous reviewers.

In alphabetical order, my thanks are due to Nick Admussen, Ian Angus, Midori Aoyogi, Nobumichi Ariga, Wendy Bacon, Anthony Barnosky, Elena Bennett, Marc Blecher, Max Boykoff, Robert Brulle, Tiago Campos, Anabela Carvalho, David Casagrande, Nigel Clark, Laura Colluci-Gray, Guareav Daga, Simon Dalby, Carl Death, Miguel Esteban, Adele Gibson, Phoebe Godfrey, James Goodman, Mike Goodman, Peter Haff, Clive Hamilton, Alf Hornborg, Cymene Howe, Marek Hrubec, Jennifer Jacquet, Keith Johnson, Peter Jacques, Giorgos Kallis, Julian Kessel, Chenyang Li, Sonia Livingstone, Heidi Norman, Leigh Martindale, Alan Mazur, Andrew Milner, Nina Moellers, Kathleen Morrison, Atsuro Morita, Tom Morton, Gisli Pálsson, Zizhe Peng, Martin Prominski, Libby Robin, Alan Rogers, Peter Rudiak-Gould, Nadine Ryan, Christian Schwägerl, David Schlesinger, Nick Simpson, Sam Solnik, Clare Saunders, Jackie Smith, Richard Stahel, Tereza Stokelova, Renata Tyszczuk, Davor Vidas, Kyle Whyte, and Andrew Zolnai. I owe a special debt of gratitude to Max Boykoff, whose pioneering work with colleagues at the University of Colorado (and elsewhere) on quantifying and analysing media reporting of climate change and global warming was a great inspiration. Thanks also to all those artists and organizations who generously gave permission to reprint images in this book, and to the team at Routledge for their work on a difficult manuscript in difficult times

My nearest and dearest know who they are and how much their love and support mean to me. My five grandchildren are too young to understand what is going on, but I want them to know, when they grow up, that being with them has given me the inner strength to finish this project. Completed during the Coronavirus pandemic, this book confirms my belief that capitalist globalization, the Anthropocene, and the pandemic are all parts of the same problem.

Stay safe, Leslie Sklair, London 2020

Abbreviations

AF	Africa
AMP	Anthropocene Media Project
AN	Asia North
AS	Asia South
AWG	Anthropocene Working Group
Cape Town	35th International Geological Congress
CC/GW	climate change/global warming
C/EE	Central/eastern Europe
ESS	Earth System Science
GHG	greenhouse gases
IGBP	International Geosphere –Biosphere Programme (1987–2015)
IPCC	Intergovernmental Panel on Climate Change
LA/C	Latin America/Caribbean
ME	Middle East
NA	North America
OC	Oceania
PNAS	Proceedings of the National Academy of Science (USA)
SIDS	Small Island Developing States
TCC	transnational capitalist class
UN SDG	Sustainable Development Goals (UN)
WE	western Europe
WWF	World Wide Fund for Nature

Part I

The Anthropocene and media

1 Editor's introduction

Leslie Sklair

There is an impressive amount of research on how climate change and global warming are reported in the media all over the world (Smith 2000, Boyce and Lewis 2009, Boykoff 2011, Eide and Kunelius 2012, Brevini and Lewis 2018, McNatt et al. 2019). However, there is very little research on how the Anthropocene, the name proposed for a new geological epoch defined in terms of human impacts on the Earth System, is reported in the media. The Anthropocene Media Project (AMP), on which this book is based, aims to fill that gap by documenting and analyzing how the media worldwide report the Anthropocene. Whereas select groups of academics, environmental professionals, social scientists, humanities scholars, and creative artists do engage actively with issues of the Anthropocene, it is likely that most people have either never heard of it or, if they have, they have no clear idea about it.

The difference between climate change (which most people seem to have heard of) and the Anthropocene, to put it simply in lay terms, is that, whereas climate affects and is affected by every ecosystem (implicating oceans, forests, soils, rocks, atmosphere, all interacting with the biosphere of living organisms, which regulates the system), the Anthropocene is a more holistic idea that directly implicates human behaviour. The environmental historian Julia Thomas (2014: 1588) expresses this very clearly, declaring that the Anthropocene: 'is admittedly a contested term, but I use it instead of "climate change" or "global warming" because they misleadingly imply that the threat is limited to atmospheric increases in methane and, especially, carbon dioxide' (expanded in Thomas 2019). Climate change seems to have become a metonym for the Anthropocene, substituting a part for the whole. This linguistic distinction matters because everyone knows that the climate (usually understood in terms of 'the weather') changes and, unless you study the science, the implications of this may be vague. Neither climate change nor global warming in themselves, as phrases, imply anything about human agency. If we explain the Anthropocene as the 'Age of Humans' (or 'of Man'), the issue of human agency is blurred. However, the term 'Anthropocene' (and the adjective 'anthropogenic') specifically implicate human behaviour. It is important to ask questions about whose interests are served by such an apparently innocuous linguistic choice. This book aims to be a resource for those who want to compare what the media write about the Anthropocene and what scholars write about it, all over the world. This is a difficult task, as the Anthropocene as a concept, within and beyond Earth System science, is open to many interpretations (see, for example, Steffen et al. 2004, Schwägerl 2014, Hamilton et al. 2015, Angus 2016a, Davies 2016, Bonneuil and Fressoz 2017, Ellis 2018, Lewis and Maslin 2018, Zalasiewicz et al. 2019).

The Anthropocene and Earth System science

The Anthropocene concept signifies both a measure of geological time and a system that is more than the sum of its parts, in which positive and negative feedbacks between ecosystems are of vital importance.[1] From these ideas, Earth System science, sometimes referred to as the Gaia hypothesis, emerged in the 1960s (see Lenton 2016, Lovelock 2016, and Steffen et al. 2004: Box 2.7). The idea of the Anthropocene is a consequence of these new ways of looking at planet Earth and the impacts of human actions on it.[2]

The consensus among environmental scientists is that the Earth System has always been changing. However, since the second half of the twentieth century, the findings of Earth System scientists strongly suggest that the rate of change, particularly for atmospheric CO_2 and other greenhouse gases implicated in the degradation of ecosystems and potentially posing existential threats to human survival, has been increasing unusually rapidly. It is argued that these changes are causing serious (possibly irreversible) damage to the viability of the Earth System as a whole, as far as humanity is concerned. This has been conceptualized as 'The Great Acceleration' (Steffen et al. 2015, McNeill and Engelke 2016, Lane 2019). On the evidence of over 6,000 peer-reviewed papers, this is more or less the conclusion reported by the Intergovernmental Panel on Climate Change, established in 1988 by the UN to monitor the situation (IPCC 2018).[3]

Scholars have been warning for decades (in some respects, centuries) that such changes may present a credible existential threat to human life on the planet.

It is interesting to compare the tone of two multi-authored scientific papers on the Anthropocene in prestigious peer-reviewed journals a few years apart. The first (Steffen et al. 2011: 862) concludes: 'Darwin's insights into our origins provoked outrage, anger, and disbelief but did not threaten the material existence of society of the time. The ultimate drivers of the Anthropocene, on the other hand, if they continue unabated through this century, may well threaten the viability of contemporary civilization and perhaps even the future existence of *Homo sapiens*'. The second (Steffen et al. 2018: 8252) concludes: 'Collective human action is required to steer the Earth System away from a potential threshold and stabilize it in a habitable, interglacial-like state. Such action entails stewardship of the entire Earth System – biosphere, climate, and societies – and could include decarbonization of the global economy, enhancement of biosphere carbon sinks, behavioural changes, technological innovations, new governance arrangements, and transformed social values'. This paper does discuss some alarming concepts: 'hothouse Earth', 'tipping points', 'planetary boundaries' and (not quite so alarming) 'safe operating spaces' for the human species. The ideas of planetary boundaries and safe operating spaces, as we will see in Part II, have attracted some media attention.[4] However, the messages contained in the conclusions to these papers seem rather different. The difference is that, in the first, the existential risk to human survival is clearly stated, whereas, in the second, the role of human agency is mobilized to create what we might (uncharitably) label an optimistic spin to the story of the Anthropocene. The following chapters document the messages mass media all around the world have been sending out in their reporting of the Anthropocene, and ask whether journalists and scientists themselves have been in any way complicit in neutralizing the perceived risks of the Anthropocene?[5]

Dilemmas of the Anthropocene

The Anthropocene was introduced as a geological concept to name and explore human impacts on the Earth System. In 2009, proponents of the distinctiveness of the Anthropocene began a process to persuade the bodies responsible for naming geological periods by establishing the Anthropocene Working Group (AWG), led by Jan Zalasiewicz of the University of Leicester.[6] This is a complicated and lengthy business, involving a mass of scientific evidence and many committees. Lewis and Maslin describe the process as 'The Messy Mechanics of Defining Time' (2018: 283–94). According to geologists, we have been in the Holocene epoch, a period of moderate temperatures and relative stability in the Earth System (especially as far as humans are concerned) for at least the past ten thousand years. It is not a foregone conclusion that the name Anthropocene will ever officially become part of the Geologic Time Scale.

A simplified Geological Anthropocene Timeline indicates some of the people, institutions and publications involved in the evolution of the Anthropocene

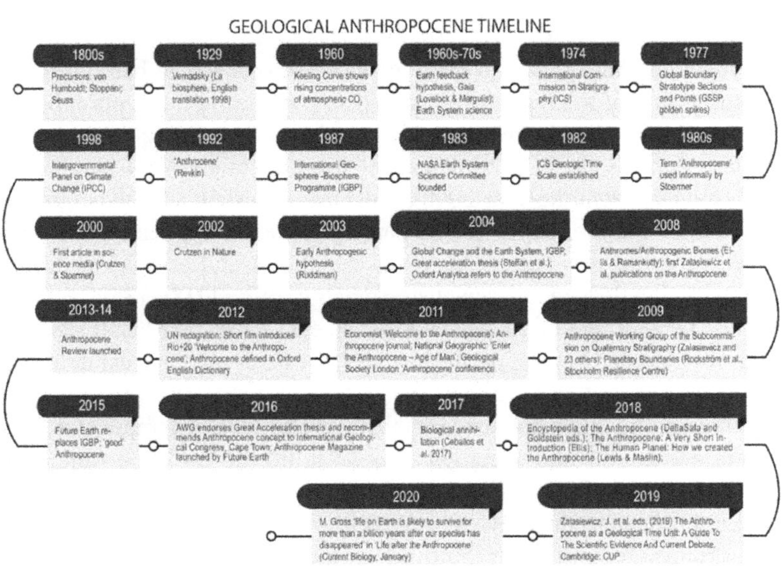

from a relatively obscure geological-stratigraphic concept into a phenomenon that impacts on everything, even 'the Anthropocene style' (Rahm 2019), and everyone.[7]

Our book is not intended as a critique of journalists or scientists, nor does it seek to apportion 'blame for the Anthropocene'. Rather, it starts from the premise that the Anthropocene puts journalists, scientists, and the rest of us in a series of impossible dilemmas created by the choices taken (knowingly or unknowingly, consciously or unconsciously, deliberately or casually) at, as yet, still-contentious historical junctures by different groups of people in different places at different times. The most important expression of these dilemmas is the debate that swirls around the effort to establish the Capitalocene (Age of Capital) as a more accurate and radical alternative name to the Anthropocene, a name that appears to portray all humanity as an undifferentiated totality.[8] This ignores the fact that all communist or socialist societies have followed similar patterns of unfettered economic growth based on fossil fuels (thus the label 'state capitalism'). While not disputing the decisive role of the capitalist system in the creation of anthropogenic ecological destruction, we need to ask if the argument is that capitalists set out from the beginning of the exploitation of fossil fuels deliberately to destroy the planet and the capitalist system for their offspring? Did Karl Marx, warming himself by the coal fire in chilly London, do the same? And today, when information about the perilous state of the planet is readily available, those of us who drive, fly, consume excessively, the lucky minorities, and the not-so-lucky billions who do not consume excessively but whose struggles to survive also impact adversely on the ecosystems that sustain them – are we all exonerated? In two previous books (Sklair 2001, 2002), I argued that the transnational capitalist class can be held responsible for ecological unsustainability because of its insatiable appetite for economic growth, and that, through the culture-ideology of consumerism, it exerts tremendous pressure on everyone on the planet towards consuming finite resources. So, those of us who chose to consume far beyond our basic needs, cause ecological damage, as do many of those who consume barely to survive. This is the reality of life that global capitalism provides in the Anthropocene. To argue that we know the few who are really to blame is as naive as saying we are all to blame equally. In the first decade of what we might label 'Anthropocene studies' (roughly from the year 2000), most Earth System scientists ignored the political and moral implications of the name (apparently implicating all humans). More recently, however, questions about capitalism and ecological justice are becoming acknowledged by scientists (notably Ellis 2018: 135ff, Lewis and Maslin 2018 and, implicitly, Zalasiewicz et al. eds. 2019: chap. 7). It seems obvious that, whereas the Anthropocene may last thousands of years, capitalism probably will not.

To intensify the dilemmas for journalists, the science is complicated, and the timescales are lengthy and uncertain, making it difficult to fit the Anthropocene into the usual media cycles. Furthermore, there is a good deal of disagreement among Earth scientists themselves over detail; for example, the terms 'uncertain' and 'uncertainty' crop up over 50 times in the first chapter of the IPCC 2018 report. However, as most scientists would explain, 'uncertain' does not mean 'we don't know', it usually refers to probability.[9] *Merchants of Doubt*, by two historians of science (Oreskes and Conway 2012), provides historical context and analytic tools to help understand the dilemmas facing scientists, journalists, and science popularizers in their efforts to communicate the ecological crisis we face (and, by implication, the Anthropocene) to the general public.[10] Oreskes and Conway introduce the argument that, wherever evidence-based research findings rub against the vested interests of big business and governments, a few scientists (often no longer engaged in research themselves) can usually be found to support the political-economic status quo.[11] Alongside the merchants of doubt are those dubbed the merchants of fear, who raise the alarm, often predicting catastrophe.[12] It is not always easy to distinguish ideologically-driven merchants of fear from working scientists in many fields, who are genuinely afraid of the consequences of the facts as they appear in the established scientific record – perhaps it is best to label them 'pessimistic realists'. What, we may legitimately ask, can researchers, journalists, and the lay public make of this?

Scholars in the social studies of science address these issues by engaging with the science-politics of the Anthropocene. Stengers (2015: 136), for example, argues that because scientists were prepared to go public on the risks of the Anthropocene before all the necessary research had been carried out and validated, this gave credence to the merchants of doubt. This is understandable, as those scientists who were already convinced of the severity of the risks, considered the situation to be urgent.[13] However, the incompleteness of research results, one of the conditions of evidence-based science, meant that doubters could keep the debate going (see van Eeden 2017a, an interesting reflection on materialism and the Anthropocene). Latour argues that science and politics are both frail human endeavours and that we need to move from 'science-versus-politics' to 'politics-with-science', however risky. Anthropocene politics 'is not a rational debate … [it is] incredibly easy to make two sides emerge even when there is only one' (Latour 2015: 147).[14] And this is why the ways in which the Anthropocene is reported in the mass media are so important. To put it most starkly, media-produced public perceptions of the severity or mildness of the risks of the Anthropocene may make the difference between human survival or the extinction of the human species on our home planet. The witty 19th-century extinction cartoon (Figure 1.1) may not seem so witty in subsequent centuries.

Adequately informed publics all over the world are more liable to change their behaviour and to pressurize governments, as are international organizations and corporations to change their behaviour, if these risks are taken seriously (Carvalho et al. 2017). This seems obvious from two meticulously researched papers in which Clémençon (2012, 2016) deconstructs UN climate change meetings over the past decades, and from Gurwitts et al. (2017), who speak of 'developed country bias' in the media. It is difficult to avoid the conclusion that, if the risks are ignored, downplayed, or otherwise not taken seriously ('neutralized'), yet turn out to be genuine, then extinction of human life on our home planet may be the result. A recent book (Grusin ed. 2019) is advertised as follows: 'the collection considers

Figure 1.1 Ichthyosaurs attending a lecture on fossilized human remains. Lithograph by Sir Henry de la Bèche, 1830, after his drawing. Creative Commons Attribution (CC BY 4.0). © Wellcome collection.

extinction as a cultural, artistic, and media event as well as a biological one'. How did we get to this perilous position?[15]

Science reporting in the media

Starting much earlier, the coverage of climate change in the mass media has always vastly exceeded coverage of the Anthropocene. Research on science in the media began in the late 1960s and that on environmental issues in the media in the 1980s (Antilla 2005: 339–40).[16] Everyone is interested in the weather (often confused with climate, especially by politicians) and the idea of anthropogenic (human-induced) ecological change has been around for centuries; for example, cutting down trees increases the likelihood of flooding, though this hasn't stopped people from cutting down trees. From the second half of the twentieth century, technological advances have made it possible to measure the effects of the various factors that cause climate and other ecological changes with increasingly greater precision. In the past few decades of the twentieth century, climate change gradually found its way onto the agenda of national governments, corporations, and international organizations, and into popular culture (Hulme ed. 2015). Howe (2014) offers a compelling argument that we need to study the Cold War to understand both the science and the politics of climate change, and this implies that the Anthropocene also needs to be considered in a wider perspective (see Lane 2019: final section). The belief of Donald Trump and others that anthropogenic climate change is a hoax invented in China (whether meant as a joke or not) suggests that history may be repeating itself, this time as farce, next time, perhaps, as tragedy.[17]

The term 'Anthropocene' emerged publicly in 2000 but was barely mentioned in the media during the following decade. From the years 2000 to 2017, our online searches of almost 2,000 newspapers, magazines, and news sites in over 140 countries brought up around 4,000 items that mentioned the Anthropocene (some items of special interest published after 2017 are discussed but not included in the numerical results). However, according to the University of Colorado research project MeCCO, which monitors around 70 media sources in 38 countries, over 400,000 items on climate change and global warming were identified between 2004 and 2018.[18] Despite methodological differences between the two surveys, it is clear that the gap is enormous. Many Anthropocene media items we identified also mentioned climate change. In 2019, a campaign 'Covering Climate Now' was launched in the United States (with over 400 media partners, notably *The Guardian* and *The New York Times*, in 50 countries). However, the word Anthropocene rarely appears in research on climate change in the media. Until now, there has been no comprehensive, systematic global survey of the extent to which and the ways in which the Anthropocene has been reported in the media. Between 2011 and 2017, Steven Corneliussen, a media analyst for the American Institute of Physics, followed Anthropocene coverage in three national newspapers in the United States, and occasionally other sources, writing up his findings in the popular science magazine *Physics Today*. Useful as these articles are, their coverage is limited. It is important to emphasise that our research is not an attempt to identify all the articles on the Anthropocene ever published in the mass media. We use the term 'mass media' in its broadest sense to mean any source of news that is generally available, freely accessible to communities at various scales (online, on the streets, at newsagents, kiosks, transport hubs, airports, and supermarkets in cities, towns, and villages).[19] The large variety of media sources identified for this project brought to the surface many interestingly different ways in which people get their news all over the world and how this facilitates the analysis of diverse coverage across communities. With a few exceptions, searches were online and sites with total paywalls were excluded.[20]

The Anthropocene Media Project

The Anthropocene Media Project (AMP) started in late 2016, when our collection of data from online searches of newspapers, magazines, and news websites from around the world began. Eventually, searches in the media of over 140 countries were carried out for mentions of the Anthropocene (only in 29 countries did the media fail to produce a single result). All the researchers were volunteers – around

45 in total, mostly graduate students, plus a few professors and others. The participants were contacted via personal networks in and around the rapidly growing scholarly community engaged in Anthropocene studies.[21] The project has no formal institutional funding. Most of the volunteer researchers are working and/or studying in the general field of environmental studies and most appear to have found out about the project either from direct calls for participants, their university teachers (the original points of contact) or through friends in the same field. Given the nature of the pool of participants, the research has had to take its place alongside coursework, finishing off doctoral dissertations, teaching, and the day-to-day demands of working in educational and research establishments. Results of media searches total over 600,000 words, including cut-and-pasted entire articles and summaries in English or translated from most of the major languages, and some researcher notes. The 4,000+ items found included long and short articles and passing references, for example, notifications of cultural events containing the word 'Anthropocene'.

Methodology

The methodology for the project is deliberately straightforward, minimizing difficulty and complexity for the volunteer researchers. We searched the websites of newspapers, magazines, and online news sites for the word 'Anthropocene' and for the names of its two most prominent scientific proponents – Paul Crutzen (an atmospheric chemist who shared the Nobel Prize in 1995 for research on the ozone layer) and Jan Zalasiewicz, a geologist from the University of Leicester, who led the AWG. Non-English language sources were mainly accessed by volunteer researchers in their mother tongues, the rest with the aid of translation software, often incorporated within online sources, with some interesting results on what Boussebaa and Brown (2017) label 'Englishization'. Sources are mostly identified by name and country, title of article, and year of publication. Whereas the word 'Anthropocene' may not appear in some of the quoted material (some editing is applied for clarity), it appears in all the articles (with the exception of a few cases, identified in the text). Most of the links cited were active in 2020.

Researching the media online is, to put it mildly, a continuing challenge. Whereas Wikipedia proved to be a good starting point for lists of online national, regional, and local media in most countries, the websites of many newspapers, magazines, and even online news sites proved to be non-existent or otherwise unavailable. For many countries that could not be characterized as media-rich, all that Wikipedia and other search engines could offer were 'stubs', incomplete or questionable entries. Sometimes, the domain names of media websites were offered for sale. Even when a link proved active, some websites had no search facility, or offered only a limited search by individual issues, by month, or year, making a proper search very laborious. Nevertheless, almost 2,000 separate searchable media sources were successfully accessed, and around forty percent of them produced at least one item with the term 'Anthropocene' in English or another language.[22] Many searches proved initially unreliable; for example, *The*

Guardian (London), a source of many articles on the Anthropocene, reprinted and/or cited in media all over the world, announced 800+ results in 2017. However, this turned out to include (apparently) all mentions of the word, items with no mentions of the word itself (though often discussing related phenomena), and repeats. Discounting all of these, *The Guardian* produced around 40 genuine items.[23] Some websites operated with restrictions, such as only a small number of whole articles being freely available. Almost all items offered titles plus two or three lines of text, often giving useful clues as to the content of the whole piece.[24] This simple methodology has theoretical significance in that it does offer a way of discovering how likely it is that casual browsers of the daily news would come across the Anthropocene. Many items were illustrated by what we can call 'Anthropocene imagery', some of which is reproduced in the following chapters. Where the word Anthropocene was found in any source, total numbers of items with the word were recorded, along with the years of first and subsequent items, and titles and summaries of articles (translated into English by native speakers and/or software embedded in the search, where available). Many media sources that devote considerable attention to climate change never or hardly ever mention the Anthropocene, and we consider this to be an important finding (*The Wall Street Journal*, with limited free access, is the most prominent example).

Though smartphone technology may be changing the situation, there is very little research available on how people in traditionally media-poor societies get information about climate change and/or the Anthropocene (or, indeed, anything else). However, in recent years, some scholars have begun to look at this question in the context of indigenous societies, a good example being the imaginative attempt to discover the links between 'media' communication and local perceptions of climate change in an indigenous society in the Bolivian Amazon (Fernández-Llamazares et al. 2015). There are suggestions in the media and scholarly literature that 'traditional' indigenous peoples might be better equipped to understand and even cope with the Anthropocene than 'moderns', discussed through the lens of International Relations by Inoue and Moreira (2016). Whyte (2017) specifically focuses on decolonizing the Anthropocene in his path-breaking studies of indigenous communities in North America (discussed in Chapter 12 of this book).

Another issue that emerges from research into media coverage of anything is whether to search simply for the presence of a specific word or phrase, or to limit searches to words in headlines or abstracts. The usual justification for the latter approach is the need to focus on genuine discussions of topics rather than passing mentions. However, in our case, this reasoning ignores the fact that (in this and possibly in other media projects) articles that include the word Anthropocene in their titles are not necessarily about the Anthropocene per se, in many cases the word occurring only in the title and a few passing mentions (or not at all; a tagged search result for Anthropocene may not actually include the word). A passing reference to the Anthropocene tells us one thing about how it is being presented, for example, where climate change is being used as a metonym for the Anthropocene (the part being substituted for the whole), whereas short or long descriptive and/

or analytical articles tell us something else. All are valuable pieces of information. In many cases, the Anthropocene appeared in cultural listings that scarcely mention the scientific origins of the term. This, indeed, is one the most important findings of our research,[25] manifesting itself in the titles of art exhibitions, film, music, poetry, dance, theatre, and literature, notably the emerging genre of cli-fi ('climate fiction'). This phenomenon connects with studies in the social sciences and environmental humanities (see, for example, Davis and Turpin 2015, Clark and Yusoff 2017, Heise et al. 2017, Robin 2017).

The Data

The AMP has produced huge amounts of text and thousands of numbers. In his review of a biography of Enrico Fermi (Nobel Prize laureate in 1938 for research on neutrons), a distinguished historian of science explains what is meant by a Fermi question: 'Don't devote more time and effort to a problem than it is worth ... [and] never make something more accurate than absolutely necessary' (Shapin 2018: 13). The upshot of this line of reasoning is that most, though not all, problems can be resolved to an accuracy of say plus or minus ten percent. This is the principle on which the numbers presented in this book are based. Unfortunately, I encountered the Fermi question after spending many fruitless hours trying to reconcile small differences between large numbers. Given the problematic nature of the data and the built-in unpredictability of online media searches, if the results are accurate to plus or minus ten percent, we consider that a satisfactory basis on which to draw conclusions about the extent and manner of mass media coverage of the Anthropocene in our survey. All the large numbers in this book are estimates (often rounded). Revisiting some websites over the course of four years sometimes gave different numbers of items that contain the word Anthropocene, varying distributions by year of publication, and availability of media sources. Averaging out these differences on the basis of common sense is bound to produce inaccuracies but, we argue, this does not compromise our results to any significant degree.

Research questions

This book poses three main research questions.

1. How likely is it that ordinary readers, with no special interest in the topic, would come across references to the Anthropocene while browsing online for the daily news and, if they did find such information, what would it be telling them?
2. How do different types of media (national/regional, big city papers/small local papers, general-interest papers/special-interest magazines, news sites) report the Anthropocene to their readers?
3. To what extent are media in different parts of the world 'provincializing' the Anthropocene by creating their own national, regional and/or ethnic narratives, in contrast with uncritically replicating Western (Eurocentric and

US-centric) narratives, and what are the implications of this for global inequalities in the Anthropocene?

Strictly speaking, none of these research questions addresses the Earth System science from which the Anthropocene idea emerged. They are questions that social scientists and environmental humanities scholars ask (see Pálsson, et al. 2013 and Castree et al. 2014). Sociologists, anthropologists, historians, and philosophers who research the Anthropocene do not normally go into the field with geological hammers, or measure ocean acidification, deforestation, the death of coral reefs, or other material consequences of anthropogenic ecosystem change. This has led some to speak about the Anthropo-scene (Castree 2015, Lorimer 2017, and Part III below). Anthropo-scene scholars do not usually do Earth System science research themselves; to a greater or lesser extent, they accept the research and try to work out what it means for their own disciplines and for human (and nonhuman) life on Earth in general. In one of the most widely discussed contributions to the Anthropo-scene debate, the historian Chakrabarty (2009) provides a framework for theorizing the connections between the Anthropocene as moments in both Earth and human history, and the reconceptualization of nature/society relationships. Building on his provocative idea that the Anthropocene represents a crisis for the human sciences, Chakrabarty began to show how provincializing the Anthropocene and critiques of its Eurocentric roots could not be separated from the experience of colonialism and imperialism – past, present, and future (see Chapter 12 of this book). As the regional chapters in Part II suggest, these issues are rarely to be found directly in the media, although indirect references from media in both colonizing and colonized countries deal with these issues, leading to ideas around what we can label 'localizing the Anthropocene'.[26] Himelboim et al. (2010) grasp the nettle of 'Old Global Hierarchies in a New Online World' in a study of 223 news websites in 73 countries, investigating the flow of news among core countries and from core countries to the periphery, concluding that news organizations use the Internet to perpetuate the existing order. Lück et al. (2016), in an imaginative study of media coverage in five countries (Brazil, Germany, India, South Africa, and USA) during four UN climate change conferences from 2010 to 2013, contrast global media frames with 'nationally coloured narratives'.[27] As we will see, media coverage of the Anthropocene follows these trends.

Despite the complexity of the science, some journalists have generally coped reasonably well in bringing both the Anthropocene and climate change to the attention of their readers responsibly, whereas others have been accused of crass sensationalism and distortion.[28] In her study of how climate change 'comes to matter' in five discursive communities (indigenous, religious, business, science journalists, and other experts) in North America, Callison (2014: chap. 2) explains, in fine ethnographic detail, how science journalists are trained (or not, as the case may be) to distinguish between 'myth busters' and 'truth tellers' in the production of stories on climate change; no easy task. However, there is no evidence that journalists have ever been formally trained to report the Anthropocene.[29] Building on theoretical and substantive advances in the study of media, researchers have mobilized one or other of several versions of framing (Lakoff 2010) and social construction of agendas to interpret patterns in reporting of climate change and/or global warming (Boykoff 2011).

The idea of 'balance as bias' was introduced to explain the surprising presence of so many climate change deniers and sceptics in the mass media in the United States.[30] In a study of coverage of global warming in the US prestige press (*The New York Times, The Washington Post, The Los Angeles Times*, and *The Wall Street Journal*) between 1988 and 2002, Boykoff and Boykoff (2004: 129) posed the question: 'is the journalistic norm of balanced reporting – telling "both" sides of the story – a mediating variable, that skewed and distorted global warming coverage? In other words, was coverage of global warming "balanced", and therefore actually informationally biased?' Their answer was a resounding 'yes'; in this case, balance produced bias. As we shall see, 'balance as bias' is also evident in coverage of the Anthropocene, but it appears to work in a subtly different way.

Building on the formidable literature on climate change in the media, this book will document the extent of media reporting of the Anthropocene and investigate how it is framed, bearing in mind how the Anthropocene emerged as a geological concept and subsequently evolved into something more, namely the Anthropo-scene. Whereas the term 'Anthropo-scene' is not yet widely used, the idea is fast becoming a defining meme for social science and environmental humanities scholars, as evidenced in edited collections rolling off the presses (for example, Godfrey and Torrres 2016, Heise et al. 2017, Oppermann and Iovino 2017, Tsing et al. 2017, Jagodzinski 2018, Arias-Maldonado and Trachtenberg 2019, Lysgaard et al. 2019, and Wilke and Johnstone 2019), plus a multitude of journal articles and books, many of which inform the discussion throughout this book.

The Anthropo-scene Timeline is based mainly on the results of our media searches.[31]

Notes

1 In scientific terms, positive and negative feedback mean the opposite to their meaning in ordinary language. Lashof (2018) has suggested that 'positive' be replaced by 'amplifying'. See also Lenton (2016: 5–7).

2 There is already a considerable literature for and against the name and concept 'Anthropocene', for example, Moore (2016), Angus (2016a), Bonneuil & Fressoz (2017), Bauer and Ellis (2018) and Bińczyk (2019).

3 For two understandably optimistic surveys of UN activity, see Biermann (2012 and 2017).

4 The nine planetary boundaries are climate change, ocean acidification, ozone depletion, nitrogen and phosphorus cycles, freshwater, land use, biodiversity, atmospheric aerosol loading, and chemical pollution (Rockström et al. 2009). Zalasiewicz and Steffen (2017) present an admirably clear summary.

5 On 'Anthropocene risk' see Keys et al. (2019). Ulrich Beck's *World Risk Society* (1999) has become the classic sociological source on risk, and, in one of his last publications (Beck 2014:171), he argued: 'Talk of the "Anthropocene" signals that geologists have now caught up with the reality of world risk society'. See also Cottle (1998) on Beck and the media.

6 Zalasiewicz et al. eds. (2019) gives the authoritative version from the AWG.

7 See Ellis (2018), Lewis and Maslin (2018), and Zalasiewicz et al. (2019: chap. 1). Oldfield (2016) explains the paradigm shift to the Anthropocene.

8 Schwägerl (2014: 65, n.132) reports that the term 'Kapitalozän' (Capitalocene) was coined by the late Elmar Altvater. The essays in Moore (2016) provide the most comprehensive survey of the Capitalocene. For a thorough critique from an eco-socialist perspective, see Angus (2016b). Also of interest here is the idea of the Plantationocene (https://edgeeffects.net/tag/plantationocene/).

9 It is notable that the first chapter in the IPCC 2018 report places climate change science within the framework of the Anthropocene, though the term is absent from the press release and most subsequent media coverage.

10 See, in this connection, Parham (2006).

11 For a sympathetic critique of Oreskes and Conway, challenging some aspects of their analysis of how science works and some sharp comments on the IPCC, see Howe (2012).

12 Many commentators link the Anthropocene with disaster and catastrophe (Clark 2014, Scranton 2015, and Barrios 2017). Truscello (2018) discusses mixed messages in documentary films, and Fava (2013) provides more sharp comments on the IPCC.

13 See Kolbert (2006: esp. chap. 3) for a poignant discussion.

14 Whereas it is open to some minor methodological criticism, the conclusions on consensus among scientists around anthropogenic climate change offered by Cook et al. (2016) are widely accepted. See also Hulme (2009) and the refreshingly polemical Maxwell and Miller (2018). For more, simply search for '97% of scientists' on the Internet.

15 For the moral case against any species extinction, see Cafaro and Primak (2014) and the online discussion that followed.

16 There is no Anthropocene equivalent anywhere of the wide-ranging comparative research on climate change in the media carried out at the Reuters Institute for the Study of Journalism (based at Oxford University); see Painter (2010) and Painter & Schäfer (2018).

17 *The Atlantic* (19 October 2016) declared 'Donald Trump is the First Demagogue of the Anthropocene'.

18 See http://sciencepolicy.colorado.edu/media_coverage. This document contains valuable brief descriptions of global media sources, many of which also appear in our study. See also the daily media reviews from *Climate Brief* by Leo Hickman and his col-

leagues. Despite this apparently massive coverage, many environmentalists and media organizations consider climate change to be under-reported; see Harkins (2019). Hong et al. (2019) found 37,670 articles (23 million words) from 84 sources in 45 countries.

19 Search engines currently produce millions of results for 'Anthropocene'. Blogs and other social media (including the substantial content of Reddit) are excluded from our results, except in a few cases explicitly indicated.

20 On the move from paywalls to free online access, see: https://www.cjr.org/the_media_today/if-readers-pay-for-your-news-youre-one-of-the-lucky-ones.php. On the decline of print media and the rise of online news, see Abernathy (2018). Though these sources are from the United States, the trends are global.

21 According to the bibliographical search engine Scopus, over 2,000 scholarly articles with the keyword 'Anthropocene' were published between 2009 and 2018.

22 For an easily searchable list of all the media accessed, including those with zero results, please contact me directly (l.sklair@lse.ac.uk). As far as I am aware, most research on climate change/global warming in the media focuses on found items and does not record sources with zero results. In our research, all searches were recorded.

23 In 2018, *The Guardian* moved to Google Custom search, which still displayed only 40 genuine results. Without doubt, *The Guardian* has published many more articles on the Anthropocene, but only 40 appear to be available online at one time. Some Google Custom searches display Anthropocene-related adverts; these were not recorded.

24 Antilla (2005: 340, n. 11) points out the importance of 'framing the lead' to any story in the media. Perhaps framing the conclusion is just as important.

25 Autin and Holbrook (2012) compare the Anthropocene as stratigraphy and as pop culture.

26 For an evocative example of what this means, see Irvine (2017).

27 A special issue of the *Journal of World-System Research* (Murphey and Sklair 2020) focuses on the connections between world-systems analysis and the Anthropocene.

28 See Jackson (2010), Shea et al. (2020). Fitzgerald (2019) and McKie (2019) mark the tenth anniversary of the notorious 'Climategate' data theft.

29 Tom Morton (2016) argues persuasively that the Anthropocene forces journalists to think beyond their usual timescales. Revkin's dot.earth blog in the *New York Times* may have served as a sort of informal Anthropocene training primer for journalists (see Chapter 4 of this book).

30 In a wide-ranging survey, Capstick et al. (2015) suggest that both climate scepticism and concern were growing in developed countries in the 2000s.

31 Like the terms 'iconic architecture' and 'starchitect', when first introduced, no doubt many consider 'Anthropo-scene' flippant. Though they do not use the term, Zalasiewicz et al. (2019: esp. sections 1.4 and 7) and many others imply it.

References

Abernathy, P.M. (2018) *The Expanding News Desert*. Chapel Hill: University of North Carolina, School of Media and Journalism.

Angus, I. (2016a) *Facing the Anthropocene: Fossil Capitalism and the Crisis of the Earth System*. New York: Monthly Review Press.

Angus, I. (2016b) 'Anthropocene or Capitalism?' misses the point (Book Review). *Climate & Capitalism* (26 September).

Antilla, L. (2005) Climate of scepticism: US newspaper coverage of the science of climate change. *Global Environmental Change* 15: 338–352.

Arias-Maldonado, M. & Z. Trachtenberg eds. (2019) *Rethinking the Environment for the Anthropocene: Political Theory and Socionatural Relations in the New Geological Epoch*. London: Routledge.

Autin, W. & J. Holbrook (2012) Is the Anthropocene an issue of stratigraphy or pop culture? *GSA Today* 22(7): 60–61.

Barrios, R.E. (2017) What does catastrophe reveal for whom? The anthropology of crises and disasters at the onset of the Anthropocene. *Annual Review of Anthropology* 46: 151–166.

Bauer, A. & E.C. Ellis (2018) The Anthropocene divide: Obscuring understanding of socio-economic change. *Current Anthropology* 59(2): 209–227.

Beck, U. (1999) *World Risk Society*. Malden, MA: Polity Press.

Beck, U. (2014) How climate change might save the world. *Development and Society* 43(2): 169–183.

Biermann, F. (2012) Greening the United Nations charter: World politics in the Anthropocene. *Environment: Science and Policy for Sustainable Development* 54(3): 6–17.

Biermann, F. et al. (2017) Global governance by goal-setting: The novel approach of the UN sustainable development goals. *Current Opinion in Environmental Sustainability* 26: 26–31.

Bińczyk, E. (2019) The most unique discussion of the 21st century? The debate on the Anthropocene pictured in seven points. *The Anthropocene Review* 6(1–2): 3–18.

Bonneuil, C. & J.B. Fressoz (2017) *Shock of the Anthropocene*. London: Verso.

Boussebaa, M. & A. Brown (2017) Englishization, identity regulation and imperialism. *Organization Studies* 38(1): 7–29.

Boyce, T. & J. Lewis eds. (2009) *Climate Change and the Media*. New York: Peter Lang.

Boykoff, M. (2011) *Who Speaks for the Climate? Making Sense of Mass Media Reporting on Climate Change*. Cambridge: Cambridge University Press.

Boykoff, M. & J. Boykoff (2004) Balance as bias: Global warming and the US prestige press. *Global Environmental Change* 14: 125–136.

Brevini, B. & J. Lewis eds. (2018) *Climate Change and the Media*, Volume 2. New York: Peter Lang.

Cafaro, P. & R. Primak (2014) Editorial: Species extinction is a great moral wrong. *Biological Conservation* 170: 1–2.

Callison, C. (2014) *How Climate Change Comes to Matter: The Communal Life of Facts*. Durham, NC: Duke University Press.

Capstick, S. et al. (2015) International trends in public perceptions of climate change over the past quarter century. *WIREs Climate Change* 6: 35–61.

Carvalho, A. et al. (2017) Communication practices and political engagement with climate change: A research agenda. *Environmental Communication* 11(1): 122–135.

Castree, N. (2015) Changing the Anthropo(s)cene: Geographers, global environmental change and the politics of knowledge. *Dialogues in Human Geography* 5(3): 301–316.

Castree, N. et al. (2014) Changing the intellectual climate. *Nature Climate Change* 4(27 August): 763–768.

Chakrabarty, D. (2009) The climate of history: Four theses. *Critical Inquiry* 35: 197–222.

Clark, N. (2014) Geo-politics and the disaster of the Anthropocene. *The Sociological Review* 62(S1): 19–37.

Clark, N. & K. Yusoff (2017) Geosocial formations and the Anthropocene. *Theory, Culture & Society* 34(2–3): 3–23.

Clémençon, R. (2012) Welcome to the Anthropocene: Rio+20 and the meaning of sustainable development. *Journal of Environment & Development* 21(3): 311–338.

Clémençon, R. (2016) The two sides of the Paris climate agreement: dismal failure or historic breakthrough? *Journal of Environment & Development* 25(1): 3–24.

Cook, J. et al. (2016) Consensus on consensus: A synthesis of consensus estimates on human-caused global warming. *Environmental Research Letters* 11: 048002.
Cottle, S. (1998) Ulrich Beck, 'Risk Society' and the media: A catastrophic view? *European Journal of Communication* 13(1): 5–32.
Davies, J. (2016) *The Birth of the Anthropocene*. Oakland, CA: University of California Press.
Davis, H. & E. Turpin eds. (2015) *Art in the Anthropocene*. London: Open Humanities Press.
Eide, E. & R. Kunelius eds. (2012) *Media Meets Climate: The Global Challenge for Journalism*. Gothenberg: Nordicom.
Ellis, E. (2018) *The Anthropocene: A Very Short Introduction*. Oxford: Oxford University Press.
Fava, S. (2013) *Environmental Apocalypse in Science and Art: Designing Nightmares*. London: Routledge.
Fernández-Llamazares, Á. et al. (2015) Links between media communication and local perceptions of climate change in an indigenous society. *Climatic Change* 131: 307–320.
Fitzgerald, B. (2019) Q&A: Michael Mann on coverage since 'Climategate'. *Columbia Journalism Review* (19 September).
Godfrey, P. & D. Torres eds. (2016) *Systemic Crises of Global Climate Change: Intersections of Race, Class and Gender*. London and New York: Routledge.
Grusin, R. ed. (2019) *After Extinction*. Minneapolis, MN: University of Minnesota Press.
Gurwitts, S. et al. (2017) Global issues, developed country bias: The Paris climate conference as covered by daily print news organizations in 13 nations. *Climatic Change* 143: 281–296.
Hamilton, C. et al. eds. (2015) *The Anthropocene and Global Environmental Crisis: Rethinking Modernity in a New Epoch*. London and New York: Routledge.
Harkins, S. (2019) Why is climate change still not top of the news agenda? *The Conversation* (19 September).
Heise, U.K. et al. eds. (2017) *The Routledge Companion to the Environmental Humanities*. London and New York: Routledge.
Himelboim, I. et al. (2010) International network of foreign news coverage: Old global hierarchies in a new online world. *Journalism and Mass Communication Quarterly* 87(2): 297–314.
Hong Tien Vua et al. (2019) Nationalizing a global phenomenon: A study of how the press in 45 countries and territories portrays climate change. *Global Environmental Change* 58 (September). https://doi.org/10.1016/j.gloenvcha.2019.101942
Howe, J. (2012) The stories we tell. *Historical Studies in the Natural Sciences* 42(3): 244–254.
Howe, J. (2014) *Behind the Curve: Science and the Politics of Global Warming*. Seattle, WA: University of Washington Press.
Hulme, M. (2009) *Why We Disagree about Climate Change: Understanding Controversy, Inaction and Opportunity*. Cambridge: Cambridge University Press.
Hulme, M. ed. (2015) *Climates and Cultures*, 6 Volumes. London: Sage.
Inoue, C.Y.A. & P.F. Moreira (2016) Many worlds, many nature(s), one planet: Indigenous knowledge in the Anthropocene. *Revista Brasileira de Política Internacional* 59(2): 1–19.
Irvine, R.D.G. (2017) Anthropocene East Anglia. *The Sociological Review Monographs* 65(1): 154–170.
IPCC (2018) https://report.ipcc.ch/sr15/

Jackson, C. (2010) Staying afloat in Paradise: Reporting climate change in the Pacific. *Reuters Institute Fellowship Paper*. Oxford: Reuters Institute.

Jagodzinski, J. ed. (2018) *Interrogating the Anthropocene: Ecology, Aesthetics, Pedagogy, and the Future in Question*. Cham: Palgrave Macmillan and Springer.

Keys, P.W. et al. (2019) Anthropocene risk. *Nature Sustainability* 2: 667–673.

Kolbert, E. (2006) *Field Notes from a Catastrophe: Man, Nature, and Climate Change*. London: Bloomsbury.

Lakoff, G. (2010) Why it matters how we frame the environment. *Environmental Communication* 4(1): 70–81.

Lane, R. (2019) The American Anthropocene. *Geoforum* 99: 11–21.

Lashof, S. (2018) Why are positive climate feedbacks so negative? https://www.greenbiz.com/article/why-are-positive-climate-feedbacks-so-negative

Latour, B. (2015) Telling friends from foes in the time of the Anthropocene. In Hamilton, C. et al. eds. *The Anthropocene and Global Environmental Crisis: Rethinking Modernity in a New Epoch*. London and New York: Routledge. chap. 12.

Lenton, T. (2016) *Earth System Science: A Very Short Introduction*. Oxford: Oxford University Press.

Lewis, S.L. & M.A. Maslin (2018) *The Human Planet: How We Created the Anthropocene*. London: Pelican.

Lorimer, J. (2017) The Anthropo-scene: A guide for the perplexed. *Social Studies of Science* 47(1): 117–142.

Lovelock, J. (2016) *Gaia: A New Look at Life on Earth*. Oxford: Oxford University Press.

Lück, J. et al. (2016) Counterbalancing global media frames with nationally colored narratives: A comparative study of news narratives and news framing in the climate change coverage of five countries. *Journalism* 17: 1–22.

Lysgaard, J.A. et al. eds. (2019) Dark Pedagogy: Education, Horror and the Anthropocene. London: Palgrave Macmillan.

Maxwell, R. & T. Miller (2018) The media produce climate change: Fake news, post-truth and journalistic pollution. In Brevini, B. & J. Lewis eds. *Climate Change and the Media*, Volume 2. New York: Peter Lang. chap. 10.

McKie, R. (2019) Climategate 10 years on: what lessons have we learned? *The Observer* (9 November).

McNatt, M.B. et al. (2019) Anthropocene communications: cultural politics and media representations of climate change. In Davoudi, S. et al. eds. *The Routledge Companion to Environmental Planning*. London: Routledge, chap 2.11.

McNeill, J. & P. Engelke (2016) *The Great Acceleration*. Cambridge: Harvard University Press.

Moore, J. ed. (2016) *Anthropocene or Capitalocene? Nature, History, and the Crisis of Capitalism*. Oakland, CA: PM Press.

Morton, T. (2016) The first draft of the future: Journalism in the 'Age of the Anthropocene'. In Marshall, J.P. & L. Connor eds. *Environmental Change and the World's Futures: Ecologies, Ontologies and Mythologies*. London: Routledge, chap 3.

Murphy, M. & L. Sklair eds. (2020) World-Systems analysis in the Anthropocene. *Journal of World-Systems Research* 26(2), whole issue.

Oldfield, F. (2016) Paradigms, projections and people. *Anthropocene Review* 3(2): 163–172.

Oppermann, S. & S. Iovino eds. (2017) *The Environmental Humanities and the Challenges of the Anthropocene*. London and New York: Rowman & Littlefield International.

Oreskes, N. & E. Conway (2012) *Merchants of Doubt – How a Handful of Scientists Obscured the Truth on Issues from Tobacco Smoke to Global Warming*. London: Bloomsbury.

Painter, J. (2010) *Summoned by Science: Reporting Climate Change at Copenhagen and Beyond*. Oxford: Reuters Institute for the Study of Journalism.

Painter, J. & M. Schäfer (2018) Global similarities and persistent differences: A survey of comparative studies on climate change communication. In Brevini, B. & J. Lewis eds. *Climate Change and the Media*, Volume 2. New York: Peter Lang. chap. 3.

Pálsson, G. et al. (2013) Reconceptualizing the 'Anthropos' in the Anthropocene: Integrating the social sciences and humanities in global environmental change research. *Environmental Science & Policy* 28: 3–13.

Parham, J. (2006) Academic values: Why environmentalists loathe the media. *Interdisciplinary Studies in Literature and Environment* 13(1): 113–133.

Rahm, P. (2019) The Anthropocene style: Towards a new decorative style. In Crespi, L. ed. *Design Innovations for Contemporary Interiors and Civic Art*. Hershey, PA: IGI Global. chap. 14.

Robin, L. (2017) Environmental humanities and climate change: understanding humans geologically and other life forms ethically. *WIREs Climate Change*. doi:10.1002/wcc.499

Rockström, J. et al. (2009) A safe operating space for humanity. *Nature* 461: 472–475.

Schwägerl, C. (2014) *The Anthropocene: The Human Era and How it Shapes Our Planet*. Santa Fe and London: Synergetic Press.

Scranton, R. (2015) *Learning to Die in the Anthropocene: Reflections on the End of a Civilization*. San Francisco, CA: City Lights.

Shapin, S. (2018) Think like a neutron. *London Review of Books* (24 May): 13–15.

Shea, M.M. et al. (2020) Representations of Pacific Islands and climate change in US, UK, and Australian newspaper reporting. *Climatic Change*. doi:10.1007/s10584-020-02674-w

Sklair, L. (2001) *The Transnational Capitalist Class*. Oxford: Blackwell.

Sklair, L. (2002) *Globalization: Capitalism and its Alternatives*. Oxford: Oxford University Press.

Sklair, L. (2017) Review article: Sleepwalking through the Anthropocene. *British Journal of Sociology* 68(4): 775–784.

Smith, J. ed. (2000) *The Daily Globe: Environmental Change, the Public and the Media*. London: Earthscan.

Steffen, W. et al. (2004) *Global Change and the Earth System: A Planet Under Pressure*. Berlin: Springer (IGBP Series).

Steffen, W. et al. (2011) The Anthropocene: Conceptual and historical perspectives. *Philosophical Transactions Royal Society London A*, 369: 842–867.

Steffen, W. et al. (2015) The trajectory of the Anthropocene: The great acceleration. *Anthropocene Review* 2(1): 81–98.

Steffen, W. et al. (2018) Trajectories of the earth system in the Anthropocene. *PNAS* 115(33): 8252–8259.

Stengers, I. (2015) Accepting the reality of Gaia. In Hamilton, C. et al. eds. *The Anthropocene and Global Environmental Crisis*: *Rethinking Modernity in a New Epoch*. London and New York: Routledge. chap. 11.

Thomas, J.A. (2014) History and biology in the Anthropocene: Problems of scale, problems of value. *The American Historical Review* 119(5): 1587–1607.

Thomas, J.A. (2019) Why the 'Anthropocene' is not 'Climate Change' and why it matters. *Asia Global Online* (20 January).

Truscello, M. (2018) Catastrophism and its critics: On the genre of environmentalist documentary film. In Jagodzinski, J. ed. *Interrogating the Anthropocene: Ecology, Aesthetics, Pedagogy, and the Future in Question.* Cham: Palgrave Macmillan and Springer. chap. 11.

Tsing, A.L. et al. eds. (2017) *Arts of Living on a Damaged Planet: Ghosts and Monsters of the Anthropocene*. Minneapolis, MN: University of Minnesota Press.

van Eeden, P. (2017a) Materializinig discourse analysis with James, Schmitt and Latour. *Palgrave Communications* 3. doi:101057-palcomms.2017.39

Wilke, S. & J. Johnstone eds. (2019) *Readings in the Anthropocene: The Environmental Humanities, German Studies, and Beyond.* London: Bloomsbury.

Whyte, K.P. (2017) Indigenous climate change studies: Indigenizing futures, decolonizing the Anthropocene. *English Language Notes* 55(1–2): 153–162.

Zalasiewicz, J. & W. Steffen (2017) Petrifying earth process: The stratigraphic imprint of key earth system parameters in the Anthropocene. *Theory, Culture & Society* 34(2–3): 83–104.

Zalasiewicz, J. et al. eds. (2019) *The Anthropocene as a Geological Time Unit: A Guide to the Scientific Evidence and Current Debate*. Cambridge: Cambridge University Press.

2 Anthropocene in the mass media

The big picture

Leslie Sklair

With so many discrete units to deal with, we decided to bundle groups of countries into nine roughly geographical regions as follows: Africa (AF), Asia South (AS), Asia North (AN), North America (NA), Latin America/Caribbean (LA/C), Central/Eastern Europe (C/EE), Western Europe (WE), Middle East (ME), and Oceania (OC).

Sources and items

Bearing in mind the discussion of the Fermi problem in the previous chapter, this chapter provides a snapshot of the main results of the project as a whole. Two provisos should be noted about all the numbers, representing results, in this book. First, this is not a record of all Anthropocene items in the mass media over the period of the study; it is restricted to those media, the online presence of which permitted freely available, reliable searches. Second, given the volatility of media websites, all these numbers should be regarded as approximate. As noted in the previous chapter, the term Anthropocene appeared in the public realm (albeit in a relatively obscure geological newsletter) for the first time in 2000 (Crutzen and Stoermer 2000), followed by a brief statement by Crutzen (2002) in the prestigious journal *Nature*. By 2006, all the regions showed at least one Anthropocene item. Also, it is noteworthy that the total results for any one year between 2000 and the end of 2017 reached 100 in only three regions – North America, Western Europe, and Central/Eastern Europe. The highest numbers of Anthropocene articles occurred in 2016 for all regions, probably due to the outreach of the Anthropocene Working Group (AWG) and global media reporting of the 35th International Geological Congress in Cape Town. Table 2.1 summarizes the main numerical findings by region.

Given that the results stretch over a period of 18 years (from 2000 to the end of 2017), the low average number of articles per source per year for each region is striking. Counting only sources with at least one Anthropocene item (non-zero sources), the arithmetic average is about five items per source over the 18-year period. This means that readers might encounter an article or item mentioning the Anthropocene only once every three to four years. Such exposure would be unlikely to signal that the Anthropocene was of great importance.

Table 2.1 Summary results by years and regions

	AF	*AS*	*AN*	*LA*	*NA*	*ME*	*WE*	*C/EE*	*OC*
FIRST[1] ITEM	2004	2006	2003	2000	2003	2006	2000	2002	2005
FIRST 100[2]	N/A	N/A	N/A	N/A	2015	N/A	2015	2016	N/A
PEAK[3]	64	65	98	95	289	64	379	106	89

[1] Year of first mention of the Anthropocene.
[2] First year in which 100+ results were recorded.
[3] Highest number of results in any year (all 2016).

Adding the zero-items sources, the figure drops to two items per source, one every nine years. In Africa, Asia South, and the Middle East, articles mentioning the Anthropocene occurred less than once per year. Even in the higher-scoring regions, namely North Asia, Latin America, North America, Western Europe, East/Central Europe, and Oceania (boosted by the large number of items from the Australian media), Anthropocene articles appeared at a rate of about three per year. Averages, of course, can be deceptive. It is probably the case that some people who normally read particularly high-yielding media sources would come across the term more frequently and be tempted to find out more about it. Exploring this in more detail, we list all the newspapers and magazines in our survey that mentioned the Anthropocene twice or more a year on average (that is, at least 36 items between 2000 and the end of 2017). This is an interesting but not very long list, less than 1% of all media searched. Six UK publications make this list, as do four from the United States, and three from Germany. The five highest-scoring publications are based in the United States, Germany, France, and Italy. This list of highest scoring publications suggests that the idea of the 'quality press' has some validity globally with respect to Anthropocene reporting, though the very popular *Daily Mail* in the UK does not usually feature in this category (see Table 2.2). Bunching of results, where a story runs for an extended period in several sources, is rare. However, even if we look at only these high-scoring publications, the results are still very modest. If we count from the year 2000, the date of publication of the first Anthropocene articles in the whole survey, only *The New York Times* (15) and *Süddeutsche Zeitung* (12) reach double figures for items per year, about one per month. Counting from the date of the first piece on the Anthropocene in each source, apart from *The New York Times* and *Süddeutsche Zeitung*, the average for these highest-yielding sources is between three and seven items per year.[1] It could also be argued that readers of most of the publications listed in Table 2.2 are more likely to have higher levels of education than the general public and may well have already come across the word 'Anthropocene'. It goes without saying that there is no guarantee that people actually see or take in what appears in the media they access, and it is quite possible that the only people to read some of the Anthropocene items we found were our researchers. It is also possible that a study of social media might give different results.

Table 2.2 Publications with 36+ Anthropocene items (beginning 2000 to end 2017)

Publication	*Total items*	*Date range of items in source*	*Items by year from first result*
New York Times (USA)	165	(2007–17)	15
Süddeutsche Zeitung (Germany)	92	(2010–17)	12
Frankfurter Allgemeine Zeitung (Germany)	90	(2000–17)	5
Le Monde (France)	84	(2004–17)	6
Focus (Germany)	62	(2007–17)	6
Wired (USA)	60	(2008–17)	6
Canberra Times[1] (Australia)	58	(2007–17)	5
Morgenbladet (Norway)	52	(2010–17)	7
Slate (USA)	52	(2009–17)	6
Telegraph (UK)	49	(2008–17)	5
Discover (USA)	49	(2004–17)	4
Daily Mail (UK)	46	(2008–17)	5
Independent (UK)	46	(2007–17)	4
Hindu (India)[2]	45	(2011–17)	6
Guardian (UK)[3]	40	(2008–17)	4

1 Restricted access excludes the *Sydney Morning Herald* and *The Age* (Melbourne) from the table.
2 Restricted to 10 free articles per month. Discounting many duplicates, from over 150 search results this is our best estimate.
3 Google Custom Search, 40 items available at any one time. Many more results were available in 2017.

Narratives and frames

Having established approximate measures of volumes of coverage by region, we can now move onto the question of how the media are reporting the Anthropocene. To do this, we explore the argument that connects journalists with environmental humanities scholars, social scientists, and eco-linguists, that stories, metaphors, and narratives about nature, climate change, and the Anthropocene in general can be effective in raising consciousness at all levels about the risks (and opportunities) that human impacts pose for the survival of humanity on planet Earth.[2] These narratives provide context for framing strategies mobilized by the reporting of ecological issues, often using climate change/global warming (CC/GW) as a metonym for the Anthropocene. A good example of this can be found in the study by Agarwal and Narain (2012), evocatively entitled 'Global Warming in an Unequal World: a case of environmental colonialism'. This is a critique of narratives emerging from what used to be called 'First World' governments, inter-governmental bodies, and environmental organizations (notably, in this case, the World Resources Institute and the UN) dealing with emissions of greenhouse gases. By implication, such narratives allocate responsibility for the potential dangers of the Anthropocene. Measuring total global emissions, the authors argue, represents 'thinking like a planet'; measuring emissions per national GDP represents 'thinking like a market', whereas measuring emissions per capita represents 'thinking like humans'. Agarwal and Narain argue that failing to compare the 'survival emissions of the poor' with the 'luxury emissions of the rich' has led to environmental colonialism.[3] Theoretical and historical studies of international climate change negotiations (Bernstein 2002, Howe 2014) provide evidence for such a conclusion. Figure 2.1 represents the best estimates we have for thinking about emissions like a planet and like humans.[4]

The regional chapters in Part II present findings from media in 'the West and the Rest', sometimes linking colonial and environmental history and current practices. This is contextualised in terms of what has come to be known as the 'provincializing' of the Anthropocene and, whereas these issues are rarely raised in precisely these terms, they can be seen bubbling away beneath the surface in global media. Our book represents (albeit tentatively) the first systematic study of the narratives that mass media in countries all over the world are mobilizing

CDIAC [1]	1751-2014		Stat. Review World Energy [2]	2015-2018
Million Tons Total CO2				
UNITED STATES	102,510,260	1	CHINA	13,302,742
CHINA	47,649,834	2	UNITED STATES	7,332,095
RUSSIA	41,256,776	3	INDIA	3,304,071
UNITED KINGDOM	20,500,813	4	RUSSIA	2,170,882
JAPAN	15,680,779	5	JAPAN	1,690,586
GERMANY	12,764,185	6	GERMANY	1,082,448
INDIA	11,385,351	7	SOUTH KOREA	970,335
FRANCE	9,697,149	8	IRAN	884,904
CANADA	8,038,299	9	SAUDI ARABIA	844,859
POLAND	6,960,097	10	CANADA	784,907
Tons CO2 per Capita				
QATAR	856.65	1	SINGAPORE	5.62
UNITED ARAB EMIRATES	472.28	2	QATAR	5.61
KUWAIT	433.40	3	UNITED ARAB EMIRATES	4.20
BRUNEI	337.02	4	KUWAIT	3.48
BAHRAIN	335.65	5	ESTONIA	2.61
UNITED STATES	322.09	6	SAUDI ARABIA	2.60
CANADA	266.19	7	LUXEMBOURG	2.56
TRINIDAD AND TOBAGO	259.10	8	AUSTRALIA	2.46
AUSTRALIA	237.58	9	UNITED STATES	2.27
SAUDI ARABIA	187.19	10	TRINIDAD AND TOBAGO	2.24

Carbon Dioxide Information Analyisis Center
https://energy.appstate.edu/research/work-areas/cdiac-appstate
Statistical Review of World Energy
https://www.bp.com/en/global/corporate/energy-economics/statistical-review-of-world-energy.html

Figure 2.1 Thinking like a planet, thinking like humans (Andrew Zolnai).

to make the Anthropocene meaningful to the general public. More than half of the 4,000+ media items on the Anthropocene identified in our research were randomly selected for classification. Three broad and sometimes overlapping narratives, extracted from the media coverage, point towards three different ways of framing the Anthropocene story.[5] These are labeled here as the **neutral frame**, the optimistic **'good' anthropocene frame**, and the pessimistic **radical change frame**. Alternative ways of interpreting what we find in the media are offered in some of the other chapters in this book, reflecting local and/or national differences in media coverage.[6] The main characteristics of these three frames are:

{1} Neutral descriptive reporting, including passing mention of the Anthropocene, disagreements among scientists, especially over starting dates,[7] and/or the presentation of the Anthropocene as a continuation of natural processes: the **neutral frame**.

{2} The optimistic **'good' anthropocene frame** comprises a broad spectrum of mutually reinforcing narratives, ranging from the moderately to the definitely alarming. All recognise that the planet and humanity itself may be in some danger, and that we cannot ignore the warning signs. However, if we are clever enough, we can save ourselves and the planet with technological fixes (notably geoengineering)[8] and other strategies that present opportunities for industry, science, and technology. The recognition of the need for change typically includes renewable energy,[9] population control, conservation, sustainability, and resilience but stops short of any radical challenge to the economic and political status quo. This master narrative provides a more-or-less sophisticated optimistic 'good' Anthropocene frame.[10]

{3} Recognition that human survival is at risk, that we cannot go on living and consuming as we do now, that we must strive to change our way of life radically, for example by bringing capitalism to an end and creating new types of global governance or state-less societies or religious/spiritual renewal, whether these are realistic or not. This is labelled the **pessimistic radical change frame**.

The fact that many articles combine two (occasionally all three) of these narratives is certainly a challenge for those tasked with categorizing media coverage of the Anthropocene. This is particularly true for {1/2}, where problems thrown up by the Anthropocene are identified but not interpreted unequivocally in terms of existential threats to human survival. Combinations of {2/3} tend to put forward a more considered view of risks and what must change, but never threatening the status quo in any radical fashion. For example, proposals to introduce electric cars might appear radical to some, but, if they do not challenge all modes of transportation, they can be considered 'business as usual', which now has its own acronym in the literature (BAU). To a greater or lesser extent, combinations of {1/2}, and {2/3} become versions of what I have labelled 'reassurance narratives' and all can be reconciled with the 'good' Anthropocene. As the regional chapters to follow show, neutralizing narratives {1} are the most common (hence the subtitle of the book). This is not surprising as many Anthropocene items are short, straightforward reports of the derivation of the term, mentions in passing in articles on other topics, and arts events that simply reference the word. Optimistic 'good' Anthropocene articles {2} occur about half as often, and these tend to be longer and more analytical. Whereas there is a good deal of pessimism in the media, genuine radical change narratives {3} are almost entirely absent everywhere. If we take narratives {1} and {2} together, the various versions of 'reassurance' account for at least 99% of all items.[11]

The 'good' Anthropocene narrative

There is an interesting backstory to the optimistic 'good' Anthropocene narrative. In 1987, the International Geosphere–Biosphere Programme (IGBP) was established to provide an institutional framework for the emerging Earth System science. Ellis (2018: 29–33) tells us that Paul Crutzen was among the first to join. It was at an IGBP meeting in Mexico in 2000 that Crutzen made his famous statement 'We are in the Anthropocene' (see Schwägerl 2014 for the 'authorized' version). That was the year in which Crutzen and Stoermer published their paper, introducing the Anthropocene to the scientific community in the IGBP newsletter. However, in 2015, the IGBP was terminated and a new organization, 'Future Earth', replaced it. The mission of Future Earth, according to its website, is as follows:

> Our organisation is rooted in the work of 20 Global Research Projects [see also Bai et al. 2016, and the more recent version in Jeanson et al. 2019]. These networks have a long history, in some cases stretching back decades, of generating research at the forefront of sustainability science. They played a critical role in forming the field of Earth system science – which considers how all of the parts of the planet, including human societies, connect to and shape each other. Our Global Research Projects also led in defining the Anthropocene, a new epoch in the geologic history of Earth marked by the influence of humans on the planet.'
>
> (www.futureearth.org)

However admirable these activities are (they include signing up to the UN's Sustainable Development Goals), the statement does not tell the whole story. Future Earth goes beyond endorsing the Anthropocene, it also propagates a specific (and, for many, highly contentious) version of it – the 'good' Anthropocene.[12] For such a recent idea, the 'good' Anthropocene has already spawned an impressive scholarly literature (see Bennett et al.2016, and its long list of references), and presence in the media. Revkin (2014), Dalby (2016), and Fremaux and Barry (2019) are illuminating guides to this debate. Most proponents of the 'good' Anthropocene sensibly advise us to prepare for the worst. Some, however, appear to use the idea in order to promote corporate interests. For example, Kareiva and Fuller (2016) argue: 'environmentalists too often reject entrepreneurial experimental approaches that could make them more relevant to policymakers, corporations, and other institutions that seek to respond more proactively to impending disruption' (ibid.: 107).[13] In contrast, narrative {3}, often characterized as 'alarmist' (see, notably, Hamilton 2017), tends to be more or less restricted to the genuinely worried anti-capitalist left.[14] I would place myself in that category (see Sklair 2001: chap. 7; 2016; 2019). The other possible narrative combinations are not considered here; they would tend to add to the 'good' Anthropocene total.

Media coverage over time

Our media searches show that the total volume of Anthropocene mentions increased slowly in the first decade of the new millennium, reaching almost 200 between 2010 and 2011. In our searches, no region had an article on the Anthropocene in each of the years between 2000 and the end of 2017. Central/Eastern Europe had only two blank years (though its overall total was relatively low), whereas all the other regions had three or more. The blank years were, unsurprisingly, concentrated in the first decade, as the term 'Anthropocene' was slowly being disseminated. Media references to the Anthropocene in North America reached double figures in 2008 and 100+ in 2014. Western Europe also reached double figures in 2008 and recorded the biggest year-on-year increases, from about 100 articles from 2014 to 2015 to almost 400 from 2015 to 2016. Results peaked in all regions in 2016 (totalling about 1250 in that year), fell back a little in 2017, and are probably similar today.

There are several obvious reasons that help to explain these changes in coverage over time, and a few less obvious reasons. Media research has some convincing explanations for changes in the volume of press coverage of CC/GW, mostly focused on extreme weather, 'hot topics' like the extinction of the dinosaurs, corporate lobbying, issue cycles, and domestic politics (see Ungar 1992, Mazur and Lee 1993, Mazur 1998, Jacques et al. 2008, Rich 2018, and Saunders et al. 2018). To explain the surges in media interest in the Anthropocene from 2010 onwards, the items in the two timelines in the previous chapter offer some clues. The formation of the AWG in 2009, major Anthropocene exhibitions in Berlin and Munich (2013–2014), and Kolbert's book *The Sixth Extinction* all featured in the mass media in many countries.[15] The AWG was mentioned almost 200 times, mostly in 2015 and 2016, often in reports on the International Geological Congress in Cape Town in 2016, where discussions on the Anthropocene were highlighted. Much of the media interest was probably due to Jan Zalasiewicz's dynamic leadership of the AWG and his strenuous efforts to publicize the Anthropocene outside the ivory towers of academia and bridge the divide between the two cultures of science and the humanities (discussed in Sklair 2020). Crutzen and Zalasiewicz are the only scientists to appear in media coverage of the Anthropocene in all nine regions.

Deniers and sceptics

Corporate and political mobilization against radical social, economic, and political action to combat CC/GW differs somewhat from what happened in the creation of Anthropocene narratives in the media. The main difference is that climate change denial, scepticism, and contrarianism were central components of an organized, well-funded strategy that garnered plenty of media coverage, especially in the United States (Brulle 2018), whereas, in contrast, media coverage of Anthropocene denial and scepticism has been relatively sparse.[16] Stanley Finney, who has occupied several senior posts in the world of professional geology (see Finney & Edwards 2016), received coverage in about a dozen newspapers for his Anthropocene scepticism. For example, *Prothom Alo*, the most popular newspaper in Bangladesh, concluded a generally neutral piece in 2016 as follows:

> There are sceptical scientists too. "The drive to officially recognise the Anthropocene may, in fact, be political rather than scientific," said Stanley Finney, a professor at California State University at Long Beach and chair of the International Commission on Stratigraphy—one of the bodies that must validate the new epoch. Finney's main objection is that not enough time has elapsed for a new epoch. "Its duration is that of an average human lifespan," he wrote in a pointed commentary, published by the Geological Society of America.[17]

Corneliussen (2016), in his review of media coverage in the US and the UK cited in the previous chapter, sums up the limited success of the Anthropocene denial, sceptic, and contrarian lobbies, noting that online searches of archives at the 'Anthropocene-promoting *New York Times*' turned up more than a hundred hits on the term, while *The Guardian* had exposed a 'whole media ecosystem' of climate-science denial. He concludes: 'Meanwhile, the Anthropocene, both word and concept, is becoming less exotic and more routine in the media'. Corneliussen gives the last word on this issue to Clive Hamilton, the prominent Australian alarmist/realist (my term), who says: 'Some scientists even write: "Welcome to the Anthropocene".' Hamilton continues: 'first, I thought they were being ironic, but now I see they are not. And that's scary. The idea of the Anthropocene is not welcoming. It should frighten us. And scientists should present it as such'. This statement is reprinted in two Indian newspapers, *The Tribune* and *The Hindu*, and in *The Belfast Telegraph* and *Physics Today* (all in 2016). 'Welcome to the Anthropocene' (ironic or not) is a central component in media coverage of the optimistic 'good' Anthropocene frame and in its emergence as a dominant reassurance narrative for the Anthropocene in recent years. We can conclude, therefore, that whereas the weight of contrarian opinion on CC/GW in media coverage focused on denial, this was not the case for the Anthropocene in the media. The majority of media representations of the Anthropocene in our results tend to conflate elements of reassurance narratives, finding a home in the 'good' Anthropocene frame, though not all in the same way.

Countries and regions

In Part II, we explore the varieties of media coverage of the Anthropocene in nine geographical regions and the countries within them. For the countries in each region, we provide, where possible, references to the enormous volume of research on climate change to give readers the opportunity to compare and contrast them with our own findings on media coverage of the Anthropocene. Also, where possible, we reference studies of media organization, by region and country, and on environmental movements around the world. There is no single template for these chapters; authors construct them in terms of the media results we found, article titles are translated into English, and light editing is applied. These chapters engage with debates around 'provincializing the Anthropocene' and its connections with controversies swirling around environmental justice between the West and the Rest. Although media coverage of the Anthropocene rarely mentions imperialism or colonialism, it appears to be an underlying theme (in both the West and the Rest), pointing to the ways in which the concept might be liberated from its white, male, Eurocentric roots. An important aspect of this is the responsibility of different countries around the world for emissions of greenhouse gases.

In 2019, the American Meteorological Association published a special report on the state of the climate globally (Blunden and Arndt 2019), with summaries of the situation, country by country, many of which appear in the following chapters. This report received the usual flurry of media attention, and then the media cycle moved on. Figure 2.2, which shows an estimate of CO_2 emissions from fuel combustion for individual countries (all of which feature in the regional chapters to follow), gives a vivid picture of the complexities of allocating blame in our unequal global system of competing states, let alone the different communities within them.

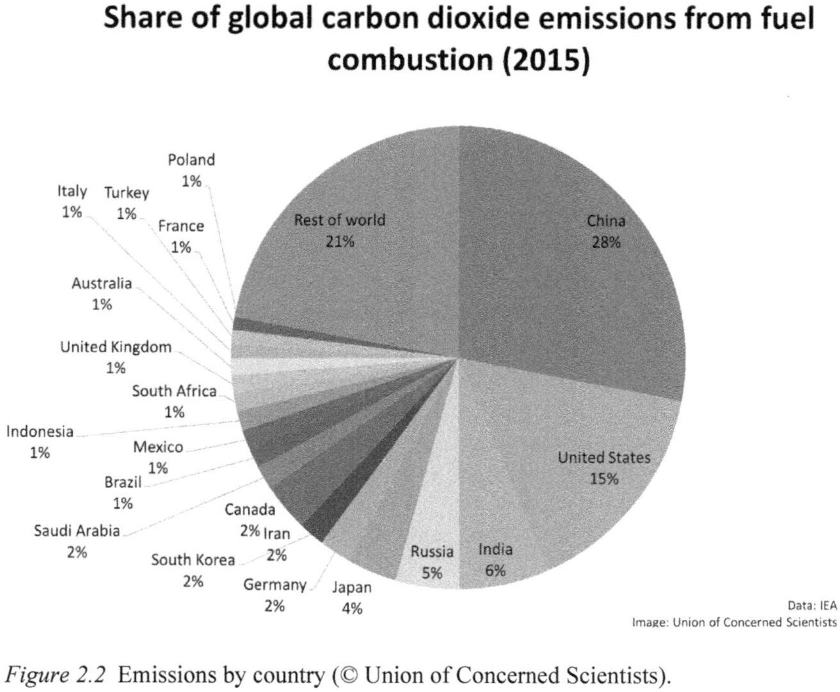

Figure 2.2 Emissions by country (© Union of Concerned Scientists).

Notes

1 These are rather artificial numbers in that they seem to penalize media that began reporting the Anthropocene early. In addition, the *New York Times* is a special case due to the impact of Andy Revkin, who wrote the highly influential dot.earth blog for the paper from 2007 to 2016 (see Chapter 4 of this book).
2 See, for example, Zolnai (2014), Stibbe (2015), Karlsson (2016), Howe and Pandian (2016), DeLoughrey (2019), and Garbutt (2017).
3 See Schmidt et al. (2013). Issues around emissions from land use, urbanization, soil degradation, and food production are rarely connected in the media but are increasingly of interest to Anthropocene scholars (Bren d'Amour, et al. 2017). The enormous carbon emissions from the production of the media themselves are also ignored (see Maxwell and Miller 2018).
4 The complexities of our era of hyper-international trade and production make it virtually impossible to compute meaningful numbers on emissions at the level of national GDP.
5 The code book of one study on media coverage of the IPCC documents over 80 discrete messages (Konelius et al. 2017). The purpose of our book, however, is to simplify the process in terms of the major research question: what is the 'takeaway' Anthropocene message for the casual reader?
6 For a critique of conventional narratives, see Kunnas (2017) and Simon (2018).
7 Global media have paid considerable attention to disagreements among scientists, usually neutralizing the risks. Obviously, the earlier the start date, the less blame attaches to our and recent generations.
8 Discussions on geoengineering range from the optimistic and technocratic (Flannery 2015, https://edgeeffects.net/kim-stanley-robinson/) to the extremely alarming, as in https://www.geoengineeringwatch.org/, a website that claimed over 35 million visits by 2020. In the mass media, the topic has been discussed mainly in Western Europe and North America, with Crutzen often identified as an advocate (Klein 2014: chap. 8).
9 Most media coverage of renewable energy naively assumes zero carbon emissions but see Di Wu et al. (2019) on 'electric' vehicles, Hornborg et al. (2019) on 'free' solar power, and Booth (2019) on biomass. For life cycle analysis (usually ignored in the media), see Hauschild et al. (2018).
10 I use the term 'master' deliberately to highlight elements of 'mankind's domination of nature' (including some women). See also the 'God Species' (Lynas 2011) inherent in some 'good' Anthropocene narratives and PERC (2016) on 'free-market environmentalism' for the Anthropocene.
11 These findings challenge the assumptions of many academics and other intellectual elites that everyone has heard of the Anthropocene, and that 'doom and gloom' scenarios prevail (see also Chapter 14 of this book).
12 See http://www.stockholmresilience.org/research/research-news/2015-09-29-seeds-of-a-good-anthropocene.html, on the Future Earth website.
13 The word 'Anthropocene' appears only twice in this article (once in the title) whereas 'climate' occurs 55 times. This linguistic choice may be seen as a risk-neutralizing strategy, consciously or not.
14 Hulme (2009) provides a powerful rejection of catastrophism.
15 As early as 2011, *Nature* has an article arguing that we might be entering the Sixth Extinction (Barnosky et al. 2011).
16 Boykoff and Olson (2013) helpfully distinguishes climate change deniers, sceptics, and contrarians. In the case of the Anthropocene, differences usually focus on the name. The geologist George Klein (2015) offers an unusually robust rejection of the Anthropocene.
17 For interesting background on Finney and the Anthropocene, see the 'long read' in *The Guardian* (30 May 2019) and my rejoinder on 3 June.

References

Agarwal, A. & S. Narain (2012) Global warming in an unequal world: A case of environmental colonialism (selected excerpts). In Dubash, K.N. ed. *Handbook of Climate Change and India*. London: Earthscan, chap. 5.

Bai, X. et al. (2016) Plausible and desirable futures in the Anthropocene: A new research agenda. *Global Environmental Change* 39: 351–362.

Barnosky, A.D. et al. (2011) Has the earth's sixth mass extinction already arrived? *Nature* 471: 51–57.

Barthel, S. et al. (2019) Global urbanization and food production in direct competition for land: Leverage places to mitigate impacts on SDG2 and on the earth system. *The Anthropocene Review* 6(1–2): 71–97.

Bennett, E. et al. (2016) Bright spots: Seeds of a good Anthropocene. *Frontiers in Ecology and the Environment* 14(8): 441–448.

Bernstein, S. (2002) *The Compromise of Liberal Environmentalism*. New York: Columbia University Press.

Blunden, J. & D.S. Arndt eds. (2019) State of the climate in 2018. Special Supplement to the *Bulletin of the American Meteorological Society* 100(9) (September).

Booth, M.S. (2019) The great biomass boondogle. *NYR Daily* (14 October).

Boykoff, M. & S. Olson (2013) 'Wise contrarians': A keystone species in contemporary climate science, politics and policy. *Celebrity Studies* 4(3): 276–291.

Bren d'Amour, C. et al. (2017) Future urban land expansion and implications for global croplands. *Proceedings of the National Academy of Sciences* 114(34): 8939–8944.

Brulle, R.J. (2018) The climate lobby: A sectoral analysis of lobbying spending on climate change in the USA, 2000 to 2016. *Climatic Change* 149: 289–303.

Corneliussen, S. (2016) Media attention increases for the term—and concept—Anthropocene. *Physics Today* (4 November).

Crutzen, P. (2002) Geology of mankind. *Nature* 415(January): 23.

Crutzen, P. & E. Stoermer (2000) The 'Anthropocene'. *IGBP Newsletter* 41: 17–18.

Dalby, S. (2016) Framing the Anthropocene: The good, the bad and the ugly. *The Anthropocene Review* 2(2): 102–106.

DeLoughrey, E.M. (2019) *Allegories of the Anthropocene*. Durham and London: Duke University Press.

Ellis, E. (2018) *The Anthropocene: A Very Short Introduction*. Oxford: Oxford University Press.

Finney, S. & L. Edwards (2016) The 'Anthropocene' epoch: Scientific decision or political statement? *GSA Today* 26(3–4): 4–10.

Flannery, T. (2015) *Atmosphere of Hope: Searching for Solutions to the Climate Crisis*. London: Penguin.

Fremaux, A. & J. Barry (2019) The 'Good Anthropocene' and green political theory: Rethinking environmentalism, resisting eco-modernism. In Biermann, F. ed. *Anthropocene Encounters: New Directions in Green Political Thinking*. Cambridge: Cambridge University Press, chap. 9.

Garbutt, R. (2017) Rear-view mirror: Vision, time, modernity and the Anthropocene. *Continuum* 31(2): 277–284.

Hamilton, C. (2016) Define the Anthropocene in terms of the whole Earth. *Nature* (17 August).

Hamilton, C. (2017) *Defiant Earth: The Fate of Humans in the Anthropocene*. Cambridge: Polity Press.

Hauschild, M.Z. et al. eds. (2018) *Life Cycle Assessment: Theory and Practice*. Cham: Springer.

Hornborg, A. et al. (2019) Has Cuba exposed the myth of 'Free' solar power? Energy, space, and justice. *Environment & Planning E* 2(4): 989–1008.

Howe, C. & A. Pandian (2016) Lexicon for an Anthropocene yet unseen: Theorizing the contemporary. *Cultural Anthropology*. https://culanth.org/fieldsights/803-lexicon-for-an-anthropocene-yet-unseen

Howe, J. (2014) *Behind the Curve: Science and the Politics of Global Warming*. Seattle, WA: University of Washington Press.

Hulme, M. (2009) *Why We Disagree about Climate Change: Understanding Controversy, Inaction and Opportunity*. Cambridge: Cambridge University Press.

Ingram, M. et al. (2015) Environmental action in the Anthropocene: The power of narrative networks. *Journal of Environmental Policy & Planning*. https://www.tandfonline.com/loi/cjoe20

Jacques, P.J. et al. (2008) The organisation of denial: Conservative think tanks and environmental scepticism. *Environmental Politics* 17(3): 349–385.

Jeanson, A.L. et al. (2019) Twenty actions for a 'good Anthropocene' — perspectives from early-career conservation professionals. *Environmental Reviews* 1–10. doi:10.1139/er-2019-0021

Kareiva, P. & E. Fuller (2016) Beyond resilience: How to better prepare for the profound disruption of the Anthropocene. *Global Policy* 7(Supplement 1): 107–118.

Karlsson, R. (2016) Three metaphors for sustainability in the Anthropocene. *The Anthropocene Review* 3(1): 23–32.

Klein, G.D. (2015) The 'Anthropocene': What is its geological utility? (Answer: It has none!). *Episodes: Journal of International Geoscience* (September): 218.

Klein, N. (2014) *This Changes Everything: Capitalism vs. the Climate*. London: Penguin.

Konelius, R. et al. (2017) *Media and Global Climate Knowledge: Journalism and the IPCC*. New York: PalgraveMacmillan.

Kunnas, J. (2017) Storytelling: From the early Anthropocene to the good or the bad Anthropocene. *Anthropocene Review* 4(2): 136–150.

Lynas, M. (2011) *The God Species. How the Planet can Survive the Age of Humans*. London: Fourth Estate.

Maxwell, R. & T. Miller (2018) The media produce climate change: Fake news, post-truth and journalistic pollution. In Brevini, B. & J. Lewis eds. *Climate Change and the Media*, Volume 2. New York: Peter Lang. chap. 10.

Mazur, A. (1998) Global environmental change in the news: 1987–90 vs. 1992–96. *International Sociology* 13: 457–472.

Mazur, A. & J. Lee (1993) Sounding the global alarm: Environmental issues in the US national news. *Social Studies of Science* 23(4): 681–720.

PERC (2016) *Environmental Policy in the Anthropocene*. Bozeman, MO: Property and Environment Research Center.

Revkin, A. (2014) Paths to a 'Good' Anthropocene. https://www.youtube.com/watch?v=VOtj3mskx5k

Rich, N. (2018) Losing earth: The decade we almost stopped climate change. *New York Times Magazine* (1 August).

Rickards, L. (2015) Metaphor and the Anthropocene: Presenting humans as a geological force. *Geographical Research* 53(3): 280–287.

Saunders, C. et al. (2018) Attention to climate change in British newspapers in three attention cycles (1997–2017). *Geoforum* 94: 94–102.

Schmidt, A. et al. (2013) Media attention for climate change around the world: A comparative analysis of newspaper coverage in 27 countries. *Global Environmental Change* 23(5): 1233–1248.

Schwägerl, C. (2014) *The Anthropocene: The Human Era and How it Shapes Our Planet.* Santa Fe and London: Synergetic Press.

Simon, Z. (2018) The limits of Anthropocene narratives. *European Journal of Social Theory*. doi:10.1177/1368431018799256

Sklair, L. (2001) *The Transnational Capitalist Class*. Oxford: Blackwell.

Sklair, L. (2016) Half-Baked. *Philosophica Critica* 2(2): 103–116.

Sklair, (2017) Review article: Sleepwalking through the Anthropocene. *British Journal of Sociology* 68(4):

Sklair, L. (2019) The corporate capture of sustainable development and its transformation into a 'good Anthropocene' historical bloc. *Civitas* 19(2): 296–314.

Sklair, L. (2020) Globalization and the challenge of the Anthropocene. In Rossi, I. ed. *New Frontiers of Globalization Research*. New York: Springer, chap. 4.

Stibbe, A. (2015) *Ecolinguistics: Language, Ecology, and the Stories We Live By*. London and New York: Routledge.

Ungar, S. (1992) The rise and (relative) decline of global warming as a social problem. *The Sociological Quarterly* 33: 483–501.

Wetts, R. (2019) Models and Morals: Elite-Oriented and Value-Neutral Discourse Dominates American Organizations' Framings of Climate Change. *Social Forces* 98(3):1–31.

Wetts, R. (2020) In climate news, statements from large businesses and opponents of climate action receive heightened visibility. *PNAS* https://doi.org/10.1073/pnas.1921526117

Wu, D. et al. (2019) Regional heterogeneity in the emissions benefits of electrified and lightweighted light-duty vehicles. *Environmental Science & Technology* https://pubs.acs.org/doi/pdf/10.1021/acs.est.9b00648?rand=jnq12e7l

Zolnai, A. (2014) Map stories can provide dynamic visualizations of the Anthropocene to broaden factually based public understanding. *Anthropocene Review* 1(3): 243–251.

Part II

Media coverage of the Anthropocene

A global survey

3 Africa's Anthropocene in the media

A kaleidoscope of contradictions[1]

Meryl McQueen and Leslie Sklair

Introduction

The story of mass media coverage of the Anthropocene in Africa is one of broadly scattered results, surprising outliers, and a complex junction of history, development, language, politics, and culture. This chapter covers 36 countries and 6 languages. There were over 200 mentions of 'Anthropocene' from 60 sources, with a further 169 sources returning no results. One source, *Conversation Africa*,[2] provided almost half of the results. Therefore, almost three-quarters of the sources searched in Africa did not refer to the Anthropocene in any context whatsoever. This is a significant finding, given that the continent is home to 1.3 billion people, many of whom are living in poverty and are likely to be adversely affected by human-driven climate change and the consequences of the Anthropocene. The inequity extends even further, given that the primary source of global climate change is industrial activity in Europe, North America, and Asia. Africa's experience of negative socio-economic and political outcomes from the advent of the Anthropocene will depend heavily on each country's ability to navigate the consequences of these changes, both predictable and unforeseen.

A sample of Anthropocene findings from media in Africa was categorized into three broad themes, which complement the frames outlined in Chapter 2:

1. Informative, defining the Anthropocene in neutral terms, reviewing start dates, etc.
2. Risk-acknowledgement, with optimism about the roles of technology (especially renewables), education and/or religion in mitigating negative effects.
3. Pessimistic view of risks, advocating radical socio-economic transformation.

We explore the diversity of mass media coverage of the Anthropocene in Africa, mainly Anglophone but with significant French- and Arabic-language news sources, Afrikaans, and indigenous-language media.[3] We identify countries with unexpectedly high numbers of mentions of the Anthropocene, such as Liberia and Zimbabwe, and highlight the patchwork nature of Anthropocene coverage in Africa. As a counterpoint to the relative scarcity of Anthropocene coverage, the chapter will also examine how one specific case of almost no Anthropocene coverage, Tanzania, contrasts with rich media narratives of 'climate change' and 'global warming' in that country. The findings demonstrate how some communities in Tanzania are successfully navigating changes to historical climate patterns using technology and education interventions that promote local control of limited resources, despite the fact that the term 'Anthropocene' is absent in most of the media discourse in the continent.[4]

Like all regions in the world, Africa faces daunting challenges resulting from anthropogenic climate change. Issues such as deforestation, population growth, water shortages,[5] food insecurity, low literacy rates, and complicated geopolitical relationships around resource exploitation suggest that the ability to understand and respond to the Anthropocene is essential to Africa's survival. In terms of media ecologies, literacy rates are particularly salient. According to the World Bank in 2016, the sub-Saharan Africa literacy rate was only 64%, compared to 99% in Europe (Central Asia scored 96%, East Asia/Pacific, 94%, and Latin America/Caribbean, 71%, with 86% overall globally). To the extent that literacy rates affect mass media consumption, this discrepancy may in part explain the relative absence of the term Anthropocene, because use of the term itself may be seen to reflect a Western-centric approach to the link between science, politics, policy, and everyday life. As is clearly demonstrated in the Tanzania example, the lack of 'Anthropocene' in the discourse does not indicate a lack of awareness of, or appreciation for counter measures against, climate change. As outlined in Chapter 1, there are important differences between climate change and the Anthropocene.

Southern Africa

The media in Mauritius, Namibia, South Africa, and Zimbabwe all produced articles on the Anthropocene. Angola, Botswana, Lesotho, Madagascar, and Malawi returned zero results. However, media in Angola, Lesotho, and Madagascar all covered climate change, sometimes extensively. This indicates that, although the mass media coverage may include elements of effects of the Anthropocene, the discourse has not (yet?) generally moved to incorporate the Anthropocene as a concept that is very important for Africa.

In Mauritius, the Anthropocene appears four times in the media. This first occurs in the popular Francophone daily, *Le Mauricien* (2016), on an art exhibition identified with the Anthropocene, with the title 'Ephemeral coast'. Curated by a Canadian, it was said to have been inspired by the degradation of the local environment due to anthropogenic climate change. A second item, also in *Le Mauricien* in 2016, expresses sober acknowledgement of the risks of the Anthropocene combined with optimism about the possibility of radical change in a report of a 'Conference on The Ecological Crisis and the Right of Nature'. This profiles a local environmental activist, Ashok Subron, who characterizes the Anthropocene in terms of 'the destructive and irreversible imprint of man on the terrestrial ecosystem', thus requiring a transformation of social and economic systems and redefinition of the human in relation to Nature in general. This is striking because of the tone of the coverage of the conference, and for its specific reference to new socio-economic systems. A third item in *Le Mauricien* (2017) takes an even more radical position, accusing the establishment on the island of 'ostrich'-like behaviour in the face of ecological crimes'. The tabloid *5 Plus Dimanche* (2016), describes a course offered by the same organization as above, quoting local ecologist Stefan Gua: 'We must rebuild our humanity to be more in harmony with nature':

> [we need to] understand the responsibility of human society and the dominant development model on the climate crisis. Today, many scientists are addressing the question of the era we are living in and describe it as the period of the Anthropocene … the course will reflect on what needs to be done in order to turn the tide of things for a fairer and more equitable exchange, in harmony with our world, for the survival of humanity and life on Earth.

These four items tell us that, in Mauritius, an active member of the Alliance of Small Island States, some elements in the local media (and apparently the government) have been taking anthropogenic climate change seriously. However, this focus appears not to have evolved into much coverage of the Anthropocene in the media.[6] Namibia's mass media coverage of the Anthropocene, eight articles from three publications with eight other publications returning no result, is divided between optimistic and pessimistic narratives. The following examples, about changing sanitation systems across the globe and, and conservation of small creatures such as insects, promote education and technology as well as new ways of thinking about how humans interact with the planet. From the Anglophone *The Namibian* (2017), we find 'Sewage Systems: Finally Facing Our Water-Loo', making a connection between the Anthropocene and British colonialism. In 'Conservation Efforts Have to Include Small Animals', *The Namibian* argues: 'Humans like to think that they rule the planet and are hardwired to do so … so the focus must be on increasing awareness among the young … the Anthropocene marches on. Putting in place strategies that conserve as many animals as possible, along with the rest of biodiversity, is not a luxury for the future'.[7]

Mass media in South Africa returned 13 articles from nine publications, with an additional four showing no results. The earliest mention is in a letter to the editor in 2005 in the magazine *Noseweek* (the South African 'Private Eye'), responding to an article supporting conservation, titled 'Frankenflora':

> Change is being forced on us globally and we need to embrace and digest some of the huge implications of modern discoveries and technological advances. We [have] to deal with a world that is verging on the Anthropocene … preservationists are tied too tightly to the now-obsolete concept of protected area conservation … too few people able to recognise the harm we are doing to the planet and who are aware of the impending disaster!

The response of the editor was rather cynical: 'Serious as the issues you raise no doubt are, humour is as essential for the survival of the species – even if it is only padkos [Afrikaans for snacks] for the journey to Doomsday. Try some. – Ed'.

A much more optimistic view of technology, an indicator of the emerging 'good' Anthropocene narrative, appears in the multi-platform *City Press* in 2011, 'The age of man need not be a disaster. Scientists … have begun calling the current geological epoch the Anthropocene … but that does not mean we inhabit an ecological hell … We can accept the reality of humanity's reshaping of the environment without giving up in despair … The Anthropocene does not represent the failure of environmentalism. It is the stage on which a new, more positive and forward-looking environmentalism can be built'. Research on how the city council in Cape Town is responding to water supply crises is of particular interest here. Simpson et al. (2019) argue that the mentality of the Anthropocene has become the city's 'New Normal'.[8]

The Afrikaans mass media engagement with the Anthropocene (*Antroposeen*) offers some intriguing insight into this subculture within the broader South African context. There were six mentions of the Anthropocene from three publications (all merged into a single website), with none from a fourth. The first is a review of a collection by a popular Afrikaans poet in the multi-platform *Rapport/ Netwerk24* (2014): 'T.T. Cloete's Latest Collection is a Fascinating Study of the Relationship Between God, Human and Nature'.The poet outlines a dystopian vision of the current socio-economic system in an attempt to reconcile science and religion, using nature as a lens through which to experience humanity's effect on the world: 'The section "Anthropocene" outlines an oblivious image of modern humans who have to sacrifice quality of life because of its rapid growth and its insatiable urge for consumption'. The daily *Beeld* (2016) offers: 'Progress Demands a Higher Toll From the Earth', focusing on the negative effects of human activity on the planet, and on the socio-economic inequalities between people across the world, ranging from numbers of people who are malnourished to the rapid rise of depression and related illnesses through the developed world over the past few decades. The last line asks, 'Is the price we paid for "progress" perhaps too high?' One final note of interest in the Afrikaans media is that, in 2016, the word of the year was '*epogwisseling*' (epoch change). This nod to the widespread use of the term indicates significant awareness among this cultural subgroup of the shift from one era to another, and therefore (by extension) to the Anthropocene. As can be seen in Figure 3.1, *Antroposeen* was another popular word that year, indicating how it jostles for attention with many other items, of potential media interest.

Rapport/ Netwerk24 (2016) explains: 'The dictionary of the Afrikaans language has chosen 'epoch change' as its word of the year. The word, which refers to the transition from one period to the next (time-shift), counts among the words that the HAT compiled which, according to them, embodies the spirit of the previous 12 months. Other words include Anthropocene – an epoch made by man'.[9]

Central Africa

Our searches of media in Central Africa covered 18 countries. Of those, Ethiopia, Gambia, Guinea, Kenya, Nigeria, Rwanda, Tanzania, and Uganda produced results on the Anthropocene, while Benin, Burkina Faso, Burundi, Cameroon, Central African Republic, Congo (Brazzaville), Congo (DRC), Ghana, Ivory Coast, and

Figure 3.1 'What do you think of HAT's word of the year?', Antroposeen in Afrikaans (2016).

Zambia did not. In Ethiopia, most results presented Anthropocene information neutrally or optimistically, with emphasis on technological solutions. However, despite a small sample size of seven instances of Anthropocene, three have elements that incorporate a discourse of fundamental socio-economic change (or at least a challenge to the status quo), 'Resistance and Alternatives'(2017) in the business weekly *Capital Ethiopia*, described as a supporter of 'free enterprise', connects modern industrialized society to an incompatibility with life itself:

> 'What should be resisted are all the various derivatives of the prevailing modern world system, as they are diametrically opposed to the above life- sustaining principles ... This view is certainly outdated in the current geological era of Anthropocene ... Consequences are already suffocating humanity ... Since modern societies are geared only to the satisfaction of the lunatic (non-sustainable) dogma of non-stop economic growth (which is imposed on all humanity by dominant powers and their institutions, at the service of transnational capital), salvation is only possible by resolutely resisting the prevailing very destructive paradigm!'

Also fulfilling its responsibility to inform the public, *Addis Fortune,* the largest English-language paper in Ethiopia, prints 'Waste not, want not' on World Environment Day 2015:

> One of the keys to productivity and decoupling environmental damage from GDP is to make prices tell the environmental truth. Again, the energy sector

> shows how important this can be. The International Monetary Fund (IMF) estimates that the total cost of public subsidy to fossil fuels amounts to more than five trillion dollars a year when direct and indirect subsidies are counted. Getting price signals right, educating consumers and making policies that foster a green economy are not only desirable, they are essential. How well we succeed will determine whether the 'Anthropocene' is an age when more than nine billion people have access to food, energy, and security, without compromising the vital life-support systems of our planet.

Gambia's single result, in *The Standard* (2017), makes a strong link between African's post-colonial period and the historical inequities that remain. This article, 'Why Africa is the Worst Victim of Climate Change' had over 300 views, suggesting a relatively large constituency for such topics, and goes on to explain at length that the complexities of provincializing the Anthropocene in former colonial territories is no simple matter. Although there is a nod to technological solutions, the clear implication is the need for radical change:

> 'Africa is still in a crucial and fragile state of transition since the colonialists physically 'left ... '.This shift in paradigm from indigenous African conception to 'Western scientific' realities of life is good ...What is bad, however, is the evil eye of the same western life-style of industrialization that continues to rob even more heavily the indigenous Africans of their typical harmony with nature. The very genesis of climate change. Droughts, food insecurity, hunger, poverty, diseases, global warming, force migration, and gross inequalities here and there are different facets of the same problem – climate change ... For the Third World, the best way forward is to adopt [the] principles of modernity on the one hand and, on the other, to pursue strategies of decolonization and of reducing dependency. It is a dual effort – to modernize and decolonize at the same time. In a sense, it is an effort to decolonize modernity.

In Kenya, over half of the results acknowledge risk, with optimism about the roles of technology, education, and/or religion in mitigating negative effects. For example, from the Kenyan daily *The Star*, 'Why Changing Eating Patterns is a Choice We Must Make' (2015) and 'Five New Principles for Post-2015 Growth,' (2013) both recommend technological and educational changes that could mitigate the effects of the Anthropocene without moving away from the current capitalist 'growth' mindset. Another article in *The Star* (2016), 'Town Can Drive Prosperity', introduces the idea that urbanization provides a significant opportunity in confronting issues facing large segments of Africa's population:

> Our epoch has been characterised as the Anthropocene. Our species, *Homo sapiens*, is capable of the same disruptive and devastating effect on our planet just like volcanoes, earthquakes, tornadoes, and tsunamis ... a fundamental characteristic of the Anthropocene is the age of urbanism. I would like to think of our kind today as *Homo sapiens urbanus* ... I dare say that Africa is

> on the warm threshold of a youthful urban moment. This changes everything … Urbanization is perhaps the most powerful game changer in our lifetime. It is full of promise but peppered with peril.[10]

Kenya's *Daily Nation*, in a book review titled 'The Big Lie About Conservation of Wildlife in Africa Over the Years' (2017), reignites the debate on colonialism and the provincializing of the Anthropocene. Although the mention of the Anthropocene in passing is accurate, it is somewhat ambivalent. The review focuses on historical and current intra- and international power struggles around conservation, race, and colonialism. The reviewer notes 'Mbaria and Ogada's [the authors] work is one of the most popular in Kenyan bookshops', and, although not uncritical, praises the book for its approach, which is clearly hostile to Western influence:

> The greatest strength of the book is its energetic polemic against mainstream conservationists. They are portrayed as a club of dishonest white people with a racist agenda against the interests of local communities. Their driving motivation, according to the book, is nothing but greed for donor funding … The book is also written in simple language. It avoids academic jargon that has become the bane of environmental studies in what has come to be called "the anthropocene" (simply, the current geological age, viewed as the period during which human activity has been the dominant influence on climate and the environment).[11]

Two other countries in Central Africa stand out in contrast to one another. Nigeria, with a population of almost 200 million people, only returned four results. Rwanda, with a population of 12 million, returned 14 results. Nigeria's results leaned toward the optimistic, technocratic, whereas Rwanda's were mainly descriptive, simply defining the Anthropocene. For example, in *Igihe* (2016): 'A new geological epoch, the Anthropocene, has begun, scientists say: We're living through one of the most extraordinary events in Earth's history — the start of a new geological epoch … the Anthropocene'.

North Africa

Media searches in North Africa covered Algeria, Libya, Mali, Morocco, Niger, Senegal, Sudan, and Tunisia.[12] We found 15 mentions of the Anthropocene from 60 sources searched, 39 of which returned no results. Algeria, Niger, and Senegal had none. Mali had one item, in *Mali Jet* (2016), an informative, if rather neutral, piece on 'Climate: The Next Ice Age Delayed Due to Warming'. Morocco's sole contribution, in *Liberation* (2016) was the passing mention of an exhibition on 'l'Atlas de l'Anthropocène'. Media in Libya mentioned the Anthropocene three times, one each in three publications, with eight more returning no results. The first, from *Libya TV* (2016), on a much publicized WWF report 'Number of Wildlife Species in Decline Due to Human Activity, About 60% Since 1970' was a neutral account of the facts, as were *Akhbar Libya* (2017) on 'Children in Gaza Tell of the Anthropocene Age at Science Festival' and *Ayn Libya* (2017) on research telling us 'Over 200 Human-Caused Minerals Supporting Evidence of Dawn of Anthropocene Age'.

Searches of 20 publications in Tunisia, three-quarters of which returned no results, brought up four articles, two about exhibitions. The Arabic-language paper *Alhayat* (2016), with 'Architecture + Other Things', reports on an installation designed to explore space and aesthetics in the Anthropocene era at the Maraya Art Centre (Sharjah, UAE), while, in the Francophone *Le Quotidien* (2015), presented an article on an art exhibition in Strasbourg (Thème de l'Anthropocène). *Tunisia TV* (2017) ran the same article on minerals that we saw in Libya above. In 2013, *Zoom Tunisia* reproduced an article by Jeffrey Sachs, former director of the Earth Institute at Columbia University, under the title: 'What Does the Future Hold for the Global Economy?' Sachs, who has become something of an ambassador of global climate change, and latterly, of the Anthropocene, projects the perfect 'good' Anthropocene narrative, combining well-informed explanations of the hazards with the optimistic belief that humanity can cope.[13] His articles are widely reproduced all over the world and turn up in most of our regions. In Sudan, the Arabic-language *Al Rakoba* has the distinction of being the only source we found in Africa to review Kolbert's book, *The Sixth Extinction*: 'Is Humanity Behind Biggest Extinction Event Since Dinosaurs?' (2014). Other African media do discuss anthropogenic extinction, but rarely entertain the prospect of human extinction.[14] With a sting in the tail, *Sudan Vision* in 2017 prints: 'Focus: Are They Reconstructing or Rather Deconstructing?' Alfatih Ziada (a Sudanese journalist reporting from Washington DC) outlines the increasing pessimism of scientists as the full effects of the Anthropocene come into focus and humanity's position becomes more precarious:

> What I'd learned by 2016 was not encouraging. In every previous cycle, the Earth had regulated itself. Then, we humans came along and started fiddling with the global thermostat. The era of climate change that began in the nineteenth century with our concerted use of fossil fuels would prove unprecedented. Scientists began to speak of our 11,700-year epoch, the Holocene, as the Anthropocene, the first period in which the actions of a particular species, our very own anthropos, changed the planet. (I used to half-jokingly call our era the Anthro-obscene.) As time itself began to telescope, many of us, in the US in particular, simply closed our eyes and pretended that species death was not staring humanity (and many other species) in the face. Geologic time would, of course, go marching on, just not for us. …

Special focus: Liberia and Zimbabwe

Separated by more than eight thousand kilometres, Zimbabwe and Liberia present intriguing cases in our Anthropocene media research. Despite their relatively small populations, sixteen and four million, respectively, the Anthropocene is relatively highly represented. In Liberia, two publications produced 24 results, with another five publications having zero results. Out of those, more than half fall under what might be labelled an informative/educational narrative, with the rest under the 'acknowledging the issue but optimist outlook' narrative or a combination of neutral and optimistic, technocratic narratives. Overwhelmingly future-oriented and positive in outlook, even the warnings about global upheaval are tinged with optimism and a sense that the solutions are just over the horizon. Many of these articles were published in the Commentary section of *The New Dawn*, a daily newspaper. Although it is impossible to document accurately, it is likely that at least some of these articles or headlines appeared on *The Daily Talk*, an English-language news medium published on a blackboard on Tubman Boulevard in the centre of the Liberian capital Monrovia. This may be the most widely read news source in the city as many Monrovians lack the money or the electricity necessary for access to the conventional mass media, creating a rather unique media ecology.

The Anthropocene as a concept is clearly identified and defined in Liberian mass media. This article in *The New Dawn* 'The Climate Pope' (2015), one of several similar items all around the world, outlines the Pope's response to the Anthropocene, explaining the science in plain language and mentioning Crutzen and Stoermer:

> Firmly rooted in science, the teaching document – the most significant from the Vatican in over a decade – recognizes the need for urgent action, as the world confronts potentially catastrophic climate change. In 2000, the scientists Paul Crutzen and Eugene Stoermer proposed that human activity, particularly in the developed world, was interfering at the planetary scale, with the fundamental forces of nature … which they called the Anthropocene … scientific evidence has reinforced the conclusion that human activity is fundamentally transforming the planet. The Vatican has already recognized this view explicitly, with the Pontifical Academy of Sciences referring to the Anthropocene in the proceedings of a meeting held in May 2014.

More details on the science of the Anthropocene are given in another item in *The New Dawn* (2012), entitled 'Our Summer of Climate Truth':

> Scientists have given a name to our era, the Anthropocene, a term built on ancient Greek roots to mean "the human-dominated epoch" – a new period of earth's history in which humanity has become the cause of global-scale environmental change. Humanity affects not only the earth's climate, but also ocean chemistry, the land and marine habitats of millions of species, the quality of air and water, and the cycles of water, nitrogen, phosphorus, and other essential components that underpin life on the planet.

Commentary from the same newspaper, 'Writing the Future' (2013), adds detail on the globalization of the specific challenges presented by the Anthropocene, with optimistic reference to the UN task force response and the contentious issue of developing versus developed countries:

> What does the future hold for the global economy? Will living standards rise worldwide, as today's poor countries leapfrog technologies to catch up with richer countries? Scientists have even given a name to our era, the Anthropocene, in which humanity ("anthropos" in Greek) is having a large-scale impact on the planet's ecosystems … The rapidly growing developing countries cannot simply follow the economic growth path that today's rich countries travelled … The greatest talents in our societies … are joining the UN's new Sustainable Development Solutions Network.

The tone remains optimistic. Yet another article from 2013 in *The New Dawn*, entitled 'Redefining Sustainable Development', raises the highly contentious question of over-population, an issue that is often reported but rarely analyzed in great depth. This article (probably syndicated) evokes Albert Einstein's problem-solving methodology and combines elements of informative, neutral definition of the Anthropocene with optimistic acknowledgement of risks:

> 'Albert Einstein once said that if he had just one hour to find a solution on which his life depended, he would spend the first 55 minutes defining the problem. Once he knew the right question to ask, he could solve the problem in less than five minutes. … Today, humanity faces such a life-threatening problem: How are we to provide adequate nutrition and a decent quality of life to a global population that is set to surpass nine billion by 2050, without irreparably damaging our planetary life-support system? To find a solution, we must start by clarifying the problem. Humans have fundamentally altered Earth's ecosystems … scientists now believe that the planet has entered a new geological epoch, dubbed the Anthropocene.

The media in Zimbabwe (13 results from nine sources, with an additional two sources yielding no results) presents a rather uniform picture, with most items describing the Anthropocene in neutral terms, and/or as incurring risks that human ingenuity could avoid. As early as 2006, there was an acknowledgement of the Anthropocene in *The Herald*, with articles adapted from other sources (mainly from Europe and the United States). For example, 'Zimbabwe: Changing Climate, Changing Language' directly engages with the new term: 'According to the Nobel Prize-winning Dutch chemist Paul J. Crutzen, we are now living in the anthropocene', and from *The Guardian* in the *Harare News* (2016), 'Welcome to the Anthropocene, the Human-Made Earth Epoch': 'The impact of humanity on the Earth is now so profound that a new geological epoch the Anthropocene – needs to be declared, according to an official expert group'. *The Herald* (in 2017) reprints 'Native Intelligence: Space, Race and Us' from *New African Magazine* by the distinguished Ugandan journalist, Kalundi Serumaga. This imaginative essay stands out for its direct engagement with Africa's post-colonial future and the consequences of continued Western hegemony. The premise – an exclusionary space exploration agenda – is focused on the possible futures of space travel and contact with other sentient species. It identifies issues of power, resource control, external exploitation, governance, and environmental destruction as pointed reminders of the stakes as the evidence for and consequences of the potentially catastrophic consequences of the Anthropocene become ever clearer. This is one of the most powerful expressions of 21st century anti-colonial sentiments, and on why all the white and richer countries are spending billions on space programmes:

> I've said it before, and I'll say it again; one day, the black and brown peoples of the world may wake up to find that the white race has abandoned this damaged planet and flown off to distant new ones. ... Planet Earth is in a dire state. For the first time in its entire history, critical changes to its (deteriorating) geological state are now being driven primarily by human activity. Science has named this the Anthropocene era.
>
> In short, there is the possibility of a long-distance, hi-tech version of the genocidal 'Wild West' culture that European settlers brought to the Americas.... if say, a Chinese space mission finds a habitable planet and settles on it, does it become Chinese territory, or does it belong to Planet Earth as a whole? ... And no discriminatory visas.[15]

Tanzania, multi-lingual Africa

Tanzania is a country of 56 million people in the African Great Lakes region. The national language of Tanzania is Swahili, although English is also widely spoken in some areas. In our research, we found no word for 'Anthropocene' in Swahili mass media and only one example in English. However, the two main Swahili dailies in Tanzania returned more than 200 results for 'climate change' and another 25 for 'global warming'.

The single reference to the Anthropocene is a 2016 *IPP Media* editorial, reprinted from *The Guardian,* entitled 'Pursuit of Social Justice Must Promote Development and Human Dignity'. The piece argues that old socio-economic models cannot continue indefinitely, and discusses links between system change and sustainability, calling for radical change but offering no concrete proposals:

> We now live in the Anthropocene era in a full world where humans are dramatically altering their ecological life-support systems. Our traditional economic concepts and models were developed in an empty world ... [of] sustainable prosperity ... we are going to need a new vision of the economy and its relationship to the rest of the world that's better adapted to the new conditions we face ... [we need] economics that respects planetary boundaries ... recognizes that the ultimate goal is real, sustainable human well-being, not merely growth of material consumption ... the economy can't grow forever on this finite planet.[16]

The Swahili media, as noted above, covered climate change extensively, often in ways that implied a recognition of the Anthropocene but never actually used the term, climate change, as a metonym for the Anthropocene (substituting the part for the whole). For example, in an article from 2017 in *Habarileo*, 'Poverty Increases Deforestation', highlighting climate change as a major driver of increasing food insecurity, poverty, and deforestation. Also notable is the positive account of the role of religious institutions and leaders in education and development work around climate change, as in *Habarileo* (2017): 'Rehabilitation Programs to Cope with Drought and Poverty'. Tensions between countries responsible for the highest greenhouse gas emissions and those with the lowest (mostly the West versus the Rest) are also highlighted in the Swahili media. In the words of *Mwananchi* (2012), this dispute makes 'the fight against climate change … increasingly difficult' and *Habarileo* (2017) states bluntly: 'Climate change is more colonialism for Africa', raising questions about 'Who is responsible for climate change?'[17] This is the most important aspect of the provincializing of the Anthropocene in Africa.

We can only speculate why the term 'Anthropocene' does not make an appearance in Swahili mass media. The focus of the media on climate change in Tanzania highlights concrete ideas of adaptation and mitigation and that might go some way to explaining the reluctance to engage directly with the Anthropocene, another rather abstract European/Western concept (though media in other parts of Africa do).The presence of the word 'Anthropocene' in mass media often indicates links among the scientific, educational, legal, and policy establishments of various countries. Although this can sometimes be a positive force, that leads to collaborative conservation efforts, all too often it results in contentious disputes over resources and blame. An example is the cross-border dispute between Kenya and Uganda outlined in an article in *The Conversation* by Christopher Rossi (University of Iowa), 'What Migingo, the World's Tiniest Disputed Island, Tells Us About International Law' (2017), illustrating how the Anthropocene's effects can spill over into international conflict:

> A series of aggressive encounters [between Kenya and Uganda] … affected the commercial interests of … Rwanda, Burundi and eastern Democratic Republic of Congo … Migingo intertwines issues of ethnicity, nationality, and politics around the competing temptations of a resource … [and] directly affects the livelihood of 30 million people … The small geographic size of the Migingo dispute belies the grave political consequences of inaction, making Migingo a metaphor for African resolve in the Anthropocene.[18]

Pan-African media

Given the difficulties of extracting an overall African narrative from media coverage of the Anthropocene, it is clear that pan-African media play a significant role in our data collection and analysis. There were over 100 mentions of the Anthropocene from six pan-African sources, with four additional sources returning no results. Almost half of the results were simply informative, defining the

Anthropocene in neutral terms and placing it in a scientific context. For example, *Africa Science News Service* provides this definition in 2015, referencing Crutzen and tracing the industralization trajectory:

> Nobel laureate Paul Crutzen characterized the current epoch as the Anthropocene. Our kind possesses the power to transform our planet on a par with volcanoes, earthquakes and glaciers. Our unprecedented success and capacity to dominate the planet, relative to other creatures, has been made possible through centuries of clever manipulation of knowledge, energy, and technology. We harnessed fossil fuels to power the industrial revolution, the agrarian revolution and, more recently, the information revolution.

Over half of the results acknowledged risks but found positive outcomes from the use of technology and/or education, approaching the 'good' Anthropocene narrative. This is encapsulated in two articles in *The Conversation.* The first (in 2017) on the 'Seeds of Good Anthropocenes' by Laura Pereira, under the title: 'Incubating Ideas on How Southern Africa Can Manage the Anthropocene':

> The Anthropocene is often associated with problems such as climate change and inequality. But there is also hope that it can come with positive change for the benefit of people. Seeds of Good Anthropocenes. … To explore this idea in practice, we convened a diverse set of people from across southern Africa. They included artists, social entrepreneurs, researchers, and policy-makers who were taken through a three-day visioning process, that was more science fiction than science. … participants will also be using the scenarios in their own work in the region and to run similar scenario processes with other groups. We will track the uptake of these scenarios to see how influential they will be in helping southern Africa to move towards a more positive future.[19]

'Climate apocalypse' in *Conversation Africa* (2016), by John Stremlau, takes the optimistic narrative one step further, arguing that Africa is 'the perfect testing ground for adapting to the Anthropocene':

> Amartya Sen … argues that environmental catastrophes are the joint products of natural and human systems. Africa could provide valuable lessons to others on how to adapt to the new realities of the Anthropocene, particularly in how to strengthen democracy. This is because it has fragile states and a history of surviving and overcoming natural and man-made deprivations. These may generate new democratic ingredients, innovatively mixed to suit its highly diverse regional population, that prove vital for sustaining democracy.[20]

The third narrative (pessimistic and urging concrete radical change) offers a bleak contrast to the discourse that 'all will be well' if only humanity works hard to adapt and mitigate negative effects of the Anthropocene. This view is comprehensively and passionately outlined by Rob Seimetz in his article, 'What Does it Mean to Live in the Age of the Anthropocene?' reproduced from the US-based leftist magazine *Counterpunch* in *Topix Africa* (2017):

> It's another day in the age of the Anthropocene, where a global game of musical chairs continues to play out. ... The only living organism that gets to see the end of the game is Mother Earth, and it will squeeze humans and most other living beings on this planet out of existence... The Age of Anthropocene will be synonymous with words like 'unprecedented', 'extinction', 'migration', 'collapse', 'ignorance', and 'greed', to name a few. It's an age where global systems implemented by humans will have global impacts on all life on Earth. ... In this age, humans will have to confront the sickness of our species. ... In the Anthropocene, humans will have to deal with the consequences of their history of violence ... Instead of wanting something to happen, we must shift our focus to understanding what is happening.

Seimetz's article does more than predict a dire future for humanity – it outlines the environmental, social, and geopolitical fallout from unstoppable forces. The second part of this article offers, if not a solution, then a path forward that incorporates the facts of the Anthropocene while extending a new way of thinking about what it means to be humans, living together on a planet with finite resources:

> When we redefine "hope" in the Anthropocene it must not include having solutions to our problems. It must include being able to live with problems we are facing that have no solutions. And that, despite the problems our species is up against, we must find a better way to live so we can have brighter days. In this redefined "hope" we will know that everything isn't going to be alright, and that we are a flawed species. Despite knowing these things, there will be a shift in thought and consciousness. This shift will be celebrating the cultivation of life and not the accumulation of possessions. It will be a shift of focusing on how hard you live and who you share your life with. This is what it will mean to live in the age of the Anthropocene.

The writer who is most often associated with the dangers of the Anthropocene and the need for radical change from a philosophical perspective is the Australian ethicist Clive Hamilton, author of several popular books on the topic. He appears six times in coverage of the Anthropocene in African media. The excerpt from this piece in *Conversation Africa* (2016) first appeared in *Nature*, and the italics go some way to explaining the impact of 'good' Anthropocene narratives.

> The idea of the Anthropocene was conceived by Earth System scientists to capture the very recent rupture in Earth history arising from the impact of human activity on the Earth System as a whole... Much of the literature on the Anthropocene ... is bedevilled by readings of the new concept through

old disciplinary lenses, readings that fail to understand the revolutionary implications of humankind taking the Earth into a new geological epoch … *In this way, the Anthropocene no longer represents a rupture in Earth history but a continuation of the kind of impact humans have had for millennia. It is thereby rendered benign* (italics added). That so many scientists, often publishing in prestigious journals, can misconstrue the basic definition of the Anthropocene as nothing more than a measure of the human footprint on the landscape is a sign of how far Earth System science has to go in changing the way many scientists still think about the Earth.

Conclusion

In his comprehensive proposal for an eco-centred approach to teaching and learning in Africa, the philosopher Bert Oliver (2018: 14) declares: 'Those of us who live in Africa are better positioned than the rest of the consumption-driven world to drive such a project of eco-awareness and conscientization, with the emphasis on critically informed action'. Nevertheless, our results suggest that there is no overall prevailing 'African Anthropocene' narrative. Given the linguistic, cultural, historical and geopolitical diversity of a continent that spans more than thirty million square kilometres and almost 20 percent of the world's population, it is unsurprising to find pockets of dense Anthropocene coverage surrounded by swaths of nil results, and hundreds of articles on climate change that might easily have mentioned the Anthropocene. Nonetheless, we can draw at least two broad conclusions, that may also be relevant elsewhere, from the rich material that media coverage of the Anthropocene presents in Africa.

First, the absence of the term does not exclude ideas of climate change adaptation and mitigation from political and media discourse. As one example of this, Tanzania's civic engagement with environmental degradation to combat decreasing yields in subsistence agriculture, using religious and educational institutions, is a clear example of science and community engagement to improve outcomes. Second, the results from Africa suggest that there may be a 'hierarchy of themes' over differing media ecologies, that emerge as the concept of the Anthropocene becomes more established in mass media discourse. This common-sense notion derives from the idea that definition precedes identification of associated problems, which leads to proposed solutions. The more radical solutions that appear in the media, albeit infrequently, rise out of a shared narrative of understanding and a frustration with the perceived inadequacy of the routes through education, technology, and incremental social change to deal with the potential risks highlighted by the Anthropocene.

Notes

1 Sarah Morris also contributed to this chapter.

2 *Conversation Africa* is an online magazine supported by many universities in Africa. It is the African wing of *The Conversation*, an international publishing organization.

3 See Adedeji (2015) on media in Africa in general; Wasserman (2010) on South Africa; Lugalambi and Nyabuga (2011) and Lublinski et al. (2014) on science journalism in Africa and Asia.
4 On environmental movements in South Africa (and probably relevant for many other countries), see Cock (2004), Brockington and Scholfield (2010), Death (2014), Fakier & Cock (2018), and Leonard (2018).
5 Turton (2013) discusses 'water wars'.The topic appears regularly in local media (Simpson et al. 2019, note 1). However, it is almost entirely absent from media reporting on the Anthropocene in Africa. Austin (2017: chaps. 4 and 5) provides a useful general summary of Africa in the Anthropocene.
6 The island is heavily dependent on the tourist trade, so the media ecology is probably sensitive to visitors as well as locals.
7 On Fridays, the paper publishes a section in the local dialect, Oshiwambo; it is not clear if these articles were available.
8 While we are not suggesting a connection, it is interesting to note that President Xi in China frequently expands on the term 'New Normal' in the context of ecology and development in his book of speeches (Xi 2017). The term has become common in the global Coronavirus pandemic
9 The online *Daily Maverick* published little on the Anthropocene until 2018–19, but this item from 2019 is certainly worth reading for its insights into journalistic dilemmas: https://www.dailymaverick.co.za/article/2019-07-30-right-of-reply-a-democratic-eco-socialist-calls-out-self-proclaimed-libertarian-ivo-vegter/.
10 There is a rapidly growing literature on the connections between urbanization and the Anthropocene, for example, Pincetl (2017) and Tyszczuk (2018).
11 Conservation is a very contentious topic, not only in African media, but in all regions. Sandler (2017) explores the issue of 'de-extinction'.
12 See also Chapter 10 of this book on media ecologies in the MENA region (Middle East and North Africa).
13 Sachs' Reith Lectures in 2007 (the second was on 'Survival in the Anthropocene') can be found at http://downloads.bbc.co.uk/rmhttp/radio4/transcripts/20070418_reith.pdf
14 For a technical survey of the economic consequences of species extinction see Sol (2019), a remarkable study of 557 regions.
15 This is one of the very few references in the media to what Baldwin and Erickson (2020) identify and document as the 'racial Anthropocene'.
16 See Fullbrook (2019) for a dismal account of how ecological issues are covered in economics textbooks.
17 For the classic statement of this critique, see Agarwal and Narain (2012). It applies to the Anthropocene as much as to climate change, as discussed in Chapter 1 of this book.
18 This is one of the few media discussions of what the Anthropocene might mean for legal systems within countries and internationally. Scholars often label this 'climate change litigation' (see Benjamin 2017, Rose et al. 2019, Setzer and Byrnes 2019), the first two specifically referencing the Anthropocene. See also Zalasiewicz et al. (2019: section 1.4.3.1).
19 Compare this well-meant, sometimes inspiring, optimistic programme with the reality check provided by Hecht (2018a), a searing indictment of the contribution of gold and uranium mining to the Anthropocene in Africa, and Hecht (2018b) on the paradoxical idea of an 'African' Anthropocene.
20 There seem to be several versions of this article. The one cited was accessed in 2017.

References

Adedeji, A.O. (2015) Analysis of use of English and indigenous languages by the press in selected African countries. *Arabian Journal of Business and Management Review (OMAN Chapter)* 4(8): 1–11.

Agarwal, A. & S. Narain (2012) Global warming in an unequal world: A case of environmental colonialism (selected excerpts). In Dubash, K.N. ed. *Handbook of Climate Change and India*. London: Earthscan, chap. 5.

Austin, G. ed. (2017) *Economic Development and Environmental History in the Anthropocene: Perspectives on Asia and Africa*. London: Bloomsbury.

Baldwin, A. & B. Erickson (2020) Introduction: Whiteness, coloniality, and the Anthropocene. *EPD: Society and Space* 38(1): 3–11.

Benjamin, L. (2017) The duty of due consideration in the Anthropocene: Climate risk and English directorial duties. *Carbon Climate Law Review* 2: 90–99.

Brockington, D. & K. Scholfield (2010) The work of conservation organisations in sub-saharan Africa. *The Journal of Modern African Studies* 48(1): 1–33.

Cock, J. (2004) *Connecting the Red, Brown and Green: The Environmental Justice Movement in South Africa*. School of Development Studies, University of KwaZulu-Natal.

Death, C. (2014) Environmental movements, climate change and consumption in South Africa. *Journal of Southern African Studies* 40(6): 1251–1269.

Fakier, K. & J. Cock (2018) Eco-feminist organizing in South Africa: Reflections on the feminist table. *Capitalism Nature Socialism* 29(1): 40–57.

Fullbrook, E. (2019) Economics 101: Dog barking, overgrazing and ecological collapse. *Real-World Economics Review* 87: 33–35.

Hecht, G. (2018a) The African Anthropocene. https://aeon.co/essays/if-we-talk-about-hurting-our-planet-who-exactly-is-the-we

Hecht, G. (2018b) Interscalar vehicles for an African Anthropocene: On waste, temporality, and violence. *Cultural Anthropology* 33(1): 109–141.

Leonard, L. (2018) Bridging social and environmental risks: The potential for an emerging environmental justice framework in South Africa. *Journal of Contemporary African Studies* 36(1): 23–38.

Lublinski, J. et al. (2014) Advances in African and Arab science journalism: Capacity building and new newsroom structures through digital peer-to-peer support. *Ecquid Novi: African Journalism Studies* 35(2): 4–22.

Lugalambi, G. & G. Nyabuga (2011) *Media Coverage of Science and Technology in Africa*. Makerere University/UNESCO.

Oliver, B. (2018) An eco-centred alternative to 'Business as usual' in teaching and learning in (South) Africa. *African Perspectives of Research in Teaching and Learning* 2(1): 6–29.

Pincetl, S. (2017) Cities in the age of the Anthropocene: Climate change agents and the potential for mitigation. *Anthropocene* 20: 74–82.

Rose, J. et al. (2019) Primal scene to Anthropocene: Narrative and myth in international environmental law. *Netherlands International Law Review*. https://doi.org/10.1007/s40802-019-00151-5

Sandler, R. (2017) De-extinction and conservation genetics in the Anthropocene, recreating the wild: De-extinction, technology, and the ethics of conservation. *Hastings Center Report* 47(4): S43–S47.

Setzer, J. & R. Byrnes (2019) *Global Trends in Climate Change Litigation*. London: Grantham Institute, London School of Economics.

Simpson, N.P. et al. (2019) When Anthropocene shocks contest conventional mentalities: A case study from Cape Town. *Climate and Development* 12(2): 163–169.

Sol, J. (2019) Economics in the anthropocene: Species extinction or steady state economics. *Ecological Economics* 165. https://doi.org/10.1016/j.ecolecon.2019.106392

Turton, A.R. (2013) Water wars in the Anthropocene: A South African perspective. *Global Dialogue* 15(2): 81–90.

Tyszczuk, R. (2018) *Provisional Cities: Cautionary Tales for the Anthropocene*. Abingdon: Routledge.

Wasserman, H. (2010) *Tabloid Journalism in South Africa*. Bloomington: Indiana University Press.

Xi, Jinping, (2017) *The Governance of China*, 2 volumes. Beijing: Foreign Languages Press.

Zalasiewicz, J. et al. eds. (2019) *The Anthropocene as a Geological Time Unit: A Guide to the Scientific Evidence and Current Debate*. Cambridge: Cambridge University Press.

4 The Anthropocene in North American media

The pursuit of the 'good' Anthropocene

Leslie Sklair, Chad Steacy, Jonathan DeVore, and Ron Wagler[1]

Searches of the media in North America delivered over 1,000 items from more than 350 sources, about two-thirds of which had zero results, meaning that coverage was concentrated in a relatively small number of newspapers, magazines, and diverse other media. Coverage was overwhelmingly from sources in the United States (the highest number of items from any country in the project). Media in Canada (English and French language) provided almost one hundred items.

Francophone Canada

Research on climate change in the Canadian media started with Young and Dugas (2011), and Corrigal-Brown (2016) provides an interesting analysis of topics that don't get covered. Media coverage of the Anthropocene in Francophone Canada is dominated by *Le Devoir*, in terms of its quality and the influence it has on public debates inside the Province of Québec. With around 20 articles on the Anthropocene, mostly between 2015 and 2017, it followed the global trend. The first, a profile of the radical environmental journalist Hervé Kempf (of *Le Monde*), 'To Save the Planet, We Must Get Out of Capitalism' (2009) is a long and well-grounded argument connecting the ecological crisis and the capitalist global system (Kempf had previously predicted the crash of 2008). However, he is advocating the end of speculative, hyper-consumerist capitalism, not capitalism per se. The year 2012 introduces the Anthropocene as an era of 'disequilibrium'. In 2015, in addition to the usual reporting of the Earth entering the Anthropocene (and some cultural events), we find a substantial article in *Le Devoir* reporting that the organization Future Earth has set up its headquarters in Montréal. This is presented as a coup for the environmental movement, especially as the local academic, Dr Paul Shrivastava, is to play a leading role. A proponent of sustainable development, he explains the Anthropocene as an era in which the Earth is transformed not only by natural processes but also by human activity. 'The Quiet Suicide of Humanity' (2015) sets a different tone. Locating the Anthropocene as the Sixth Extinction, the literary critic Michel Lapierre concludes: 'Neither Adam Smith, the first theorist of liberal capitalism, nor Karl Marx, father of scientific communism, could have foreseen that the industrial revolution would have given rise to such a serious problem. To know the truth, it was necessary to answer the brilliant questions of curious minds, like Elizabeth Kolbert, followers of the experimental sciences, after having patiently listened to a great neglected voice: that of nature'. The work of another local academic, Elena Bennett, is enthusiastically reported in 'One Hundred Inspiring Projects For the Planet' (2016), giving examples of her 'Seeds of a Good Anthropocene' project, endorsed by Future Earth. *La Presse,* the most popular Francophone newspaper in Canada, provides about ten items. The first in 2010, 'Did the Extinction of Large Herbivores Cool the Planet?' is a short technical article, followed by another in 2012 on 'glaciation'. Geoengineering is introduced in 'Climate Manipulation: it is 'Urgent' to Supervise Research' (2013), citing Clive Hamilton, Bill Gates, and Paul Crutzen, but coming to no firm conclusions. 'Anthropocene: The Opinion of Experts' (2016) is a brief but informative summary of the *Sixth Extinction* sourced from two prominent magazines, *New Scientist* and *Science*, quoting a palaeobiologist from the University of Alberta. It is presented as a purely scientific question. 2016 also brings 'Environment: Science Helps Optimism' on the 'seeds' project of Professor Bennett (similar to the item in *Le Devoir* referred to above). In 2018,[2] *La Presse* published a controversial article, 'To Save the Planet, Do Not Have Children'. The journalist Marissa Groguhé reports conversations with half a dozen people who, out of conviction, no longer wish to give birth to children, most feeling it would be unfair to bring a child into a world in danger. Groguhé tells us: 'Anthropocene – The Human Era', a documentary that arrived two weeks earlier on Montréal screens, is fuelling the debate. A Facebook group, 'The Happy Collapso' (students of 'collapsology'), is reported to have 4,000 members. The article explores the options (no kids, adoption, reducing consumption). Whereas many items on the Anthropocene in global media reference the 'population problem', very few explore the implications for families. More typical is the free independent Vancouver weekly *The Georgia Straight* (2015): 'How seven billion humans are collectively warming the planet is an invisible but devastating example of what scientists increasingly agree is the beginning of the Anthropocene: a proposed epoch defined by *Homo sapiens* overtaking nature as the dominant force on Earth'. The popular tabloid, *Journal de Montréal*, has one article on the Anthropocene, the conventional 'Man Enters the Planet in a New Epoch, According to Scientists' and the free *Journal Metro* repeats 'Montréal Hosts the Seat of Future Earth' (2015), the story in *Le Devoir* above. The only magazine in French Canada we found with anything on the Anthropocene was *L'Actualité* (said to sell a million copies); the first item on the Anthropocene in 2015 also reported the move of Future Earth to Montréal, with the second, in 2017, 'Health and Science', reporting a common story of health risks in the Anthropocene. *Sélection du Reader's Digest* gave no results.

Anglophone Canada

Coverage of Anthropocene-related news in Anglophone media in Canada was more voluminous than that in the Francophone media. *The Toronto Globe and*

Mail is said to be Canada's most widely read newspaper for the elite, and its coverage of the Anthropocene is relatively extensive. Beginning in 2003, with an article labelled 'SOCIAL STUDIES', [sic] Paul Crutzen tells us that the Anthropocene Era can be said to have started in the latter part of the 18th century, as a product of the industrial revolution (credited to *The Boston Globe*, the article was updated and republished in 2017). Also in 2003 (and also republished in 2017, as are several articles in this paper), 'Can Enviro-Optimists Save the Movement From Itself', by the controversial conservative columnist Margaret Wente, is an attack on the renowned environmentalist Bill McKibben (also controversial) for his activism. Condemned by Wente as 'the firebrand leader of the crusade to kill the Keystone XL oil pipeline, [he] is losing the battle for hearts and minds', the article speaks of splits in the green movement between purists (pessimists) and pragmatists (optimists), the latter being personified by Peter Kareiva, at the time senior scientist for Nature Conservancy. Wente argues that techno-optimists, like Kareiva, are 'the key to save the movement from terminal irrelevance'. She concludes: 'Meantime, the planet indeed may be more resilient than we thought. So, on Earth Day, please do something to improve your corner of it. And cheer up – the Anthropocene Age might be better than you think'.[3] In 2015 'Scientists Call for New Era in Earth's History to Recognize Human Impact' repeats the usual story at length with a local reference, 'International Scientists, Including One Canadian [Professor Alex Wolfe, University of Alberta, co-author of a paper in *Science*, on which the article is based] Say that Humans Have Changed the Earth So Much That it's Time to Recognize We Have Created a New Era in the Planet's History'. However, a letter to the editor in 2017, '100 or Older?' dramatizes the story, arguing that 'largely because we are entering the Anthropocene, when Earth systems are being stressed to the point of collapse … chances are, life expectancy will be much reduced in the 21st century'. Another Canadian professor reinforces this sense of urgency, 'Mark Jaccard on Responsible Growth' (2014) identifies carbon pollution as the problem. Jaccard concludes: 'if you look farther afield to examples like California [he also cites Québec and British Columbia], with its regulations on fuels, vehicles, and electricity, you see there are things that can be done'.[4] 2014 brings 'The Sixth Extinction: The World Has Endured Five Mass Extinctions. Can We Survive a Next One of Our Own Making?', a historical analysis of Cuvier's seminal insight that extinction has always been a natural process, expertly summarizing Kolbert's *Sixth Extinction,* and making clear human responsibility for the current, accelerating rates of extinction among animals.[5] Whereas coverage in *The Toronto Globe and Mail* is informative and sometimes provocative, it rarely indicates the potential risks of the Anthropocene to humanity in detail.

The Toronto Star, a popular broadsheet daily, also offers a range of Anthropocene coverage, beginning with two eye-catching titles in 2010. The first, by Stephen Scharper from the Centre for Environment, University of Toronto, 'We All Lose in the War against Nature. We Could Be on the Verge of an Age of Mass Extinction'. Scharper explains: 'I remember as a child visiting New York's Bronx Zoo. After marvelling at the human-eating lions and fierce-toothed bears, I was confronted by a mirror. The sign above boldly stated, "You are looking at the most dangerous animal in the world"'. Citing an article in the journal *Environmental Science & Technology*, Scharper introduces the reflections of the late Canadian eco-theologian and Roman Catholic priest Thomas Berry. This theme is reinforced with 'New Epoch to Usher in Cataclysmic Extinction. Thousands of Species to Disappear as New Age Dawns, Scientists Say'. Zalasiewicz is quoted: 'the Anthropocene represents a new phase in the history of both humankind and of the Earth, when natural forces and human forces became intertwined, so that the fate of one determines the fate of the other'. 2011 brings 'A Deep Look at the Earth's Future' arguing the necessity to consider alternative energy sources 'Otherwise, we're going to have a massive crisis as the cheap petroleum runs out' (not exactly the point, one might argue). The article quotes palaeoecologist Curt Stager (author of *Deep Future*) ending on a reassuring note: 'Humans are just so resilient … there will be a lot of suffering … But I have no doubt that we will be here for the entire story of this long carbon legacy'. Differences between central government and provincial policies are exposed in 'Surviving the Anthropocene: Ottawa Plans Cuts to Climate Programs' (2014). 'A "Good" Anthropocene? Diane Ackerman's Pretty, False Future' by science and technology reporter Kate Allen, is a devastating review of the popular science book, *The Human Age*. Allen describes the book as: 'a lyrically wrought romp through some of the innovative solutions, adaptations, and modifications our species has created for our broken planet and our increasingly hot and uncomfortable place in it. We just have to get motivated. And there is absolutely no way to do that if all that we are confronted with is doom and gloom'. Allen highlights the critical question that Ackerman fails to answer, namely who are 'we'?, citing the work of Clive Hamilton and Rob Nixon, who ask the same question.

As we show at various points throughout this book, issues of blame, inequality, and Eurocentrism have become central for the ways in which social scientists and humanities scholars (and even some Earth scientists and a few journalists) approach the Anthropocene. *The Toronto Star* with 'Earth Pushed Beyond Four of Nine Planetary Boundaries, Scientists Warn' (reprinting an article in 2015 by Joel Achenbach in *The Washington Post*), is one of several reports in the global media of a potentially alarming paper by Will Steffen and 17 other scientists in *Science*. However, Earth System researcher Ray Pierrehumbert sounds a note of caution: 'The boundary is not like the edge of the cliff'. This is a very long, nuanced, technical discussion. The free tabloid *Metro* (part of *The Toronto Star* group) picked up the Anthropocene story early in 2008 with 'Our World is Changed: Scientists Say New Era in Earth's History Needs Recognition', repeating the view that the Anthropocene epoch began in 1945 with the first nuclear test explosion (originally published in *The Globe and Mail*). It was 2016 before *Metro* revisited the topic with a few more articles. For example, 'We've Officially Made Our Mark' by David Suzuki, described as 'one of the most visible ecologists in the Canadian public sphere'. Suzuki (a controversial figure in Canada) concludes 'So there you have it, the case for the Anthropocene. We've done it. We've written our name on the wall. We're the king of the hill, lord of the sandbox. We're now the most powerful force of change on the planet, so much that we actually get our own epoch. A pretty big responsibility for a naked ape that emerged on the plains of Africa only 150,000 years ago; and so what now, little human, what now?' *Metro* follows this with a story on a local author, 'The New Anthropocene Epoch has Dawned and It's Very Worrisome' (2016), citing, rather marginally, Ian Angus, 'Ontario-based editor of the online journal *Climate & Capitalism*' and writer of a new [eco-socialist] book on the Anthropocene (see Angus 2016a).

The conservative *National Post*[6] does better, with seven articles, the most informative of which is 'Human Activity Helping Push State of the Earth System into Unknown "Danger Zone"' (2015). This article, by Margaret Munro of Postmedia News (dubbed 'the Fox News of Canada'), reports a paper from *Science,* concluding that humans are the prime driver in the deterioration of the 'Earth System', thus the Anthropocene. The tone is gloomy rather than alarmist, focusing on technical fixes. 'Fallacy of Wilderness: There's No Landscape on Earth Untouched by Humans, Scientists Say' (2016) is reprinted from *The Washington Post.* The article challenges, at great length and with scientific authority, conservationist assumptions that nature was ever 'pristine' and raises questions around Anthropocene imaginaries. The reference here is a paper from *PNAS.* Melinda Zeder (co-author) is quoted: 'Today isn't a radical right turn, and that all of a sudden we're having these impacts, which is sometimes the attitude, but it is part of the progression of this continuing ramping-up of scope and impact that today is taken to ridiculous levels'. Also quoted is William Ruddiman, whose early Anthropocene hypothesis has attracted much attention: 'This emerging evidence confirms large early agricultural effects on landscapes and on greenhouse gas emissions'. The last word goes to archaeologist Jon Erlandson (another co-author of the *PNAS* paper): 'There's a story of ingenuity built into this deeper time perspective. Humans have over and over again met challenges in different parts of the world, and they've found ways to meet those challenges. Every generation has its challenges. But we have found ways to solve them ...No matter how bad we screw things up'. This well-crafted article could be seen as a perfect reassurance narrative, suggesting that the present and recent generations of humans have no specific responsibility for any current ecological emergency.

In *The Canadian Geographic*, 'Creative Cartography: David Thomas Smith's Anthropocene' (2013) discusses 'a sense of the scale on which the world operates; the power that mankind has at its finger tips, and then, hopefully, they begin to question how that power is used'. 2016 brings 'John Smol on the Anthropocene: The Lake Expert and Palaeolimnology (Study of Inland Waters) Pioneer Addresses the Details and Evidence of a New Epoch: 'There are still many out there who say "Climate change is a natural process", and it's time to give up on that. The climate change we are experiencing is not natural'. Reporting on the efforts of the AWG in Cape Town, 'Anthropocene: Geologists Urge Global Recognition of New, Human-Influenced Epoch' is the first of several Anthropocene stories in *Maclean's* magazine. 'Plan B for Global Warming. A Bold Scheme to Cool Earth Almost Overnight. But Are We Ready?' (2009) by the magazine's senior writer, Jonathon Gatehouse, is a long profile of David Keith,[7] whose 2004 paper, 'The Influence of Large-Scale Wind Power on Global Climate' signalled a transition from thinking about renewables to thinking about geoengineering. Gatehouse tells us: 'Verboten in mainstream forums, the topic was only debated at secret NASA and White House-organized confabs' but is now [2009] being taken more seriously, despite uncertainties surrounding it. Still extremely controversial, the urgency for action appears to be forcing scientists to see geoengineering as an option (see Chapter 2 of this book, note 8, and Pearce 2019). 'Good News, Bad News' (2014) reports the 'dubious distinction' of the Anthropocene, its hundreds of academic papers, and the debates around it. We also find two interesting Anthropocene arts stories in MacLeans's magazine, both in 2016. 'Postcards From the Edge of the Earth' announces a new photography exhibition addressing the Anthropocene. The curator is quoted: 'I don't want to be apocalyptic'. Previewing a widely acclaimed 2018 exhibition, 'The unjudging eye of Ed Burtynsky's camera' (see Chapter 13 of this book), *The Globe and Mail Report on Business* magazine prints two articles by David Suzuki: 'An Economy is a Means, Not an End' (2012), in which he uses the concept of the Anthropocene to argue for a systemic change in our economies, criticizing the obsession with growth and the measure of welfare through GDP. He concludes 'From the abolition of slavery to the Bretton Woods system of monetary management, we have found ways to tailor the economy to our needs, not the other way around. Our current economic systems and thinking are not in tune with today's reality. It's time to find ways to change, before we're forced into it', updated in Suzuki's 'There's No Big Data Without Intelligent Interface' (2016). He is often a target for conservative columnists, for example, Peter Foster in *The Financial Post Magazine*, 'Man as Killer Asteroid' (2014) and 'The Left's Enemies List' (2015).[8]

The provincial media mostly ignore or give scant attention to the Anthropocene in contrast to its coverage of climate change issues. However, *The Calgary Herald* (Postmedia) in 2017, prints 'Five Things to Do This Week in Calgary' with a bizarre announcement: 'Adults-Only Night: The Human Age' at a local theatre, explaining 'There's a movement to name the current geological age after humans' because of 'our effect on water, the air and a look at what kind of artefacts current humans would leave behind for future beings to discover'. *The Edmonton Journal* offers 'Techno Fossils Signal New Epoch, University of Alberta Scientist Says' (2016), a lengthy and well-informed introduction to the familiar story of the Anthropocene. 2017 brings more of a cutting edge with 'Opinion: Why Do We Love Wildlife to Death?' by a law professor at the University of Alberta, on an encounter with a mother moose and her calves: 'Ours is the age of the Anthropocene; a time of mass extinction. Humans are the dominant force responsible for significantly altering the natural world, and we are losing wilderness and wild animals at an alarming rate. Reversing these trends requires a larger discussion about sustainable co-existence'. *The Kelowna Capital News* reports in 2017: 'Scientists Call the Interval Since the Industrial Revolution the "Anthropocene" … Empowered by fossil fuel–driven technologies, a rapidly growing human population and an insatiable demand for constant growth in consumption and the global economy, our species is responsible for the calamitous consequences'.

Occasionally, media connect the Anthropocene and the Sixth Mass Extinction. It is tragically ironic that *Fort McMurray Today* (Postmedia) publishes some uncomfortable items on anthropogenic risks.[9] For example, in 2014, an item sourced from Reuters, 'Is Humanity Behind Biggest Extinction Event Since Dinosaurs?', offers an interview with Elizabeth Kolbert on *The Sixth Extinction.* Kolbert is asked 'What species is most likely to survive the 'Anthropocene'? She replies: 'People are looking at cockroaches, which have survived relatively unchanged for quite a long time. And we know that cockroaches do really well with human disturbance. So, I think if I wanted to pass on my DNA, I would probably choose to be a cockroach ... If you want to pass on your genetic material, do not be a flightless bird'. On that rather chilling advice, we turn to how the media in the United States report the Anthropocene.

The Anthropocene in the US media

Searches in the United States were carried out in six broad geographical areas (partially overlapping) – East Coast, Appalachia, Rustbelt, North, South, and West Coast. Coverage of the Anthropocene in the US media tends to be more repetitive than elsewhere, perhaps a function of the increasing dominance of large media groups in recent decades (Abernathy 2018). This explains why the length of this section does not reflect the fact that we found over ten times more items in US media coverage than in Canada, where the coverage is more diverse. The central role of *The New York Times* and, in particular, the 'dot.earth' column by Andy Revkin, in the coverage of the Anthropocene in the United States and abroad is difficult to exaggerate.

The Revkin effect

In 'Embracing the Anthropocene' (2011) in *The New York Times*, Andy Revkin announces 'Room for Debate', an invitation to share his space in the *Times* (usually seen as the most influential media outlet, at least for the better educated, in the United States).[10] Revkin engages with others who want to argue or agree with him (often, both), giving him the opportunity to name check many other Anthropocene scholars (testifying to his open-minded approach to the topic). He concludes:

> Taking full ownership of the Anthropocene won't be easy. The necessary feeling is a queasy mix of excitement and unease. I've compared it to waking up in the first car on the first run of a new roller coaster that hasn't been examined fully by engineers. That's a very different sensation than, say, mourning the end of nature. It's more a celebration, in a way – a deeper acceptance of our place on the planet, with all of our synthetic trappings, and our faults, as fundamentally natural. In fact, in the broadest sense we have to embrace the characteristics, good and bad, that make humans such a rare thing – a species that has become a planet-scale force'.

Referencing Alan Weisman's remarkable book, *The World Without Us*, Revkin declares: 'We're stuck with *The World With Us*'. Revkin is also responsible for privileging the emerging 'good' Anthropocene narrative. Six articles from the *Times* give a flavour of his contribution and his unrivalled qualifications to explain the Anthropocene to the public. The first, a long essay published in 2011 with the title 'Confronting the Anthropocene' explains: 'I'm participating in a one-day meeting at the Geological Society of London, exploring the evidence for, and meaning of, the Anthropocene'.[11] Revkin name-checks the Australian scientist Will Steffen for a TEDx video presentation, items in *National Geographic* and the BBC, Paul Crutzen and Christian Schwägerl, all key sources in the Anthropocene (and Anthropo-scene) story. He also highlights what was, at the time, a much-neglected issue, still not adequately addressed: 'Students in school are still thinking that we are living in the Holocene; teaching students that we are living in the Anthropocene, the Age of Men [sic], could be of great help. Rather than representing yet another sign of human hubris, this name change would stress the enormity of humanity's responsibility as stewards of the Earth. It would highlight the immense power of our intellect and our creativity, and the opportunities they offer for shaping the future'. Revkin's presence at the meeting is 'because of a quirky role I played almost 20 years ago in laying the groundwork for this concept of humans as a geological force'. Quoting the biologist Eugene Stoermer, who said 'I began using the term "anthropocene" in the 1980s, but never formalized it until Paul [Crutzen] contacted me … a popular book about Global Warming, published in 1992 by Andrew C. Revkin, contained the following prophetic words: "Perhaps earth scientists of the future will name this new post-Holocene period for its causative element – for us. We are entering an age that might someday be referred to as, say, the Anthrocene" [sic]'. Revkin goes on to applaud the role of Erle Ellis, an ecologist at the University of Maryland, at the meeting: 'Ellis's presentation was a mesmerizing tour of the planet's profoundly humanized ecosystems, which he said would be better described as "anthromes" and arguing that the Great Acceleration, described by Steffen and others, is already well under way'.[12] Revkin continues, 'it's entirely possible for humans to design their future, at least in a soft way, boosting the odds that the geological record will have two phases – perhaps a "lesser" and a "greater" Anthropocene'. Revkin's articles are usually adorned with links, and this one takes us to a rather sceptical view of how 'the facts' become like religious faith, a rather ambivalent message in the context of US-based climate change denial (see Brulle 2018). 2014 brings Revkin's 'Exploring Academia's Role in Charting Paths to a "Good" Anthropocene'. The image chosen to illustrate this article in the *Times* is of an art installation at Edge Hill University near Liverpool by British artist Robyn Woolston, 'Welcome to the Fabulous Anthropocene Era' (see Figure 4.1), the artist explains:

> In its entirety "Habitus" (2013– ongoing) comprises of 12 factual signs showing geological eras, epochs and eons that span millions of years. Words like Silurian, Devonian, and Jurassic can be seen on signposts located within the trees on the campus at Edge Hill University. When you look more closely, you

Figure 4.1 Fabulous Anthropocene (2013) © Robyn Woolston www.robynwoolston.com.

> realize they are all pointing towards a Vegas-style sign. This is because they all detail histories prior to the one we are currently living within. Everything that has gone before has led us to our current situation. Whether financial or environmental, the state of the world has been built on layer upon layer of knowledge, rather like sedimentary rock. And so, the centre-point of the installation is the sign that reads in gushing words: "Welcome to the Fabulous Anthropocene Era". A time perhaps unlike no other, where we're faced with signs that reflect back to us the care with which we've chosen to look after the planet…or not' (Robyn Woolston, personal communication, 2019).

In 'Association for Environmental Studies and Sciences' (2014), Revkin writes: 'The theme of this year's meeting, hosted by Pace University (my academic home), was "Welcome to the Anthropocene": From Global Challenge to Planetary Stewardship'. Anthropocene (he explains) 'of course, is the shorthand term that is increasingly being used to describe an era in Earth history named for its causative element – us'. Applauding the growing interest of scholars from many disciplines in the Anthropocene, Revkin states: 'Any effort to bridge the "two cultures" divide in a way that recognizes both human agency and material reality is a good thing' ... It's time to grasp that uncomfortable, but ultimately hopeful, idea'. 'Rx [medical prescription] for the Anthropocene? A Dose of "Anthropophilia" (2016)' emerges from another TEDx talk, '[the novelist of cyberpunk fame] William Gibson's proposal that the future is here now – just not evenly distributed. So far, misanthropy has dominated. Woe is me; shame on us.

But our exasperating diversity of perception, experience, and character may be our salvation in the end'. This links to a summary of the talk, to Revkin's 2011 post (see above), and his article 'Anthropocene Journey' in *Anthropocene Magazine* in 2016, which takes the form of a qualified apology for his previous naivety on various aspects of Anthropocene studies, including his failure to appreciate the Capitalocene and its critique of the 'we' in discussions of culpability. 'After 16 years of percolation and debate, anthropocene has become the closest thing there is to common shorthand for this turbulent, momentous, unpredictable, hopeless, hopeful time—duration and scope still unknown'. He concludes: 'What has your Anthropocene journey been like, and where is it going? Read on for mine, and weigh in. For a soundtrack while you read, I recommend "Anthrocene", the new song by the Australian musician Nick Cave (and bandmate Warren Ellis), using a spelling I proposed way back in 1992' (Revkin himself has another career, as a musician, on which see Wodak 2018). Revkin links to Elizabeth Kolbert's twitter intervention 'Two words that probably should not be used in sequence: "good" & "Anthropocene"', and Clive Hamilton's critique of Revkin's own 'good' Anthropocene narrative.[13] And finally, in 'Building a "Good" Anthropocene from the Bottom Up' (2016), Revkin highlights the work of several academics, notably Elena Bennett (see section on Canada above). Bennett is quoted: 'It feels like we're bombarded, almost daily, with negative visions of the future. That negativity about the future of the environment drags us down, makes it difficult for us to talk to our kids about the future and the environment, and generally threatens to keep us from doing the work we need to do to make the planet safer, better, healthier, and more just'. Supported by Future Earth (see Hamilton 2017: passim), her project aims to explore positive visions of futures that are socially and ecologically desirable, just and sustainable, with another link to many inspiring examples of bottom-up ecological projects from all over the world. This message is reinforced in 'TED's Science Curator Sees Hope in Earth's Anthropocene Age' (2016), where David Biello explains why he sees 'hints of hope' in Earth's emerging Anthropocene 'age of us'. This, obviously, was encouraging news for proponents of the emerging 'good' Anthropocene narrative.

Revkin was not the only one writing in and to the *Times* about the Anthropocene. In 2013, the Vietnam veteran (and now academic and public intellectual) Roy Scranton wrote a long and passionate article: 'Learning to Die in the Anthropocene'. Scranton writes: 'this civilization is *already dead*...We can continue acting as if tomorrow will be just like yesterday, growing less and less prepared for each new disaster as it comes, and more and more desperately invested in a life we can't sustain. Or we can learn to see each day as the death of what came before, freeing ourselves to deal with whatever problems the present offers without attachment or fear. If we want to learn to live in the Anthropocene, we must first learn how to die'. An influential book (Scranton 2015) followed. But apart from a few mentions of Scranton's visits to campuses in the United States, his pessimistic message rarely appears in global media. Most reporting of the Anthropocene in *The New York Times* (as elsewhere) is a mixture of passing mentions, neutral recitation of the geophysical facts about the phenomenon, with occasional discussion of possible consequences, and/or acknowledgements of potential risks with optimistic predictions that humans (or sometimes just scientists and/or entrepreneurs) are clever enough to find ways of adapting to or mitigating. In 2014, *The New York Times* has an important article by the environmental journalist Jim Robbins, 'Building an Ark For the Anthropocene: We are Barrelling into the Anthropocene, the Sixth Mass Extinction in the History of the Planet'. Robbins references a paper in the journal *Science*. 'The world's species are disappearing as much as 1,000 times faster than the rate at which species naturally go extinct ... By 2100, researchers say, one-third to one-half of all Earth's species could be wiped out'.[14] An appropriately poignant epitaph for *The New York Times* coverage appears in 2014: 'Memorials ABBEY, EDWARD 1927-1989. As we hurtle to the end of the Anthropocene, thank G-d you are in heaven keeping its environment safe. Love, Ingoo'.

The two best-selling tabloids in New York City covered extinction, with *The Daily News* (only one Anthropocene item compared with 3,000+ on climate change) proclaiming in 2016: 'We Will Lose 2/3 of Our Wild Animal Populations by 2020 Due to Human Impact on the Earth, New Study Finds', citing the WWF Living Planet report. The article concludes: 'Earth has transitioned into a new geological epoch, the Anthropocene ... but there are still grounds for optimism'. *The New York Post* does better with five results, reporting a celebrity's book choice in 2014 (Kolbert's *Sixth Extinction*) with the comment: 'It's alarming that geologists are renaming this period the Anthropocene, whereas politicians are doing nothing about it'; 'Everything is quietly dying all around us: As humans continue to swallow up the globe, we do so at the expense of the planet's biodiversity as animal populations continue to decline at an alarming rate' (2016); and in 2017: 'Humans are Changing the Climate 170 times Faster than Nature: Study'.

Anthropocene themes in the US media

Many of those who featured in Revkin's *New York Times* columns were either professional Earth System scientists or environmental journalists, but most journalists who write about the Anthropocene are not. How, then, do they deal with the science? The most common scientific Anthropocene-related themes in the media in North America were the idea of a proposed new epoch, climate change (and its association with global warming, ocean acidification, coral reef destruction, rising sea levels, greenhouse gases, fossil fuels, loss of habitat, and pollution), human population growth and consumption, humans as the dominant global force in nature, and species extinction and loss of biodiversity. In general, as the typical excerpts reproduced below illustrate, North American mass media accurately reported the basic scientific information associated with the Anthropocene, though often in simplified forms. All these topics appear in multiple sources. Whereas it is acknowledged in some of the articles that scepticism exists in North America toward certain topics associated with the Anthropocene (notably CC/GW), these references are brief and minimal in number, which suggests differences in media treatment of climate change denial (of which there is a significant amount reported in the US media) and Anthropocene denial (very little reported).[15] Beginning with the Anthropocene as a new proposed geological epoch, in 2011, the eminent conservationist of the American West, the late Tom Bell, wrote in *High Country News* (the influential non-profit publication based in Colorado, which he founded in 1970): 'Mankind's course through history has been marked by epochal milestones. We are now entered into another of those epochs. It is to be marked by huge change and great upheaval, brought about by global climate change. It has already been noted by geoscientists, who have proposed calling it the Anthropocene, or human-caused epoch'. In 2016, *The Washington Post* publishes a wide-ranging piece by its science reporter, Sarah Kaplan: 'Earth entered the epoch of humans … At least that's the consensus of the Working Group on the Anthropocene, the team of 35 scientists tasked with figuring out whether humans have left enough of a mark to qualify for our own unit of geologic time'. *The Albuquerque Journal* has a long, meticulously researched, but neutralizing article: 'Why 1950 May Be When Earth Entered New Epoch' (2016), which provides a clear description of the efforts of the AWG, led by the British scientist Jan Zalaisiewicz, to have the name officially recognized.

Obviously, climate change appears in Anthropocene coverage. *American Scientist* magazine states, in 2015, 'We are now advancing into a geologic era defined by the environmental effects of human activities, named the Anthropocene. Our energy system appears to have put us on a course toward a permanently warmer planet, with uncertain consequences – that is, unless we make some dramatic changes'. The idea that the whole of humanity is exerting a tremendously negative environmental force on Earth, and this force is now greater than all the natural forces combined, is a common theme in mass media and scholarly literature. *The Casper Star Tribune* reports 'Geoscientists have proposed official recognition of a new geological time period, the "Anthropocene" (human-caused) epoch. It would mark the period when humans became the predominant force over the Earth's geophysical environment' (2008), while, from *Discover Magazine* (2016):

> It's hard to deny that humans are shaping planet Earth: from atmospheric composition to urban heat islands and widespread habitat loss, "before" and "after" comparisons reveal pervasive changes. These alterations are among the justifications for the proposed "Anthropocene" geologic epoch, which recognizes the dominant role of human activity on a number of planetary parameters. And, whereas the most dramatic changes have occurred since the Industrial Revolution, a new study suggests that our ancestors' activities were having dramatic effects on biodiversity thousands of years earlier, long before coal-fired power plants kicked the Anthropocene into overdrive.[16]

This not uncommon sentiment may help to explain why there is so little serious coverage of renewable energy in our results. Generally speaking, journalists in North America (and elsewhere) rarely connect renewables with the

Anthropocene, despite a considerable amount of research on the topic, notably the work of Jacobson (see Figure 4.2).[17]

Whereas many believe that dramatic increases in human population and consumption of natural resources drive the Anthropocene, in-depth coverage is slight. In 2013, *The San Antonio Current* warns: 'Earth Nears Tipping Point, Due to Excessive Population'. John Harte (UC Berkeley, ecosystem scientist) is quoted, 'the combination of climate change and nine billion people to me is one that is just

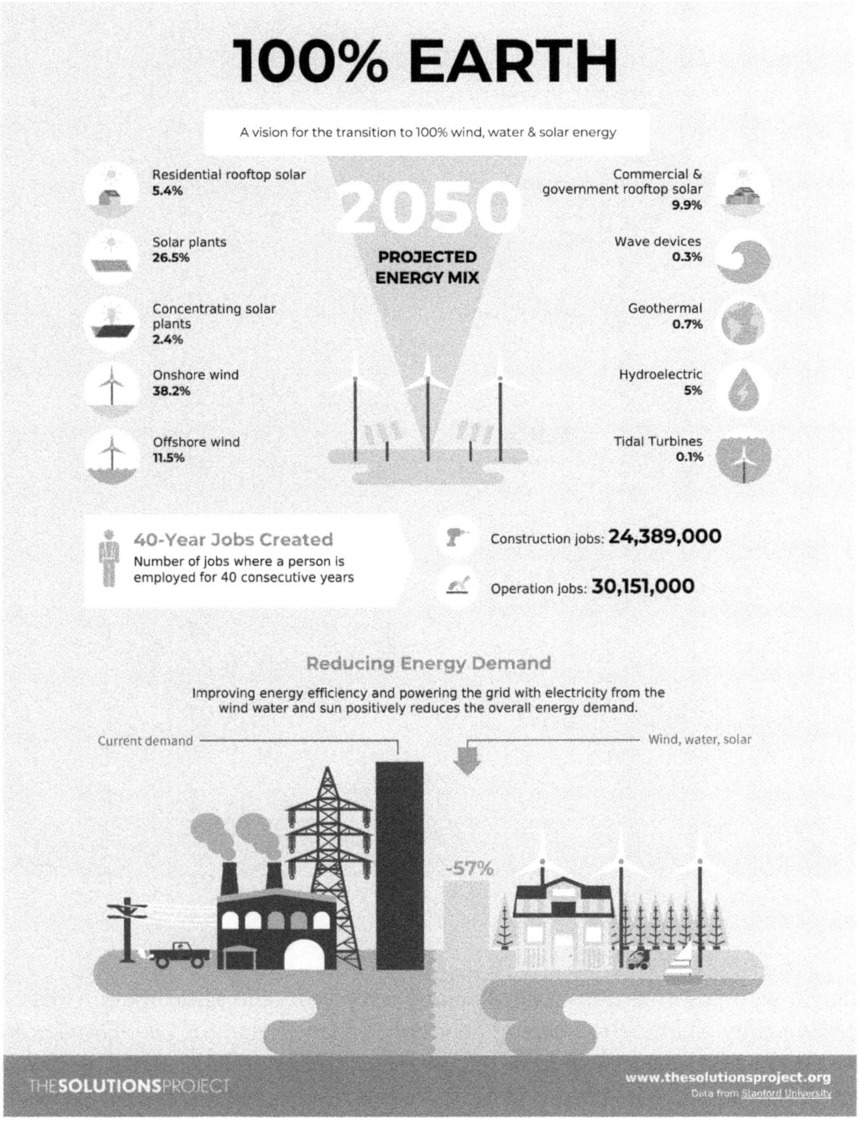

Figure 4.2 Infographic for 143 Countries (Mark Jacobson). www.thesolutionsproject.org.

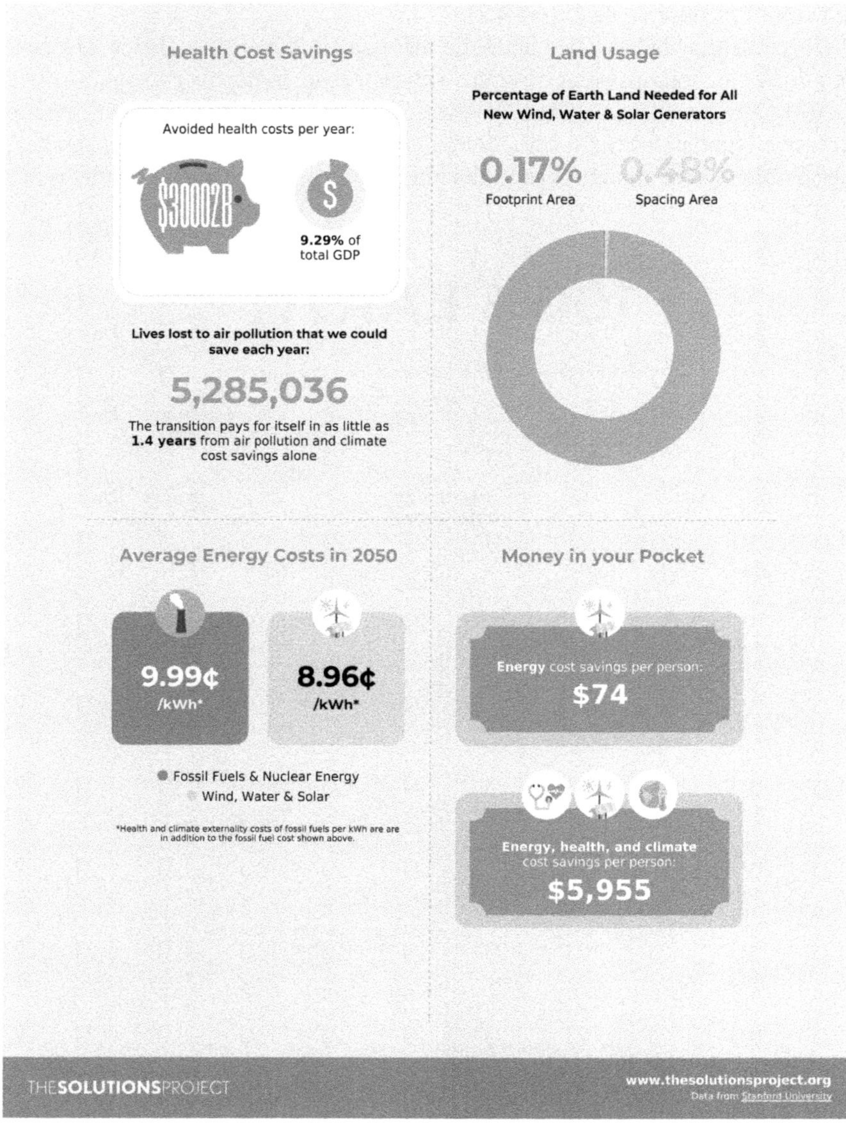

Figure 4.2 Continued

fraught with potential catastrophes', citing a report by the Millennium Alliance for Humanity and the Biosphere at Stanford. Adrian Higgins of *The Washington Post* summarizes the gravity of the biodiversity problem in 2017, when he states: 'the Anthropocene goes beyond just climate and weather and marks a period when we have altered the landscape for our needs, from building highways, expanding cities, converting forest and savanna to agricultural use, constructing dams and flooding river valleys, and all the rest. There is an environmental price for all this progress, in the loss of habitats, species and biodiversity'. This is an especially important statement insofar as it clearly shows that there are wider issues than climate change at stake. Illustrating the use local media frequently make of local peer-reviewed research, *The Santa Barbara Independent* (2014) reports on a study from UCSB published in *Science*: 'Overexploitation and Human Destruction of Habitat are the Key Drivers of Wildlife Population Decline and Extinction'. A researcher is quoted: 'this paper was an alarm call'.

In 2012, *The Boston Globe*'s coverage of the Anthropocene is dominated by the installation 'Anthropocene Extinction' (not to be confused with the widely reported musical phenomenon of the same name) by the street artist known as Swoon at the Institute of Contemporary Art. However, as was suggested in Chapter 2 of this book, most articles linking the Anthropocene (however obliquely) with species extinction do not explore the possibility of human extinction. There are, however, some notable exceptions to reassuring narratives around extinction and biodiversity loss. For example, in 2014, *The Austin Daily Texan* reviews Kolbert's *Sixth Extinction* at length, also quoting from her earlier and more alarming book *Field Notes from a Catastrophe*, where she warns that we are destroying ourselves. *Newsday* also reports rising extinction rates, 'Earth's Wildlife is Slipping Away: Humans Can Do Amazing Things. And Yet, We Can't Stop Killing Our Wildlife' (2016), explaining that 'some scientists say we're now living in the Anthropocene age'. Notably, the article goes on to elaborate: 'when we lose diversity, we start to lose life itself. Because all species – including humans – depend on other species for their survival … Saving the Earth is more complicated'. But the ending is optimistic, arguing that cooperation between nations is 'a tough lift, but the climate change pact [in Paris] from last year offers hope'. *The Tallahassee Democrat* in 2014 also writes on extinction: 'anthropocene defaunation' meaning human-caused animal decline, giving no indication that humans are at risk as well (also in *News-Press FL* and elsewhere). It is notable that science-based topics were covered in the US and Canadian prestige media, media in big cities and small towns. Most of the coverage of extinction of flora and fauna regards it as regrettable (or simply 'natural'). However, on the risks to human life, most of the Anthropocene coverage in North America (and all other regions) simply passes no judgements, leaving it to readers to work out for themselves the implications of 'the new age of man', disputes among scientists over naming the Anthropocene, and starting dates. Examples of this come from media all over North America. This is evident from the preponderance of articles mentioning the Anthropocene in passing, for example, *The Arizona Daily Star* (2016): 'Lectures' by Sara Brown in the Art and Culture section; 'Arizona Archaeological and Historical Society Lecture Series, The Archaeology of Native American Depopulation, Reforestation, and the Dawn of the Anthropocene'; *Orlando Sentinel* (2017), 'You are living in a unique time on planet Earth – mineralogically speaking … In a paper published this week in *American Mineralogist*, a team of researchers argues that human activities have helped create a distinct geological era. They call it the Anthropocene Epoch'; *The Charleston Post & Courier*, 'Editorials: Cut Black Carbon, Warming'…

Geologists are thinking about naming the current climate era the Anthropocene, to highlight mankind's influence on the planet. Not a bad idea' (2017). The Salt Lake City newspaper *Deseret News* gives nine results, all in 2017, of which seven refer to the same art installation: 'Community Calendar … Anthropocene' by Luke Watson; *Christian Science Monitor* (2016):'Geologists Say That We are Now Living in the Age of Humans … Geologists Recommend That a New Geological Epoch, Characterized by Profound Human Impact, Be Formally Recognized'.[18] Most of these items are short, but many long, well-researched articles also avoid making judgements about the possible consequences of the Anthropocene. A good example is the work of the veteran science journalist Ron Meader in *The Minneapolis Post*, notably his essay 'From Climate Change to "Technofossils", We're Revising Earth's Geologic History' (2016), as good an introduction to the key facets of the Anthropocene as you can find outside the academic literature.[19] However, there is little consideration of the implications for human survival. *Anchorage Daily News*, borrowing from Sarah Kaplan's *Washington Post* article cited above, focuses exclusively on the formal recommendation for having the Anthropocene accepted into the geological nomenclature. Jan Zalasiewicz is quoted: 'if we're happy with [the formal recommendation], we hope it will also have a chance of convincing our colleagues'. Similarly, in the Spanish-language *Diario Las Américas* (Miami), we find 'The Earth Has Entered a New Geological Epoch, According to Scientists' (2016), which covers Anthropocene debates, causes that led to the 'great acceleration' of the mid-20th century, and Zalasiewicz (again) on different markers that show the reality of the Anthropocene transition, accurate but neutralizing.

Nevertheless, we do find more nuanced narratives of the type: 'Yes, We Have Problems, But Welcome to the Anthropocene'. In 2016 *Slate* magazine (slogan: 'A Steady Stream of Hair-Raising Headlines') offers 'The Anthropocene: Great Marketing, Wrong Product' reminding us that, in 2011, *The Economist* welcomed its readers to the Anthropocene, and that humans had changed the way the world works: 'It might even make some people look around and think about how they propose to live on a terraformed planet. But it is profoundly misleading and runs the risk of making us far too comfortable, for it suggests long time periods of stability'.[20] This position runs against the grain of most Anthropocene reporting in North America, as, for example, an article in *The San Diego Tribune* from 2011, titled, rather light-heartedly, 'Goodbye Holocene, Hello Anthropocene'. Two articles in *The Los Angeles Times* send the same type of neutral message. In the first (2015), the environmental editor Geoffrey Mohan writes about the evolving adoption of the term in 'Anthropocene Epoch for Earth Etches Human Impacts in Stone'; and Scott Martelle writes, in 2016: 'Welcome to the Dawning of the Age of Anthropocene: How Much has Human Technological Progress Affected the Planet? Enough, Some Scientists Say'.[21] *The Philadelphia Inquirer* in 2017: 'The Dark Cloud Reports That Some Climatologists have Started to Call our Current Age the "Anthropocene", in Which Conditions on the Planet have been Dramatically Altered by Man. We have to Take Responsibility for What We've Done and Take Charge of Our Future'. *The Washington Post* covers the

Anthropocene in many articles, mostly acknowledging the risks in an optimistic manner, as in 2016, 'We're at the Controls on Planet Earth, But We're Not in Control. The Proposed Name for This New Epoch is the "Anthropocene" or the Age of Humanity' reporting the views of the eminent American astrobiologist David Grinspoon, another proponent of the 'good' Anthropocene. In 2017, in *The Washington Times*, 'The Future isn't All Bleak for the Natural World' reviews *Inheritors of the Earth: How Nature Is Thriving in an Age of Extinction* by the British ecologist Chris Thomas. As his title suggests, Thomas goes further than the usual risk/optimistic Anthropocene scenario. In 2014, a group of Harvard students sued the university to force divestment from fossil fuels. While *The Washington Times* article on this does not mention the Anthropocene, seven comments (freely available on the website) did. For example, one correspondent wrote 'based on what's happening in this Holocene–Anthropocene transition …the question [of the drastic effects of fossil fuels] is even more fundamentally appropriate in light of denial that there is even an Anthropocene... that an anthropogenic forcing is even happening'.[22] Coverage in *The Baltimore Sun* is mixed, but, in some respects, somewhat more alarming than the norm. In 2014, we find 'Op-Eds: Calling All Baltimoreans to Citizen Science', pointing out that in the Anthropocene: 'The idea of a single microbial world means that an infectious agent in one place in the world can be in any other place quickly', quoting Sharon Hrynkow, president of the Global Virus Network, on a threat only highlighted in the media when outbreaks occur, though much-discussed in the medical literature.[23] 'Scientists: Human Activity has Pushed Earth Beyond 4 of 9 Planetary Boundaries' (2015) explains: 'the Earth in the coming decades could cease to be a safe perating pace for human beings.' (This article summarizes a paper from the journal *Science*, and does raise some alarm.) *The Tampa Bay Times* (2016) is one of several sources that review the book *Half-Earth* by the eminent American sociobiologist Edward Wilson, summarized (rather abstractly) in conclusion that the Anthropocene is a clear call to save the planet, 'But there is time to change course'.

Perhaps surprising to some followers of climate change debates, *Fox News* offers seven items explaining the Anthropocene. The first, in 2015: 'Researchers Argue "Age of Man" Began in 1610' is a generally accurate discussion of debates around start dates for the Anthropocene; 'Planet Earth. All Human-Made Objects on Earth Amount to 30 Trillion Tons' (2016) reports research by Zalasiewicz and others on the immensity of the technosphere, another borrowing from the journal *Anthropocene.* 'Have Humans Caused a New Geological Era?' (2015) gives its support to the name Anthropocene. Quoting the British scientist Anthony Brown, it concludes: 'We have to combine the anthropogenic with the natural variability in the climate system. It is a question that geologists never had to face before'. On balance, the website of *Fox News*, while posing questions about the extent of 'natural causes', seems to suggest that anthropogenic changes to ecosystems are also important. *USA Today* (2013), in a long and informative piece: 'Climate Change Dates Back to Dawn of First Farmers', is one among many reports of the findings of 'early' Anthropocene proponent William Ruddiman: 'the Anthropocene era, the time of humans making a mark on the planet more striking than natural forces, extends not just to the beginning of the industrial era but to the dawn of agriculture' and, the paper adds, 'more climate scientists are agreeing with him on that today'. A consequence of this research (no doubt unintended) is that blame for the dangers of the Anthropocene falls as much on our distant ancestors as on us.[24]

Provincializing and localizing

Although 'provincializing the Anthropocene', explicitly criticising Eurocentric narratives, does not appear even once in our results from media in the United States, there is a good deal of what we might label 'localizing' the Anthropocene, connecting anthropogenic impacts (usually destructive) on ecosystems, with what Ellis calls 'anthromes' (see Sklair and Boykoff 2019). For example, *The New York Times* ran a story under the title 'The Southwest in the Anthropocene, Say Hello to the Anthropocene, a New Era When People's Actions Alter Natural Conditions on Earth' (2008), featuring the work of William de Buys, author, conservationist, and film-maker, picking up on an article he wrote for the journal *Rangelands*. In two items from *The Boston Globe* (both in 2014), a lyrical but also alarming piece connects the local (Provincetown Harbour) and the global in the context of ocean food chains: 'Off the coast of the Cape, a time capsule ... We live in a new age, not of Aquarius, but of the Anthropocene – the era, which may have begun with the Industrial Revolution, or, as some scientists believe, with the atomic half-life of the last century, around 1950. Within that half-century of half-life, the world has changed exponentially'. Even more explicit is 'How New Englanders Invented the Wilderness: The Birth of the Movement to Protect Places "Untrammelled By Man" ...Their willingness to adapt and think anew guide us as we enter what many scientists consider a new geological epoch-– the Anthropocene – and confront an environment that is changing more quickly than we are'. Two items from Florida, both in *The Tampa Bay Times*, also supply graphic local detail. 'Biologists in Florida! Barreling into the Anthropocene!' (2014) tells us: 'To thwart something called "coastal squeeze"; a network of "migratory greenways" is envisioned so that species can move on their own away from rising seas to new habitat', reporting the research of Reed Noss, a professor of conservation biology at the University of Central Florida; and in 2016: 'What John Muir's Walk Through Wild Florida Can Teach Us Today. We are in the Age of the Anthropocene ... And with global climate change looming, thoughts of the natural world today tend toward fear and control rather than peacefulness and preservation. In revisiting the Thousand-Mile Walk, can we learn anything from Muir's intimate connection to the harsh yet richly verdant South?' Perhaps this sentiment helps explain the dense media coverage of Justin Brice Guariglia's photographic exhibition in the Norton Museum in Palm Beach (see Chapter 13 of this book).

Some of the 'localizing' coverage does raise the alarm. For example, reprinting an item from *The Washington Post* in 2015, *The Tampa Bay Times* with 'Protect One Species, Endanger Another?' highlights a vital aspect of the extinction debate that is often fudged. More dramatically, localizing brings stories of disasters, as in *The New York Times* archive of tropical storms and hurricanes which have caused billions of dollars in damage and cost many thousands of lives over the last century, and in Florida where rising sea levels threaten thousands of homes. Similarly, *Palm Beach Daily News*, 'Holiday Heat Expected to Bake East Coast Next Week' (2014), prints a wide-ranging report covering local lightning strikes and 'lessons from China' but referencing the Anthropocene only in passing. Elsewhere, *The Democrat-Gazette* (Arkansas) reports from further afield, 'Commentary: Fires of the Anthropocene ... the astonishing fire at Fort McMurray evokes the legend of the monster destroying its maker, reveals the looming outlines of the age of human domination called by many geologists the "Anthropocene", and should be an awakening for us ... droughts, floods, and fires of the Anthropocene will soon destroy most of what we hold dear?' (2014). Likewise, through syndication via sources, such as the AP and AFP, many local newspapers in the United States alert readers located far from places typically represented as 'vulnerable' to climate change, such as coasts and fire-prone landscapes. An article in 2014 by the science journalist Seth Borenstein (AP) and carried by *The Utica* (New York) *Observer-Dispatch*, *The Scranton* (Pennsylvania) *Times-Tribune*, *The Charleston* (West Virginia) *Gazette-Mail*, *The Chattanooga* (Tennessee) *Times-Free Press*, and *The Marietta* (Georgia) *Daily Journal*, warns residents of greater Appalachia: 'People are changing Earth so much, warming and polluting it, that many scientists are turning to a new way to describe the time we live in. They're calling it the Anthropocene – the age of humans ... the way humans and their industries are altering the planet, especially its climate, has caused an increasing number of scientists to use the word Anthropocene to better describe when and where we are'. Quoting the astronaut John Grunsfeld, who said that when he looked down from orbit, there was no place he could see on the planet that didn't have the mark of man, so he uses the term Anthropocene 'because we're intelligent enough to recognize it'. Such localized reportage is significant in that it breaks down the concept of the Anthropocene into 'plain' language for non-science and, perhaps, non-politically left-leaning audiences. It characterizes anthropogenic environmental change in terms such as 'pollution', thus assimilating the Anthropocene to common environmentalist discourse, as well as suggesting the 'problem' to be of a type that has often been addressed by the modern state strategy of applying tighter controls to industry. It further links local effects to a universal process by way of the astronaut Grunsfeld's literally global perspective, which portrays a world where 'no place... [doesn't] have the mark of man'. Just the same, because such coverage is often syndicated for a geographically dispersed audience, the specifically local manifestations, contours, and causes of anthropogenic environmental change often remain unseen.[25] Similarly, Zac Smith's *Santa Barbara Independent* article, 'Welcome to the Anthropocene: Encased in Smoke, an Air Purifier Is the Only Reprieve' (2017), offers a stark reminder of the human cost of environmental degradation in California. Smith writes: 'It is up to us to shape the world we want to be surrounded by. Will it be one spent indoors with an air purifier in every room, or will we summon the creativity necessary to restore natural beauty alongside human comforts? Either way, it will be called the Anthropocene. Whether or not this damned smoke ever clears, let's all work to create an Anthropocene that exists in accordance with our noblest human traits – compassion, kindness, forgiveness – and against the arrogance and selfishness that allowed this fire to get so out of control'.

There is one story that goes some way to help understand the dilemmas of reporting on ecological issues in the United States, and perhaps elsewhere. In *Public Source*, a non-profit digital-first news organization dedicated to delivering public-service reporting and analysis in Pittsburgh, we find 'Pittsburgh has Glaring Environmental Problems. So, Why the Greenwashing?' (2016) by Donna Roberts (an educator, formerly of Pittsburgh). She surveys events addressing environment and sustainability in the city, including the 'Natural History Museums in the Age of Humanity' conference at the Carnegie and the opening of its 'We Are Nature: Living in the Anthropocene' exhibition about life in the time of climate change. Of special interest is the Society of Environmental Journalists (SEJ) annual conference. Roberts comments: 'by the end of these extraordinary gatherings, however, I kept wondering whether the truth about our region's environment is getting lost. Why were our region's very serious existing and potential environmental and health issues not being prominently acknowledged and recognized during these events? Were conference-goers being greenwashed?' The SEJ conference, sponsored by University of Pittsburgh's School of Engineering, welcomed attendees with the message that Pittsburgh had become 'America's Most Liveable City'. Roberts' assessment of this is withering.

> I studied a related thematic panel display about our region's air. Of course, I thought it would have comparisons of Pittsburgh's air quality past and present, maybe some data about toxic industrial pollution that plagues our region, perhaps a mention of the controversial Shell ethane cracker plant in Beaver? I doubted it would mention the 6,700 air pollution violations by U.S. Steel's Clairton Coke Works in three and a half years, but you never know. This was a conference of environmental journalists, after all. But no, none of it was mentioned. ... One point of redemption was an environmental justice panel on the conference's opening night. The panel included Braddock Mayor John Fetterman who reminded us: 'Your zip code determines your destiny', while referring to his city as a 'zone of sacrifice'. In addition, one of nine associated field tours – on the sidelines of the conference – focused on the cracker plant and fracking'.

This long and passionate article documents the environmental risks that residents in the Pittsburgh area face on a daily basis and the dilemmas that environmental journalists face. In a revealing project, Battistoli et al. (2017) searched *The New York Times* and The *Los Angeles Times*, and two local newspapers, *The Houston Chronicle* for coverage of Hurricane Harvey and *The Tampa Bay Times* for coverage of Hurricane Irma. They found 630 news articles from these four sources, of which only 23 mentioned 'climate change' or 'global warming' or both. So, if media fail to connect climate change and global warming with local ecosystems, it is no wonder that local disasters are rarely directly connected with the Anthropocene.

Appalachia and the Rustbelt

Appalachia is considered to be the poorest and most deprived region in the United States, largely rural and chronically under-developed. The Rust Belt is the name given to the region around the Great Lakes and the Midwest, once a prosperous set of industrial hubs, but, since the 1980s, caught in a spiral of de-industrialization. While overlapping geographically to some extent, in terms of culture and political economy, they seem very different. This section compares media coverage of the Anthropocene in these two regions.

Anthropocene coverage in the media of greater Appalachia is, perhaps unsurprisingly, geographically uneven. Out of 134 daily and weekly newspapers in 101 communities searched in the Appalachian region of 12 states,[26] 60 percent of the articles occurred in only 14 sources between 2000 and the end of 2017. Furthermore, of these 14 sources (all newspapers), five were college or university publications. However, distribution was spread relatively evenly throughout the region. The four media sources that produced the most mentions of the Anthropocene were *The Pittsburgh Post-Gazette*, *Gazette-Mail* (Charleston, West Virginia), *The Roanoke Times* (Virginia), and *The Athens Banner-Herald* (Georgia). Though we found many original articles, syndicated content was an important contributor to the discussion in much of Appalachia. Two articles, 'As Coral Reefs Die, Huge Swaths of the Ocean Floor are Eroding' (2017) and 'What the Earth Will be Like in 10,000 Years According to Scientists' (2016), were distributed by *The Washington Post*, and appeared in ten and eight smaller local newspapers, respectively, across the region. These articles portrayed the Anthropocene as a global threat as well as focusing on climate change as a key feature of the proposed new era. Some syndicated material also took regional understanding of the Anthropocene beyond just climate change. A version of the article by Jim Robbins (cited above), 'Building an Ark for the Anthropocene', was carried by *The Lexington Herald-Leader* (Kentucky) in September 2014, depicting the Anthropocene as an era defined by a mass extinction of animal species; similarly 'Humans have Caused an Explosion of Never-Before-Seen Minerals All Over Earth' in *The Charleston Gazette-Mail* (2017), syndicated by *The Washington Post*. The focus of the latter article, however, hewed much closer to the natural sciences notion of the Anthropocene as an era physically embodied in a new layer in the geologic record. This portrayal was likewise featured in syndicated content from AP reprinted in *The Pittsburgh Post-Gazette* 'One Word: Plastics' (2016), sardonically depicting the Anthropocene as 'the Slobocene era, or the Trashiferous period', where the sheer volume of plastic and concrete waste is infiltrating the Earth's many ecosystems to such a degree it is depositing an 'artificial sedimentary' layer that will be detectable to natural scientists for centuries, if not longer. Perhaps ironically, this article departed from the agnostic (if also pessimistic) presentation of the era and its possible amelioration, common to most regional coverage, by communicating a techno-optimist view of the future, pointing to a possible technical solution in the production of 'a renewable plastic-like material from carbon dioxide and waste plant fibre [whose] widespread adoption… could radically change the view from the future'.

Locally authored articles often appear in media where scientific research, public events, or museum exhibitions are located. The aforementioned Pittsburgh *Gazette-Post* led the region in coverage (with 17 articles) in large part due to an exhibition held by the city's Carnegie Museum of Natural History in 2017, 'We are Nature: Living in the Anthropocene' (noted above). Because of the educational nature of this exhibition, reportage was often substantive, repeatedly familiarizing its readership with the Anthropocene as 'a recently proposed epoch marked by humanity's impact on the environment rather than geological shifts', as 'Carnegie Museum of Natural History Evolves, Director Strives to Ensure Relevance' (2017) put it. A series of public talks at the University of Georgia in Athens, likewise, put the Anthropocene into regular circulation during the autumn of 2014 via the city's daily newspaper, *The Banner-Herald*, where issues as diverse as industrial food production, land-use conversion, energy consumption, and increasingly frequent tropical storms are presented as features of the new Anthropogenic era. Coverage of the Anthropocene in Charleston (West Virginia) focused on a series of book reviews in 2016 in the local paper, *The Gazette-Mail*. 'New Releases Explore Collision of Environment, Culture' discusses books by Stephanie Foote, Bruce Jennings, and Brian Black, which present the Anthropocene as a phenomenon based in, but also extending beyond, climate change. The reviews provide substantive discussions of book content – and therefore the features and meaning of the proclaimed new era.

Letters to the Editor and guest editorials also figure prominently in Appalachian media coverage of the Anthropocene, though to a lesser degree than in the Rust Belt. However, news and event coverage are varied enough that 'letters and eds' were not as important a source of heterodox perspectives as in the Rust Belt media either. These types of items did fill some coverage gaps. For example, Waynesville's *Smokey Mountain News* featured a guest editorial in December 2014 by local conservation activist, Bill McLarney, that depicts the Anthropocene as a looming crisis driven by the actions 'of one species – us' and *The Winchester Star* (Virginia) introduced its readers to the AWG. Such items extended the concept's reach into small communities served by media outlets without the resources to rely on syndicated or more famous expert-authored coverage. Given the region's relative lack of large cities, and its historic and current dependence on energy resource extraction, the relatively high amount of non-sceptical coverage of the Anthropocene is impressive. Long-inhabited by native peoples, the region was settled by European-descended immigrants beginning in the first half of the 19th century, after the urbanization of the eastern seaboard. And, despite its proximity to the urban centres, and its partial overlap with the urbanized Steel/Rust Belt, Appalachia is marked by its relative economic and cultural isolation from the rest of the country. Media portrayals of the Anthropocene are, however, present if not exactly comprehensive in select communities.

The media in 'Rust Belt' states reflect most of the national trends we have seen, but local platforms especially find dissenting and heterodox voices in opinion pieces and letters to the editor. The Rust Best substantially overlaps with Appalachia, and includes all or parts of Illinois, Indiana, Kentucky, Michigan, Ohio, Pennsylvania, West Virginia, and Wisconsin. The media we searched for Anthropocene items ranged from the nationally significant, such as *The Chicago Tribune* to the regionally notable, such as *The Cleveland Plain Dealer* and *The Dayton Daily News*, and minor local newspapers, such as *The Repository* (Canton, Ohio). These local newspapers constituted the majority of media searched. By October 21 2017, most of these media websites turned up zero results for the term 'Anthropocene'. By December 10 2019, results had increased in several of the larger newspapers – almost trebled for *The Chicago Tribune* (from five to 14 hits) and increased fourfold for *The Cleveland Plain Dealer* (from two to eight), while *The Cincinnati Enquirer* moved from zero to one. Meanwhile, results in some regional papers mysteriously decreased, such as *The Vindicator* (Youngstown, Ohio), where search results dropped from two to zero. Much Rust Belt media coverage of the Anthropocene consists of reprinted AP articles, reproduced elsewhere in North American media, for example, 'With Their Mark on Earth, Humans May Name Era, Too' (2014) in *The Daily Herald* (Arlington Heights, Illinois). One recurrent topic relates to the idea that any problems resulting from the Anthropocene will be solved through human ingenuity. For example, 'Obama Looking at Cooling Air to Fight Warming' (2009) in *The Dispatch-Argus* (Quad Cities, Illinois) was widely posted throughout the US. The article cites geoengineering by injecting particulates into the atmosphere as a 'potential emergency option', reporting that even the conservative American Enterprise Institute views geoengineering as an economically 'feasible and cost-effective' approach. Several local and regional newspapers mention previous research by Crutzen, one of the main proponents of the 'Anthropocene' concept, as one of three scientists to win a 1995 Nobel Prize in Chemistry for their work on helping to fix the hole in the ozone layer. His name appears in *The Journal Times* (St. Racine, Wisconsin), *Madison.com*, *The Pittsburgh Post-Gazette*, *The Morning Call* (Allentown, Pennsylvania), *The Toledo Blade* and *The Northwest Indiana Times* (Munster). Crutzen's previous success with the ozone layer lends credence to the idea that other atmospheric problems (hallmarks of the Anthropocene) will be similarly solved through technical fixes and tweaks. Indeed, even Crutzen himself promotes geoengineering in 'Scientists: Man May Need to Dirty Skies' (2006), in *The Pantagraph* (Bloomington, Illinois).[27] In short, geoengineering appears to constitute a common-sense and sensible approach to the Anthropocene, with relatively few authors citing the potential for unforeseen consequences. These perspectives relate to a view of human ingenuity and inventiveness that found expressions in some of the reassurance narratives cited earlier in this chapter. Rust Belt media did raise serious issues, such as mass extinction, also discussed above. However, some of these items read like trivia, such as 'Scientist Says Rats Could Someday be Size of Sheep' (2014) in *The Dayton Daily News* (Ohio), which ponders how ecological niches for sheep-sized rats may open up as a result of mass extinctions. 'Not impressed?' the author playfully asks, 'How about the size of cows?' *The Chicago Tribune*, 'Planet Earth, the Half-Empty Zoo' (2015) defines the Anthropocene as a mere 'geological age when people crowd out wildlife', while offering little discussion of climate change and environmental degradation.

If some Rust Belt media appears to downplay, if not trivialize, the idea and consequences of human dominance, then they are merely reflecting national trends. However, a handful of letters to the editors in local newspapers challenges conventional ideas about the Anthropocene. For example, a letter to *The Herald-Times* (Bloomington, Indiana), entitled 'The Deadzone' (2012), unleashes a diatribe against the manicured green lawn – a cultural paradigm of the Midwest – urging the city to promote 'less mowing and water use' in order to 'curtail greenhouse gas emissions and encourage true biodiversity instead of the dead zone we now call a lawn' while another to *The Pittsburgh Post-Gazette* (Pennsylvania) in 2015 argues: 'We should see deer and other animals as "other nations" rather than as "underlings"'. *The Leader-Telegram* (Eau Claire, Wisconsin), in 'Voice of the People' (2016), suggests that the Anthropocene concept brings 'greater awareness of our profound, disrupting influence', and poignantly concludes: 'The Anthropocene epoch will terrify and challenge our humanity as well as our technology and communal instincts'. Also in 2016, an opinion piece in *The Dispatch Argus* (Quad Cities, Illinois) by a former state senator of Illinois, 'Whatever We Call it, Age We're Living in is Marked By Destruction', addresses multiple consequences of climate change, and asks if we 'find it so hard to accept diversity now, how will we adapt to a crowded, overheated planet with everyone competing for a limited food supply?' It is telling that some of the most heterodox views in Rust Belt media are not ideas simply borrowed from news feeds, but rather voices of people from a region of the country that has already witnessed the loss of much of the world it has previously known.

Conclusion

The results reported above may suggest a large amount of Anthropocene coverage in North America; however, a cloud map of environmental stories in the United States puts this into perspective (see Figure 4.3).

The word cloud created by Boykoff et al. (2018) shows the frequency of words (four letters or more) invoked in media coverage of CC/GW in *The Los Angeles Times, The New York Times, USA Today, The Washington Post,* and *The Wall Street Journal* in the United States in 2018. 'Anthropocene' with fewer than four mentions fails to make the cut. The special significance of the Anthropocene for historic and contemporary 'natural' phenomena in the United States (less so in Canada) could be seen as presenting localizing, rather than provincializing, counterweights to the somewhat more Eurocentric narratives.[28] Whereas the tone of North American media coverage is generally 'reassuring' and/or 'neutralizing',

Figure 4.3 Word cloud USA, Boykoff et al. (2018).

one article stands out. This appears in the Montréal-based bi-monthly business publication, *Les Affaires*, under the title 'It is Your Greatest Risk But You Refuse to See It' (2016). The journalist Francois Normand documents both the risks of the Anthropocene and some important institutional recognition of them. His conclusion (illustrated with an unambiguous graphic of three business types with their heads in the sand) is worth quoting:

> The struggle against climate change will represent many opportunities for investors, notably in green energies or technologies that enable the reduction of GHG emissions. That being said, the global ecological crisis in which we have entered is at a major point, where the risks exceed by far the financial opportunities, according to some analysts. The impact of our way of life on the planet is so huge (we are transforming its geology) that scientists are stating that Earth has entered in the Anthropocene, a new geological era affected by humans. And to confront this unprecedented challenge in the history of

humanity, and manage the numerous risks associated to it, we must not bury our head in the sand.

This is the only Anthropocene-related item we found in *Les Affaires*, and there was no follow-up to it. In some respects, it represents one of the most important single articles in the media of North America, possibly in our global survey.

Notes

1 Vivienne Bennett, Julian Kessel, Meryl McQueen, Annette Siemssen, and Billal Tab also contributed to this chapter.
2 As explained in Chapter 2 of this book, some media items of exceptional interest published after the end of 2017 are noted, but not counted in the numerical results.
3 For a persuasive analysis of the complexities of the ways in which the 'carbon industries' in Canada fuel climate change denial and obfuscation, see Carroll et al. (2018) and Daub and Carroll (2018).
4 This is a common sentiment in much of the reporting and, while welcoming such initiatives, it is also important to pay attention to research that provides a reality check on emissions embedded in alternative sources of energy (see Chapter 9 of this book, note 2).
5 Lind (2018) offers a critical analysis of 'Extinction' language.
6 This publication and over 100 others in Canada were owned by the Postmedia Group. For a withering critique see David Olive, 'The Problem with Postmedia' (in *The Star*, 30 January 2016).
7 Keith, a professor at Harvard, founded the multi-millions dollar private company Carbon Engineering in 2009, with investment from major fossil fuel corporations and Bill Gates.
8 See Monaghan and Walby (2017) on environmental movements as a 'security risk' in Canada.
9 Fort McMurray was at the centre of the fires that devastated northern Alberta in 2016. Note the differences between coverage in various Postmedia-owned papers.
10 Dot.earth started as a blog in 2007 and moved to the opinion pages of *The New York Times* in 2010, where it ran until 2016. Revkin is a member of the AWG.
11 See Figure 9.2 for the poster of this meeting.
12 An early enthusiast for the term 'Anthropocene', Ellis now appears to have turned against it (Bauer and Ellis 2018).
13 The full essay can be found at https://www.anthropocenemagazine.org/anthropocenejourney/ See also Dalby (2016).
14 In this widely cited article, Robbins refers to the Intergovernmental Platform on Biodiversity and Ecosystem Services (formed in 2012 by the governments of 121 countries to promote the UN SDG), highlighting what bees do for us. For an interesting critique of the idea of ecosystem services in capitalist society, see Kosoy and Corbera (2010).
15 On the related issue of environmental movements in the United States, see Carmichael et al. (2012) and Jenkins et al. (2018) on their funding. These movements rarely highlight the Anthropocene.
16 On coal mining communities in the United States, see Bell and York (2010) and Brown et al. (2016).
17 See Jacobson (2020). For the data on which this graphic is based, see http://web.stanford.edu/group/efmh/jacobson/Articles/I/143WWSCountries.pdf
18 *The Monitor* produced 16 results, most of which were more informative than versions of this much-reproduced headline.

19 Meador's article is based on the classic paper by Waters et al. (2016). On technofossils (also mentioned but barely discussed in *Corriere della Sera, The Age, Vokrug sveta,* and *The Financial Times*, among other media sources), see also Taffel (2016).
20 On terraforming, rarely seen in the media, see Sklair (forthcoming).
21 The links in this article take us to a report of the AWG, and a piece by Elizabeth Kolbert, neither of which expresses much alarm.
22 At the time of writing, while the Divest Harvard campaign is gaining support, the university authorities have resisted taking decisive action. See https://nobodywins2019.home.blog/ The campus divestment movement in many countries is spreading rapidly.
23 From the perspective of life mid-2020, this looks horribly prescient. On Coronaviruses, see 'Bats in the Anthropocene: Conservation of Bats in a Changing World' at https://ebrary.net/12536/environment/coronaviruses.
24 On the complexities of blame in the Anthropocene, see Chapter 14 of this book. Houston (2013) addresses environmental justice, rarely mentioned in our results.
25 It is surprising that media in North America apparently have so little to say about the Arctic in terms of the Anthropocene, given its strategic and ecological importance (Stoddart and Smith 2016). On the symbolism of the Arctic, see, for example, Anderson (2013), and on the geopolitics, Paglia (2016).
26 New York, Pennsylvania, Ohio, West Virginia, Maryland, Kentucky, Virginia, Tennessee, North Carolina, South Carolina, Alabama, and Georgia.
27 However, the 'ozone problem' was relatively easy to fix (see Angus 2016a: 82–88).
28 Lane (2019) offers an alternative and equally convincing view of the 'American' Anthropocene.

References

Abernathy, P.M. (2018) *The Expanding News Desert*. Chapel Hill: University of North Carolina, School of Media and Journalism.

Anderson, A. (2013) 'Together we can save the arctic': Celebrity advocacy and the Rio Earth Summit 2012. *Celebrity Studies* 4(3): 339–352.

Angus, I. (2016a) *Facing the Anthropocene: Fossil Capitalism and the Crisis of the Earth System*. New York: Monthly Review Press.

Battistoli, B.F. et al. (2017) Voices in the storm: The lost discourse of climate change in hurricanes Harvey and Irma. *International Journal of Crisis Communication* 1: 72–78.

Bauer, A. & E.C. Ellis (2018) The Anthropocene divide: Obscuring understanding of socio-economic change. *Current Anthropology* 59(2): 209–227.

Bell, S.E. & R. York (2010) Community economic identity: The coal industry and ideology construction in West Virginia. *Rural Sociology* 75(1): 11–143.

Boykoff, M. et al. (2018) *A Year-End Review of Media Coverage of Climate Change and Global Warming in 2018*. Boulder: Media and Climate Change Observatory, University of Colorado.

Brown, K.L. et al. (2016) Ruin's progeny: Race, environment, and Appalachia's coal camp blacks. *Du Bois Review* 13(2): 327–344.

Brulle, R.J. (2018) The climate lobby: A sectoral analysis of lobbying spending on climate change in the United States, 2000 to 2016. *Climatic Change* 149: 289–303.

Carmichael, J.T. et al. (2012) Building environmentalism: The founding of environmental movement organizations in the United States, 1900–2000. *The Sociological Quarterly* 53(3): 422–453.

Carroll, W. et al. (2018) The corporate elite and the architecture of climate change denial: A network analysis of carbon capital's reach into civil society. *Canadian Review of Sociology* 55(3): 425–450.

Corrigall-Brown, C. (2016) What gets covered? An examination of media coverage of the environmental movement in Canada. *Canadian Review of Sociology* 53(1): 72–93.

Dalby, S. (2016) Framing the Anthropocene: The good, the bad and the ugly. *The Anthropocene Review* 2(2): 102–106.

Daub, S. & W. Carroll (2018) Why is the CEO of a big Canadian bank giving speeches about climate change and Pipelines? *Policy Note* (7 October).

Hamilton, C. (2017) *Defiant Earth: The Fate of Humans in the Anthropocene*. Cambridge: Polity Press.

Houston, D. (2013) Crisis is where we live: Environmental justice for the Anthropocene. *Globalizations* 10(3): 439–450.

Jacobson, M.J. (2020) *100 Per Cent Clean, Renewable Energy and Storage for Everything*. Cambridge: Cambridge University Press.

Jenkins, J.C. et al. (2018) Foundation funding of the environmental movement. *American Behavioral Scientist* 61(13): 1640–1657.

Kosoy, N. & E. Corbera (2010) Payments for ecosystem services as commodity fetishism. *Ecological Economics* 69: 1228–1236.

Lane, R. (2019) The American Anthropocene. *Geoforum* 99: 11–21.

Lind, K. (2018) *The Unbearable Loss of Beings: Curating, Documenting, and Resisting Anthropogenic Mass Extinction*. PhD dissertation, University of Indiana.

Monaghan, J. & K. Walby (2017) Surveillance of environmental movements in Canada: Critical infrastructure protection and the petro-security apparatus. *Contemporary Justice Review* 20(1): 51–70.

Paglia, E. (2016) *The Northward Course of the Anthropocene Transformation, Temporality and Telecoupling in a Time of Environmental Crisis*. Stockholm: CRISMART.

Scranton, R. (2015) *Learning to Die in the Anthropocene: Reflections on the End of a Civilization*. San Francisco, CA: City Lights.

Sklair, L. (forthcoming) Terraforming: The Anthropo-scene confronts architecture. In Jones, P. ed. *Sociology & Architecture*. London: Routledge.

Sklair, L. & M. Boykoff (2020). Mass Media Representations of Anthromes. In: Goldstein, M.I., DellaSala, D.A. (Eds.), *Encyclopedia of the World's Biomes*, vol. 5. Oxford: Elsevier, pp. 407–414.

Stoddart, M.C. & J. Smith (2016) The endangered arctic, the arctic as resource frontier: Canadian news media narratives of climate change and the north. *Canadian Review of Sociology* 53(3): 316–336.

Taffel, S. (2016) Technofossils of the Anthropocene: Media, geology, and plastics. *Cultural Politics* 12(3): 355–375.

Waters, C.N. et al. (2016) The Anthropocene is functionally and stratigraphically distinct from the Holocene. *Science* 351: 137–147.

Wodak, J. (2018) Shifting baselines: Conveying climate change in popular music. *Environmental Communication* 12(1): 58–70.

Young, N. & E. Dugas (2011) Representations of climate change in Canadian national print media: The banalization of global warming. *Canadian Review of Sociology* 48(1): 1–22.

5 Challenges and ideas of representations of the Anthropocene in Latin American and Caribbean media

Viviane Riegel, Sofia Ávila, and Jerico Fiestas-Flores[1]

Introduction

This chapter analyzes the results of media searches in Latin America and the Caribbean. It covers 25 countries (Argentina, Bahamas, Bermuda, Bolivia, Brazil, Chile, Colombia, Costa Rica, Cuba, Dominican Republic, Ecuador, Grenada, Guyana, Haiti, Honduras, Jamaica, Mexico, Nicaragua, Panama, Paraguay, Peru, Puerto Rico, Trinidad and Tobago, Uruguay, and Venezuela). Only the Bahamas, Bermuda, and Guyana returned zero results. In total, over 300 mentions of 'Anthropocene' were found in over 80 sources, written in four languages (Spanish, Portuguese, English, and French). A total of 160 other sources returned no results.

Research on the relationship between mediated messages on the one hand and both social representations and behaviours on the other hand, points to many interconnected factors and a complex network of influences. This chapter reports on work carried out within the framework of social representations, which aims at understanding common knowledge, grounded in language and in daily life, and explaining how scientific concepts are integrated into everyday thinking and action. Moscovici defines social representations as 'a set of concepts, statements, and explanations originating in daily life in the course of inter-individual communications' (1984: 181). These are influenced by both informal communication and the media. Rudy and White (2014: 129) assert that Crutzen and Stoermer's claim, that our current geological era is best understood as the 'Anthropocene', has strong resonances with discussions of hybridity in the social sciences. They stress that the concept of the Anthropocene captures two complementary ideas. First, the impacts of modern society, through agriculture, urbanization, population growth and CO_2 emissions, are so great that humans can now be viewed as a geological force on the planet. Second, with the widespread and uneven social transformation of the planet, there is no nature that is in any straightforward way 'natural' (ibid). Rudy and White also point out that dualist worldviews are not cultural universals. Thus, a 'worldist' (Ling 2014) way to approach environmental studies and global environmental governance, conceptually, empirically, and as a political project, would be to consider different ways of seeing nature and conceiving society–nature relations in actual ontological and epistemological parity with the Western world.

The ways of understanding nature of indigenous people may differ substantially from Western ways. This is not an essentialist claim about the value of indigenous knowledge, but a call to appreciate the ways of knowing of indigenous peoples. Wilson (2004), for example, claims that indigenous traditional knowledge offers a potential basis for rebuilding indigenous communities, and their recovery could also offer a basis for re-thinking global environmental studies. Bearing this in mind for the analysis of the results from Latin American media, we first discuss the three narratives offered in Chapter 2, and then add a fourth found in the media in Latin America and the Caribbean which, combined, illustrate the structures of media ecologies in Latin America. Based on this analysis, we identify a discussion that problematizes the process of provincializing the Anthropocene, focusing on the different perspectives of Latin American scholars, which present a local perspective, connected to indigenous ways of thinking the universe. As a way of contributing to the discussion of the representations of Anthropocene in Latin American media, and the challenges of provincializing the term to the region, we present some projects and alternative ideas that represent the Anthropocene locally.[2]

Categories of the Anthropocene within Latin American/Caribbean media ecologies

Over 300 articles with mention of the term 'Anthropocene' were published in Latin American/Caribbean by the beginning of 2000 until the end of 2017, with no articles published before 2004 and the majority of items concentrated between 2011 and 2017. In the early years, only media in Mexico and Argentina published items on the Anthropocene (13 articles). Between 2011 and 2013, there was a first wave of publication with 18, 12, and 23 articles, respectively. This first moment of dissemination can be connected first, in 2011, to the publication of the articles in *The Economist* (UK) 'Welcome to the Anthropocene', and in *National Geographic* (USA) 'Enter the Anthropocene – Age of Man', as well as the 'El Anthropocene' End of the World Biennale in Argentina, with works by artists from all five continents (13 articles mention Anthropocene in the Argentine media in 2011). In 2012, the UN Rio+20 conference was introduced by a short film, 'Welcome to the Anthropocene'. The second wave of media coverage took place between 2014 and 2015, with 40 and 42 articles, respectively (about one-quarter of the total), with a possible association with the inclusion of the term in *The Oxford English Dictionary* in 2014, and the exhibition 'Anthropocene' at *Museo Subacuático de Arte* in Mexico in 2015 (with 16 articles in the Mexican media between 2014 and 2015). The third and most recent wave, between 2016 and 2017, is the most important one, with 99 and 80 articles, respectively, on the Anthropocene in the region (over half of the total), boosted by media coverage of the opening of the Anthropocene-themed Museum of Tomorrow in Rio de Janeiro in December 2015, and with the subsequent increased visibility of the city due to the hosting of the Olympic games in 2016 (the Brazilian media published 17 articles on the Anthropocene between 2016 and 2017).[3]

The articles on the Anthropocene in Latin American and Caribbean media were written in four languages, mostly in Spanish (about 85%), about 10% in Portuguese (these being the two main languages spoken in the region), and the rest in English and French. No searches were conducted in other languages that are spoken in the region (Quechua, Guarani, Aymara, Creole, or Papiamento).[4] Considering the representations of the Anthropocene in the media, the main narrative in Latin America is neutral, which provides evidence for the 'balance as bias' interpretation found in coverage of climate change, as discussed in Chapter 1 of this book. Many articles focus on describing the term, as well as the controversies around starting dates of the Anthropocene era. Specifically, most articles can be seen as neutralizing (around 40%) or optimistic (around 35%) or otherwise reassuring. Pessimistic radical change narratives occurred in very few articles. These three narratives are also sometimes overlapping. The remaining articles connected the word with the geological (scientific) concept and/or referenced its use in a cultural context.

Examples of neutralizing articles[5]

'Earth Has Entered into a New Geological Era' (2014), in the influential Peruvian daily *El Comercio*, describing the reason why we have entered a new geological era, is based on a paper in the US journal *Science* by a team led by Zalasiewicz, chairman of the Anthropocene Working Group (AWG). The source of the article in Peru was the agency EFE (a Spanish international media agency, the fourth-largest in the world and the largest Spanish-language agency) under the original (translated) title: 'Anthropocene, an Era That Would be Marked by the Rapid Environmental Changes Caused by Human Intervention'; 'Let´s Think About the Ecological Crisis' (2017), in *La Nación* (Argentina), describes the meaning of the term Anthropocene in both catastrophic and opportunistic perspectives, with reference to Bruno Latour and the discussion of agency; *El Telégrafo* (Ecuador) published 'At the end of the century there will be 30,000 million tons of plastic (2016)', reporting the beginning of a new geological era, quoting Zalasiewicz and his work on the geological significance of plastic for the Anthropocene; 'Humanity has Permanently Altered the Earth's Geological Future' (2015), in the Brazilian *Revista Época*, explains how the term Anthropocene appeared and describes its subsequent development. It affirms that it is premature to formalize the Anthropocene, even if there is evidence of human-induced changes. Finally, 'The Construction of Cities – With Their Immense Cement, Glass and Metal Structures – is Leaving a Visible Mark on the Strata of the Earth' in *Granma* (Cuba, 2014) is a long, informative article (borrowing from the BBC), quoting Crutzen and Zalasiewicz in discussions around the transition from the Holocene to the Anthropocene, and starting dates, but giving no indication that there is anything to be worried about: 'The only thing experts seem to agree on is the name: Anthropocene ... Many believe that the name of the epoch in which we live should reflect what is happening on the planet...that the Earth is changing rapidly as a result of human activity'. Others believe that, in addition to adjusting the scientific discourse to reality, the demarcation of a new era could have practical consequences, even legal (hence the presence of a lawyer in the working group)'.[6]

Optimistic narratives

The second, optimistic narrative is found in a broad spectrum of articles, alarming to a greater or lesser extent. The recognition of the need for change typically includes conservation and renewable energy but rarely advocates radical challenges to the economic or political status quo. This narrative, often presented as a sophisticated 'good' Anthropocene frame, was found, in combination with neutral descriptions, as the second-most-common narrative (in over 120 articles). The countries with most results in this category (in countries with more than five articles) are Nicaragua, Colombia, Chile, Brazil (all more than 50%), Ecuador, Mexico, and Argentina (around one-quarter). In the case of Nicaragua and Colombia, this was the main narrative in all the articles, with a similar story in Chile, Brazil, and Ecuador. The media in these countries are less 'neutralizing' than others in the region. In Mexico, optimistic narratives were also predominant. In Argentina (the country that provided a quarter of all the results from Latin America),[7] optimism was less commonly found than neutral narratives. The following are examples of optimistic articles with respect to the prospects for humanity in the Anthropocene. In Mexico, the nominally non-partisan but often labelled 'right-wing' newspaper *Reforma* offers 'Writing the Future' in 2013. This is one of many syndicated articles written by Jeffrey Sachs (the North American economist and climate change/Anthropocene researcher). As an invited contributor, Sachs argues that his formula is different from 'business as usual'. However, his proposal is limited to changes in the energy and technological matrix. Economic growth, in his view, is still possible and desirable if we change these two aspects of the equation. The 'good' Anthropocene narrative can be found in the idea that the Anthropocene, while challenging, is a 'great opportunity' for humanity, science and technology, geoengineering, and business in general.[8] 'North Pole or Zero Pole?' published in the Nicaraguan *La Voz del Sandinismo* (Christian, Socialist, Solidarity) in 2016, emphasizes that the most striking finding in the Anthropocene for the scientists is that humans have proven to be the greatest force of nature, able to change, by their actions, the behaviour of the planet. 'Pacifists and Hatred' (2015), in the Colombian paper *La Patria*, illustrates the dilemma of the world today, facing terrible challenges that are not only political, religious and ideological, but also represent threats to our own Earth, devastated by the Anthropocene. However, *La Patria* concludes that men have the power to change this, if they learn from history. 'Anthropocene, the New Geological Epoch That Gathers Force Among Scientists' in Chile's *La Tercera* (no date) emphasizes the importance of the term to scientists as well as environmental groups and activists. Considering that the human impact on the planet is excessive, we have to modify our behaviour as a species. It opens the way to changes in behaviour. 'Fresh Water Will be Scarce in One or Two Generations' in the Mexican paper *El Universal* (2013) mentions a conference 'Water in the Anthropocene' which was taking place in Germany, arguing that fresh water will soon be scarce. The article develops the idea that the planet and humanity are in danger, but we can find technological fixes.[9]

Pessimistic, radical change narratives

The third and by far the least-mentioned narrative, pessimistic but proposing a concrete, radical change frame, was clearly found in only five of the countries (Argentina, Mexico, Chile, Grenada, and Brazil). For example, there is media coverage of the radical ideas of Latin American writers Maristella Svampa, Juan Villoro, Eduardo Viveiros de Castro, and Déborah Danowski (see Chapter 12 of this book). *The New Today* (Grenada, 2016) reprints 'Let It Decay: Cultural Precarity in the Anthropocene' from the radical political arts magazine *Miami Rail*. This is a wide-ranging, thoughtful essay by the artist and educator Christopher Lee Kennedy. He concludes: 'So, perhaps we don't need anything new, we don't need more layers of "innovation", or more bureaucracy, but to rather simply make-do with what's being left behind. To let things decay, and to find value in the shifting terrains now emerging in the ruins of disaster capitalism. Here, there may be something far more valuable in the cracks and rubble than one may think'.

A fourth narrative

In addition to the three narratives (neutral, optimistic 'good' Anthropocene, and pessimistic, advocating radical change), we observe a trend in Latin American/Caribbean media that might be considered to be a fourth narrative, namely descriptive articles with inadequate historical or political contexts for understanding of the Anthropocene. This category can be explained in terms of the main regional cultural systems, which are still relatively dependent on international media, tending towards a more descriptive orientation. The articles that occupy this category fail to connect the historical relations between modern industrial societies and the Anthropocene, as well as ignoring scientific studies, proving the human transformation of the Earth and climate. Such articles dilute the need to address and transform current ways of producing, consuming, and living (hence de-politicizing the debate). For example, 'Global Warming, Since Prehistory' (2010), an article in one of the most important Mexican newspapers, *El Universal*, generally ignores the links between the Anthropocene and modern industrial societies, and explains the concept from a very distant historical perspective. 'The Planet Entered into a New Era', in Chile's newspaper of record *El Mercurio* (2015), discusses how human activities have triggered the extinction of ecosystems around the world, but with no specific position or historical context.[10] 'Five Things We Learnt in 2014', in Peru's *El Comercio*, places 'Anthropocene' as number five, discussing neither the reasons nor its consequences. Finally, '2016 Will be the Year of the Anthropocene' in Brazil's *Revista Época*, tells us that 2016 marks the year when the start of the new age of the *antropoceno* will enter the language. It does not present any discussion, regarding sources or context for this event.

Media ecologies

The results of the media searches in Latin America and the Caribbean show the preponderance of the reproduction of narratives and news from international media, with a neutral, decontextualized and/or optimistic perspective. Anthropocene articles in Latin American/Caribbean media were found from about 80 sources and these can be divided into four main groups: international media (over half), big national media (about one-third), with small national media and online-only media making up the rest. With respect to the media ecologies in Latin America, big international media corporations (for example, EFE, *El País*, and the BBC) and national media groups tend to reproduce contents in a homogeneous way. Using the concept of media ecology, as developed by Postman (see Gencarelli 2000), we analyze media as interacting systems, and highlight the media strategies that are important, as ways of describing their symbolic agency in Latin American and Caribbean countries. Taking into account the fourth narrative, in media that avoid historical and political context, discussing media coverage of the Anthropocene in the region presents many challenges. This is also related to the reliance on scholars and writers from the Global North and the absence of local perspectives, with few exceptions, and the high proportion of sources (around two-thirds) which never mention the Anthropocene.

Is the Anthropocene being provincialized in Latin America and the Caribbean?

Chakrabarty (2009, 2018) and Morrison (2018) argue that there is a Eurocentric bias in most Anthropocene discussions.[11] To counteract this, there is a need to acknowledge that the Global North (Western Europe and the United States, mainly) is only one part of the Earth, and not its centre or its main part, and that the concepts that are created within this region are not the only ones that should be valued. To a significant extent, the notion of the Anthropocene, and much of the analytical apparatus surrounding it, represents an effort to impose (rather homogenized) Western–Northern historical experiences, frameworks, and chronologies onto the rest of the world. This can be illustrated with reference to results from the Caribbean, a group of islands particularly vulnerable to increasingly extreme weather events. Coverage of the Anthropocene in Caribbean newspapers, while limited in volume, is significant in content.[12] Media in Cuba produced only three items, however, they were original and substantial articles (in *Juventud Rebelde*, *Granma*, and the regional paper *Ahora*). For example, *Juventud Rebelde* (2012) in '¿Del Holoceno al Antropoceno?' explains: 'It is no secret to anyone that, from the depths of the oceans to the atmosphere, the Earth is undergoing enormous physical and environmental transformations. But did you know that humanity has impacted the planet so significantly that many scientists dare to say that we have created a new geological period in Earth's history?' The article concludes: 'However, the founder of the hypothesis, Paul Crutzen, believes that the essence does not lie in the revision of the texts of geology, but in a purpose of greater magnitude. In his opinion, the most significant thing is to draw attention to the consequences of our collective action, and the possible ways to avoid the worst of the endings. What I hope is that the term "anthropocene" is a warning to the world'. The sole result from Honduras is an article in *La Prensa* (2016), translated as 'Beginning of the Anthropocene: The Earth Transformed by Human Beings'. Based on an article in the journal *Science,* and referencing the AWG, it concludes: 'Humans have changed the Earth's system enough to produce a wide range of signals in the sediments and the ice, and these are singular enough to justify the recognition of an Anthropocene age on the Geological Time Scale'. Whereas this is accurate, it does not tell the whole story; for example, it fails to identify (or even discuss) which humans and societies bear the greatest responsibility for our current dilemma. *Diario Libre* (Dominican Republic, no date) reprints an article from EFE: 'Anthropocene, the New Geological Epoch That Gathers Strength Among Scientists'. The Anthropocene is presented as a thesis, resulting from years of research by a group of experts from different countries, with reference to the Geology conference held in 2016 in Cape Town. Various members of the AWG, including Zalasiewicz, are quoted: 'Wagreich (University of Vienna) declares himself optimistic about the possibilities, and sees, in the growing sensitivity about global change that can already be observed in the geological records, a phenomenon without reversal. Alejandro Cearreta (University of the Basque Country) declares: 'The formal approval [of the term "Anthropocene"] would be a boost to all the people and environmental groups that consider that the human impact on the planet is excessive and that we have to modify our behaviour as a species'. Coverage in the English-language Caribbean media was sparse, though a few English-language articles appear in *The Jamaica Gleaner*, for example, 'Welcome to the Anthropocene' (2016), written by Gwynne Dyer, a much-syndicated freelance journalist based in London, which discuss the AWG at Cape Town, and the work of the scientists Crutzen and Lovelock (the latter best known for the Gaia thesis). Dyer's articles place him in both the neutral and 'good' Anthropocene camps: 'The Epoch When Human Beings are Reshaping the Earth', mentions the risks of the Anthropocene but rings no alarm bells. *The Jamaica Observer* republished a few wire service articles mentioning the Anthropocene. One item, published in 2017, is notable for linking the Anthropocene and extreme weather events, not in Jamaica but in the nearby Dominican Republic. Under the title 'Hurricane-Battered Dominica Celebrates 39 years of Independence', *The Jamaica Observer* writes:

> Prime Minister Roosevelt Skerrit [is] urging citizens not to be daunted by the tasking of rebuilding the island … We do not lightly choose to rebuild better or to build the first climate-resilient nation in the Anthropocene. The message has found us. We understand the task. We shall rebuild so as to bring a new hope into being for humanity. We must. We will. Our tryst with humanity's

> destiny is not a chance thing. But he also acknowledged that the task before the island is 'formidable', adding we have lost an estimated 200 per cent of our GDP and our goal is to rebuild better, not just to replace what was lost.

The Jamaica Gleaner connects with the international community in '1.5 [referencing global warming targets] To Stay Alive Campaign Hailed a Success by CARICOM' (2016), celebrating the multi-media activities of Panos Caribbean and the Caribbean contribution to the Paris climate talks, illustrated by a powerful image by a local artist (see Figure 5.1). The article concludes: 'What remains is to have these things actioned'. Some would argue that we are still waiting.

The Trinidad Guardian, with six articles, stood out for its relatively high coverage of the Anthropocene in the Caribbean, but these are mostly neutral reports,

Figure 5.1 Campaign poster '1.5 to stay alive' © Jonathan Gladding.

reproducing what is usually found in the international media. For example, 'Enter the Anthropocene' (2013) tells us: 'Anthropocene. The word is so new that spell-check doesn't recognise it. "Anthropo-" for "man", and "-cene" for new. Dutch Nobel laureate Paul Crutzen first used it in print in 2000. It's the proposed word for a new geological epoch to replace the Holocene. The Holocene began 11,700 years ago; it embraces all of humankind's recorded history, from our origins as hunter-gatherers to agriculturists to the world of cities we live in today'. However, we find some less-reassuring items. An unusually radical (and potentially provincializing) article on the rise of the Anthropocene appeared in *Le Nouveliste*, the newspaper of record in Haiti, in 2009. This was written by members of an environmentalist NGO, based in Port-au-Prince. They warn that, under the world capitalist system, there would be no end in sight to global climate change:

> This ecological disaster is the fruit of a potentially deadly developmental model. Let's say it bluntly, the productivist West, in pursuit of 'infinite' quantities of production to consume more and more, has wasted the non-renewable material resources of the Earth, has broken the global ecological balance and destroyed the fundamental principle of equity... The world is changing irreversibly. The West, too, must change in its own ways of functioning, in its modes of industrial production and also in its relations with other civilizations and with the global environment ... The whole problem lies in the functioning mechanics of the global capitalist system. And we do not believe that there could be a green capitalism. Asking the rich countries of the North to reduce their greenhouse gas and CO_2 emissions to a reasonable level is tantamount to asking them to abandon the imperialist capitalist system and the model of productivist development that they practice.

This is clearly a powerful and fatalistic message, a call for radical change but, by its own admission, with little prospect of success. It should be noted that, in contrast to the relatively small coverage of the Anthropocene in the Caribbean media, climate change featured very regularly in many of their sources over the years, sometimes with thousands of articles, notably in *Juventud Rebelde, Listin Diario, Diario Libre, el Caribe, Hoy,* and *The Jamaican Observer.*

As can be seen from these examples of media reporting from Latin American and the Caribbean, many articles are direct references to problems, that are focused only on Western–Northern countries, related mainly to the ideas of industrialization and development. The environmental debt of the Global North is rarely discussed locally as an issue that should be argued by the region's authorities in international forums. As noted above, the challenges of provincializing the Anthropocene can also be connected to the relative absence of representation by local researchers and thinkers in Latin American or Caribbean media coverage of the Anthropocene. For example, the work of the Mexican Juan Villoro, identified in the local media as 'one of the most prominent writers of Mexico today' is represented by only one article in Mexico and one in Chile. Similarly, little attention is paid in the media to the Argentine sociologist Maristella Svampa (one article in

Argentina), and two distinguished Brazilian scholars, the anthropologist Eduardo Viveiros de Castro and the philosopher Débora Danowski, who are referenced in only one article in Brazil and another in Peru.

Juan Villoro, the Mexican public intellectual, advocates green communism in *El Reforma* (2014), a position that is as unfashionable in Latin America as it is in the rest of the world, though the inspiration he finds in the events in Chiapas some decades ago suggests a provincializing tendency. Informed discussion of the 'provincializing' of the Anthropocene in Latin America would necessitate further discussion of its causes and consequences for the countries in the region, and a commitment of resources to projects that can have an influence on this question. The global debate around the Anthropocene could also widen its perspectives to include discussion from the East and from the South, regarding mainly the contexts of social inequalities that exist within these regions, as well as consideration of different metaphysical systems that can present other possibilities of discussing the presence of humankind on Earth (see Chapter 12). With this in mind, it is a cause for concern that there is no coverage of the Amazon in the context of the Anthropocene in the media of Latin America, but perhaps not so surprising (see Tollefson 2015 and, for a historical survey, Roosevelt 2013).

Media representations of the Anthropocene as a cultural phenomenon in Latin America and the Caribbean

Media coverage of the Anthropocene as a cultural phenomenon in Latin America appeared in more than 50 articles from the whole region. These covered art exhibitions, literature, film documentaries, and other art forms. Examples of these are media representations of two important Anthropocene-related cultural events, the 'End of the World Biennale' in 2011 and the opening of the Museum of Tomorrow in Rio de Janeiro in December 2015. Consuelo Ciscar Casabán, the curator of the Third 'End of the World Biennale' introduces it in an article in *La Prensa* (2010) 'Una Geografía Diversificada', informing us that the Biennale will be held in Ushuaia at the end of October 2011, and the theme will be 'Welcome to the Anthropocene'. The article explains:

> The title of the Biennale refers to ideas of the Nobel Prize-winning scientist Paul Crutzen. He was the first to suggest that we are living in the 'Anthropocene', a new geological epoch in which human dominance of biological, chemical, and geological processes on Earth have altered the planet. Crutzen's term was controversially discussed more than a decade ago. Now, as more potential consequences of human activity – such as global climate change and sharp increases in plant and animal extinctions – have emerged, the term has gained support. Currently, the worldwide geological community is formally considering whether the Anthropocene should join the Jurassic, Cambrian and other, more familiar units, on the Geological Time Scale. The exhibition focuses on the triad of art, society, and environment. Several works deal with the four elements: Earth, Wind, Fire, and Water.

Four more articles covering this cultural event were found. 'Art in the Anthropocene', in the Argentine paper, *Extra El Clarín* (2011), presents the Biennale as 'a visual and participatory event for the end of the world'. The curator´s perspective is to present art in a transversal way with new laboratories of ideas. 'Quincho´s Conversations', in Argentina's *Ambito Financiero* (2011), describes the opening of the Biennale, highlighting the fact that members of the local government participated in the event, which aims to bring contributions of art to reverse the ecological imbalance. 'Recommendations for the Weekend' published in both Argentinian and Ecuadorian editions of *Ambito Financiero* (2011), describes the origins of the artists participating in the cultural event. It explains that the artists are attempting to contribute by calling attention to the need to save the planet. 'Artists Will Call to Save the Planet in a Biennale', in *El Universal* (Mexico, 2011), states that the event is 'a great opportunity to demonstrate that another world is possible if compromise and common sense guide these changes'.[13]

The rather spectacular website of the Museum of Tomorrow offers an institutional description for the experience they call 'Anthropocene':

> After we learn where we come from (the Cosmos), and about how we fit into the environment of the Earth, it is time to ponder where we, the human species, are now. Studies show that the world's population will reach 10 billion by the year 2060. The amount of change likely to occur over the next 50 years will be greater than that of the past 10,000 years. The impact of Man's activities on the planet and on geological processes is shown in the Anthropocene, the era of the human, the central experience of the Museum of Tomorrow, both conceptually and architectonically. The exhibition aims to show the significant ways in which mankind is affecting the planet, and that whereas Man is causing long-lasting changes, the choices we make today can shape the future. (https://museudoamanha.org.br/en/anthropocene)

The following four articles summarize how the media in Latin America report the museum and its exhibition. 'Museum of Tomorrow Invites You to Think About Man's Impact on Earth' appears in *Revista Época* (2015), naming the eminent climatologist Carlos Nobre, the only Brazilian who participated in the AWG, and his emphasis on the important task of raising public awareness around discussions of the Anthropocene. The Museum of Tomorrow has an important role to play in this task. 'José Roberto Marinho: It Will Always be a Museum of Tomorrow', also published in *Revista Época* (2015) is an interview with the former president of the major Brazilian media corporation, Globo, which supports the museum through his Foundation. Marinho highlights the idea that the central theme of the museum revolves around the concept of the Anthropocene and that the creation of a space that enables dialogue towards this concept is very important. An article published by the Argentine *Extra El Clarin* in 2016 describes The Museum of Tomorrow as: 'The Most Modern and Impressive Museum in Brazil', noting the contribution of the Argentine, Andrés Clerici, a specialist in creating 'third generation' museums, as the creative director of the Museum. He says that the museum 'is a humanist museum'. Finally, 'A Flower for the Future: Museum of Tomorrow', in Mexico's *El Reforma* (2016), highlights the new 'iconic building' of Rio de Janeiro, created by the Spanish architect Santiago Calatrava, who is quoted: 'The Museum of Tomorrow represents the headquarters for the science of the future, based on the questions that humans have been asking about since time immemorial: where do we come from? who are we? where are we? where are we going? The choices we make today will drive into a series of possible tomorrows'. The museum display is notable for quantifying anthropogenic impacts (see Figure 5.2).

Most of these articles show neutral or optimistic narratives with respect to the Anthropocene, as well as providing, in some cases, useful short introductions to the science of the Anthropocene and shedding light on the growing attention that the term is attracting beyond scientific spheres. These representations of the Anthropocene´s global discourses in local cultural events or spaces may give materiality to the discussion on the lives of individuals, with the participation of local artists and audiences, as well as indications of new thinking from the region, such as in the case of The Museum of Tomorrow, with the final space of

Figure 5.2 Anthropocene display (2019) © Museum of Tomorrow, Rio.

the exhibition providing an indigenous home for discussion of the possibilities for the future of humankind.

Conclusion

As can be seen from these results, coverage of the Anthropocene in the media of Latin America and the Caribbean is varied, offering some different interpretations of our basic three narratives. Also notable is the engagement, sometimes in depth, sometimes not, with the work of social scientists and environmental scholars, who have reflected on the Anthropocene. However, most of the newspapers and magazines, whether national or local, mainly report what other larger media systems, especially international media, have published previously. This means that most of the material reproduced in media from the countries of Latin America and the Caribbean present little additional local discussion, with almost no difference between popular or quality newspapers. Yet, when analyzing alternative and specialist media, targeted to small groups, there are other strategies. For example, the volume and depth of coverage of Anthropocene-related creative arts events is impressive (see Chapter 13).

Notes

1 Renato Mader and Jeb Sprague also contributed media searches for this chapter.

2 For a thorough survey of how the media in Brazil report climate change, see Carvalho and Loose (2018). See also Lycarião and Wozniak (2017) on how media in Brazil reported the 2009 UN climate conference.

3 The focus in Issberner and Léna eds. (2017) on 'predatory development' as a feature of the Anthropocene in Brazil contrasts with its absence from our results.

4 See Salazar (2017) on indigenous media In Latin America. For an introduction to the work of Phillip Budka, a social and cultural anthropologist from Vienna, on the anthropology of indigenous media, see http://www.philbu.net/blog/category/media/indigenous-media/

5 As in all other chapters of this book, we have applied some editing to most of the excerpts.

6 This was prescient, as litigation on ecological issues has become a serious problem (see Kotzé 2014, 2019). The lawyer is Davor Vidas, a specialist on the law of the sea and a founding member of the AWG (see Zalasiewicz et al. eds. 2019: Section 1.4.3.1).

7 Mercado (2012) offers a detailed analysis of how the quality press in Argentina reports climate change, more 'sustainability' than 'catastrophe'.

8 Howe (2015) discusses 'energy transitions' in Latin America and the Caribbean.

9 Major Anthropocene projects in Berlin (and Munich) can be seen as laying the foundations of sensibly optimistic 'good' Anthropocene narratives to balance out other, more alarming, stories (see Chapter 13 of this book).

10 Dotson et al. (2012) provide an interesting comparison of liberal and conservative media coverage of climate change in Chile.

11 Chapter 12 of this book discusses their work in more detail.

12 For example, while the Cayman Islands did show several media sources that discussed climate change, not one mentioned the Anthropocene. However, see the study by the distinguished ecological historian, the late J.D. Hughes (2017).

13 Many newspapers and websites all over the world were recommending Anthropocene-related cultural events in their news and entertainment sections, as demonstrated in Chapter 13 of this book.

References

Carvalho, A. & E.B. Loose (2018) Climate change in Brazilian media. In Brevini, B. & J. Lewis eds. *Climate Change and the Media*, Volume 2. New York: Peter Lang, chap. 5.

Chakrabarty, D. (2009) The climate of history: Four theses. *Critical Inquiry* 35: 197–222.

Chakrabarty, D. (2018) Anthropocene time. *History and Theory* 57(1): 5–32.

Dotson, D. et al. (2012) Media coverage of climate change in chile: A content analysis of conservative and liberal newspapers. *Environmental Communication* 6(1): 64–81.

Gencarelli, T. (2000) The intellectual roots of media ecology in the work and thought of neil postman. *The New Jersey Journal of Communication* 8(1): 91–103.

Howe, C. (2015) Latin America in the Anthropocene: Energy transitions and climate change mitigations. *The Journal of Latin American and Caribbean Anthropology* 20(2): 231–241.

Hughes, J.D. (2017) Grand Cayman environmental history: A case study of the Anthropocene. *Ekonomska-I Ekohistorija* XIII(13): 88–95.

Issberner, L.-I. & P. Léna eds (2017) *Brazil in the Anthropocene: Conflicts Between Predatory Development and Environmental Policies*. New York: Routledge.

Kotzé, L. (2014) Rethinking global environmental law and governance in the Anthropocene. *Journal of Energy & Natural Resources Law* 32(2): 121–156.

Kotzé, L. (2019) Coloniality, neoliberalism and the Anthropocene. *Journal of Human Rights and the Environment* 10(1): 1–6.

Ling, L.H.M. (2014) *The Dao of World Politics: Towards a Post-Westphalian, Worldist International Relations*. London: Routledge.

Lycarião, D. & A. Wozniak (2017) The prism of the public sphere: The COP15 coverage by the Brazilian media system. *Journal of Public Deliberation* 13(1). www.publicdeliberation.net/jpd/vol13/iss1/art8

Mercado, M.T. (2012) Media representations of climate change in the Argentinean press. *Journalism Studies* 13(2): 193–209. https://doi.org/10.1080/1461670X.2011.646397

Morrison, K.D. (2018) Provincializing the Anthropocene: Eurocentrism in the earth system. In Cederlöf, G. & M. Rangarajan eds. *At Nature's Edge: The Global Present and Long-term History*. New Delhi: Oxford University Press, chap.1.

Moscovici, S. (1984) The phenomenon of social representations. In Farr, R. & S. Moscovici eds. *Social Representations*. Cambridge: Cambridge University Press.

Roosevelt, A.C. (2013) The amazon and the Anthropocene: 13,000 years of human influence in a tropical rainforest. *Anthropocene* 4: 69–87.

Rudy, A.P. & D. White (2014) Hybridity. In Death, C. ed. *Critical Environmental Politics – Interventions*. London and New York: Routledge, chap. 13.

Salazar, J.F. (2017) Indigenous media in Latin America. *Miami Rail*. https://miamirail.org/performing-arts/indigenous-media-in-latin-america/

Tollefson, J. (2015) International media spotlight on the amazon roams, but rarely enlightens. *Elementa Science of the Anthropocene* 3. http://doi.org/10.12952/journal.elementa.000058

Wilson, A.C. (2004) Introduction: Indigenous knowledge recovery is indigenous empowerment. *American Indian Quarterly* 28(3&4): 359–372.

Zalasiewicz, J. et al. eds. (2019) *The Anthropocene as a Geological Time Unit: A Guide To The Scientific Evidence and Current Debate*. Cambridge: Cambridge University Press.

6 The Anthropocene in the media of North Asia

Leslie Sklair, Ka Ho Mok, and Yuyang Kang[1]

This chapter reviews Anthropocene coverage in the mass media of six countries in North Asia (China, Japan, Taiwan, South Korea, Afghanistan, and Mongolia). Most of the results came from the media in China. We compare differences and similarities in media coverage of the Anthropocene from these six countries. Japan, South Korea, and Taiwan also contributed significant results.

China

China (with over 80 items) was divided into five broad geographical areas (Northeast, Beijing/Tianjin, Shanghai, Hong Kong/South, and West), plus the multi-lingual transnational coverage of *People's Daily* (the official Party paper), *China Daily* (the organ of the State Council information office) and *Global Times* (a Party tabloid daily). Differences between these media systems are noted, as are issues of a 'Chinese Anthropocene', raised by the legally constituted policy of 'ecological civilization' and the relationship between the Anthropocene and Confucianism Anthropocene translates in Chinese as *renleishi* (era of human beings).[2] Most of the media in Mainland China are owned by the government, and target readerships that could be classified as educated elites or general public. Our searches show that publications for the general public yield very little on the Anthropocene. In publications for party cadres, most items on the Anthropocene fall into neutral or optimistic narrative frames, with few proposals for radical change to cope with associated risks. Adaptation and mitigation strategies were most common. An early study, sponsored by the United Nations Development Programme (UNDP) on environmental reporting in China (Tolan 2007/08), quotes Chinese journalists: 'You can't criticize the government about policy … If we criticize political issues, we must criticize a company or a factory … not the government'. However, respected journalists agreed that 'a shift had taken place in China's policy toward climate change, and therefore the media coverage', citing the statement made at the National People's Congress that climate change represented a 'significant problem for China … It's the first time that a senior official mentioned climate change as a problem in such a forum … That means the central government wants to do something about climate change and that, inevitably, is being reflected in the changing coverage' (ibid.: 9). It seems likely that this change of policy prevails to some extent today, although the situation with regard to emissions and fossil fuels is even more complex (see De Burgh & Zeng Rong 2012).[3]

Coverage of the Anthropocene in China is dominated by media in coastal areas (southern and eastern China).[4] Media in the politically and economically less important west of the country produced only four results. As an exotic word, Anthropocene usually appears in connection with global issues in Chinese media articles, very few of which address local issues. The portrayal of the Anthropocene does not vary much between different media outlets. Although there are many reassuring narratives relating to the Anthropocene, few deny the seriousness of the threats that humans are facing in the Anthropocene era. The difference lies in whether or not humanity can resolve the environmental challenges caused by industrialization and excessive consumption. An historical review of the processes of environmental change in South China from 1550 to1850 (Marks 1996) shows that Lingnan (South China) could achieve sustainable farming, at least in contrast to the destructive and unsustainable monocropping commonly found in North America. Marks argues that the eighteenth-century economy in South China was neither capitalist nor becoming capitalist, succinctly characterized as 'commercialization without capitalism'. Nevertheless, the commercialization taking place in contemporary China, especially after 40 years of 'economic success' since 1979, has produced negative socio-economic and environmental consequences, requiring a balance between economic growth and sustainable development, according to Neuweg and Stern (2019).[5]

The problems associated with the idea of sustainable development are addressed directly, but rarely critically, in media coverage of the Anthropocene in China. *People's Daily,* mentions the Anthropocene three times between 2000 and 2017. The first, 'Microbiologist: Mankind May be Extinct Within 100 Years' (2010) quotes the eminent Australian scientist, Frank Fenner: 'I spoke of the end of the world to try to draw people's attention and let them know what a bad situation it really is'. Explaining that Paul Crutzen introduced the term, the article concludes on a more optimistic note, 'And we talk of other things than the end of the world, and luckily so, because we'll be around a lot longer yet, if we keep our heads - and our hope'. This sets the tone for most, though not all, Anthropocene coverage. The other two articles, both in 2016, present the usual neutralizing narrative: 'Scientists Declare the Dawn of the Human-Influenced Anthropocene Era'; and 'Dawn of the Anthropocene: Humans Have Tipped the Earth into a New Geological Period – and Now Experts Believe it Started Around 1950'. The views of Fenner are reported by several newspapers in China around 2010. His warnings about human extinction appear, for example, in *The Yangtze Evening News* and the daily *Yang Chen Wan Bao*. The latter reports: "Frank Fenner, a 95-year-old famous scholar in Australia, recently pointed out that human beings could be extinct in 100 years, due to population explosion, waste of resources, and global warming. He further points out that 'not only our children, but other animals will disappear'. The Party daily, *China Youth News,* also reports Fenner's warning in 2010, along with the criticism: 'science and technology would help mankind to resolve the possible crisis'. A book review, reprinted from *The Guardian,* in the magazine for party youth, *Elite Reference*, 'Humans Stand on the Verge of their Extinction' (2011), praises Melanie Challenger's book, *Extinction.* The review concludes: 'Sadly, most seers do not penetrate into the depths of the numb human spirit. Mankind is in charge of the well-being of the entire planet. We need to know ourselves better and need to redefine how we get along with other creatures'. The magazine *Sanlian Life Week* in 2012 is also pessimistic: '7 Billion Population and 2012 Prediction – How Dangerous is the Earth', warns 'If the ecosystem really has serious imbalances, the earth will find a balance the same as in the past after a long evolution, but what about humanity?'

In our results, Jan Zalasiewicz (the leader of the Anthropocene Working Group, AWG) and Loren Rieseberg (see below) are the most cited scientists. Will Steffen's name occurs mostly in international editions of Chinese media (he is the author of many influential Anthropocene papers in scientific journals). The first instances of Zalasiewicz's name appear in two southern newspapers, *Shenzhen Du Shi Bao* and *Guangzhou Daily*, both in 2012. Zalasiewicz is quoted for his article in *Anthropocene Magazine*, where he argues the planet is deeply polluted by plastic. He explains that some plastics which sank into the sea have 'fossilized' and become 'geologic indicators'. His name also appears in 2016 in *Nan Fang Du Shi Bao* and *Global Times*, when he led the official proposal for recognition of the Anthropocene at the International Geological Conference, an event reported by media all over the world. *Shenzhen Shang Bao* reports a lecture by Loren Rieseberg (winner of the prestigious Darwin Wallace Medal), 'Cooperation Among Nature Reserves Should Be Enhanced' (2017). Rieseberg argues that species extinction is happening at a much greater pace than the natural extinction rate before the presence of humans. He also highlights the importance of promoting science education for children, an important theme that attracts little attention in the global media (see Chapter 14 of this book).

China Youth Daily, in 'Human Activities Accelerate Geological Evolution? – Scientific Knowledge' (2010), provides an excellent summary sourced from the *National Geographic* magazine website, quoting the latest research results of an international team of geologists, explaining the Anthropocene as a new era. The paper adds: 'This news immediately attracted widespread attention. The US *Daily Science* website, *Scientific American* magazine, and the British *New Scientist* magazine all reported. However, some experts say that human influence on geological evolution has not yet been fully established'. This is a long article, clear and with much detail. Chinese media review many popular science books that discuss the Anthropocene and climate change, mostly by foreign scholars, for example Elizabeth Kolbert's *Sixth Extinction* in the Beijing daily *Xi Jin Bao* and *Shenzhen Te Qu Bao* (2016 and 2017, respectively). Kolbert is quoted in the latter: 'the surprising characteristic of the Anthropocene is the terrible damage on the local biosphere based on the geographic distribution, which is totally different from the fifth extinction. Rapid damage and destruction by human beings creates countless irreversible disasters'. *Jin Bao* reviews Diane Ackerman's book *Human Age* in 2016, which argues that, in the Anthropocene era, nature is not 'natural' anymore, and, in 2017, *Shenzhen Te Qu Bao* reviews Ackerman's, *The Moon by the Whale Light*, discussing human impact on animals. We find a rare local reference in *Yang Cheng Wan Bao*, 'If This World Turns into the Grave of Fireflies' (2015), reviewing a remarkable book, *The World Without Us*, by Alan Weisman. The reviewer warns that, in a few hundred years in the Anthropocene, millions of creatures became extinct or endangered because of humans. Citing local folklore that considers every firefly as a human soul, if we continue like this, the world will become 'the grave of the fireflies'. *Guangzhou Daily* reviews *This Fleeting World* by David Christian in 2017 and *Xinhua Daily* also mentions this book, arguing that, in the past 250 years, the Earth has faced severe environmental destruction and experienced much greater rates of species extinction than at any time in the past 6,500 years. 'This is why scholars say that we are in a new era, the Anthropocene, a chaotic era that is full of unpredictable and drastic changes'.

Only a small number of Anthropocene-related publications discussed in the media are by Chinese writers. These scholars tend to focus upon philosophical discussions on how humans should stay in harmony with nature. Two articles in the influential *Guangming Daily* exemplify this: 'Confucian Civilization and the Contemporary World' (2015) profiles the research of Chenyang Li, from the Department of Philosophy at Nanyang Technological University in Singapore, who examines the 'cultural roots of the Anthropocene's new humanism' from the perspective of Confucianism. Li argues that Confucianism is of great significance to neo-humanism, and that the Confucianist concept of harmony between heaven, earth and people foresaw the Anthropocene. *Guangming Daily* prints 'New Trend of Contemporary Western Historical Theory' (2017), giving a Chinese interpretation of the new historiography of the 1970s. 'Concept creation has become the internal driving force of academic imagination. Concepts, such as the Anthropocene, the small Ice Age, the Eastern era, the late ancient period, the early modern period, the long 20th century, the deep history and the saddle period, prompted researchers to pay more attention to concept definition and its theoretical research … the new method is changing with each passing day, constantly breaking the limits of human's understanding of history'. This appears to suggest that Chinese scholarship is in a catching-up mode and that should involve the reinterpretation of the classics (*Guangming Daily* is also translated as 'Enlightenment' Daily).

The investigative local Party-owned *Orient Morning Post* in Shanghai ceased publication at the end of 2016, and it appears to have been replaced by the online site *Surging News*, which printed some Anthropocene articles in 2015. 'New Earth Geology, Anthropocene will Start Next Year?' reports that some scientists had chosen 1610 for the start date, and that the AWG is working to establish this. A more extended and challenging article, 'The Soil Science of Life', argues that 'soil pollution and soil environmental standards are undoubtedly the best indicators of humanity's entry into the Anthropocene'. It is notable that the writer acknowledges geographical differences, and that new soil environmental quality standards in China are being revised. 'Whether the new standard can effectively protect the national health, we will wait and see'. *The Lancet's* influential

'Human Health Protection Report on the Anthropocene Era' is welcomed, though the challenges are said to be great.[6] *Surging News*, with 'The Snoring of the Hummingbird' (2017), reflects on an art exhibition, interpreted as an ironic symbol for anthropocentrism, and *Hangzhou Daily* (2014) features 'Taipei Biennale, There is a Big World in the Small Play', explaining the name of this exhibition 'Anthropocene', and that the works of 52 artists from around the world are on display. Three business papers publish one item each. *First Financial Daily*, 'Reflect and Mature in the Haze'(2013), reports the modernization of the national governance system 'to establish a complete system of the ecological civilization system, which let us see the dawn in the thick haze'[7] (one of only two mentions of ecological civilization, the other being in *Global Times* in 2018, a contribution to China's claim to 'green' leadership). In *The Economic Daily*, 'Hiring a Plastic Worker in the Plastic Kingdom' (2014), reports on the 'successful' campaign against plastic pollution, and *The 21st Century Business Herald* reprints an article from *Global Times*, 'Experts: Human Nuclear Test has Destroyed the Earth and Made Geological Era Enter into the Anthropocene' (2016).[8]

In 2016, the major Shanghai-based newspaper *Wenhui Daily* offers 'How to Talk About "Resilience of City and Countryside" When Technology Progresses with Regressive Wisdom?', a long report on an initiative from 200 Chinese and foreign scholars that became known as the Tongji declaration. Making an analogy with the Anthropocene, (citing an article published in the US journal *Science*), these scientists adopted what is termed 'teleological wisdom, analogical practice'. The original intention of this declaration was to call for more people in industry to return to traditional ecological wisdom and reverse the ecological crisis. 'Please Remember, You Live in the "Anthropocene" Era' reinforces this message of the need for change. This is explained as follows: 'Human activities have had a tremendous impact on the Earth's atmosphere, geological layers and ecosystems. We need to pursue economic development rather than speed, develop new energy, take global large-scale and more strategic cooperation to replace the continuous exploration and repeated trial of the old way. Now it's our turn to write our own story in the Anthropocene era'. These contemporary environmental ideas in China have roots in the 1980s, according to a pioneering study on the creation of 'Beautiful China' by Maurizio Marinelli (2018).[9] Tracing the history of new environmental thinking in China, he tells us that the 13th Five-Year Plan (2016–20) enshrined the concept of 'ecological civilization' in law. In his widely distributed volumes of speeches and writings, President Xi Jinping reinforced these themes with essays on 'Ecological Civilization' (Xi 2017, volume 2, pp. 421–431) and 'Complete a Moderately Prosperous Society and Realize the Chinese Dream' (ibid. pp. 62–67). At this time, Xi was winning approval all over the world for his green credentials whereas Trump was being condemned for his lack of eco-awareness.[10]

Despite extensive searches through media in China's western provinces, only three out of 24 sources mentioned the Anthropocene, all with neutralizing frames. In 2010, *Chinese Business Newspaper* (Xian) published a version of the plastics story, 'Transformation of the Earth in Six Ways by Humans: Annual Output of 60

Billion Tons of Plastic Products', referencing the Holocene–Anthropocene transition and quoting Erle Ellis: 'The impact of humans on the Earth now outweighs the geological effects and the effects of natural climate on the planet'. In *Gansu Daily* 'A New Study Proposed the Starting Point of Anthropocene was the Middle of the Last Century' (2016) reports the work of the AWG, and 'Truth is Eternal, There is No Need to Speak it, and the Writing of the Article Doesn't Need to Cater to Others' (2017), quotes a Tang Dynasty nature poem by Qiji. Referencing the suggestion in the journal *Nature* that geologists formally consider accepting the Anthropocene, the article discusses a Chinese novel, with an interesting reflection on the idea of a Chinese Anthropocene: 'As a Chinese writer, I am also very happy to accept the term and the full meaning it represents'. Zhao Defa's *Anthropocene* 'explores the frailty of human nature and the possibility of human salvation under the guidance of religion and philosophy'. The novel, published in 2016, perhaps the first-ever published with that title, is also discussed in *Qilu Weekly* (2016). *Guangming Daily* reviews it enthusiastically in 'Changes of Life and Adjustments of Literature' (2017), highlighting the fact that whereas novels on the theme of the city were predominant in recent years, Zhao Defa 'distilled the expressionism of realism and modernism in a furnace … Placing the construction of coastal cities against the background of globalization, he used the latest geological concept of the Anthropocene, showing that we should be in awe of nature and cherish it, through the narrative of the protagonist Shi Jiao's commitment to environmental protection' The cover (see Figure 6.1) translates as '*The Anthropocene* is due to be remembered in history: it tells the story of consuming cars and living in luxurious houses at the expense of air, food, and water necessary for human subsistence; of abandoning oneself in sex at the sacrifice of health; of inventing airships and nuclear weapons while begging for compassion and love from God'.[11]

The Hong Kong Economic Times reports the opening of the Museum of Tomorrow in Rio de Janeiro in 2017, noting that it has a gallery on the Anthropocene, 'showing the visitors how human activities have affected the earth since the industrial revolution … changes and damage in the past and forecasting the future, encouraging visitors to think about what we can do to make a change' (see Figure 5.2). Whereas the media in Hong Kong pay considerable attention to problems of climate change, the Anthropocene is rarely mentioned. Two items from *The South China Morning Post*, set the tone: 'Raw and Rustic Styles to be Next Big Thing, Fashion Forecaster Says' (2013), arguing that 'designers are drawing inspiration from the 'new geological age' … to create clothes featuring a rustic texture and ethnic adornments'. A local designer, Teo, is quoted, 'The earth is no longer made up of dirt, rock, and minerals, but plastic, garbage, and cement' connecting this to the Anthropocene. 'Anthropocene Dawn: Scientists Say Humans Have Brought on a New Geologic Epoch' (2016), sourced from Reuters, focuses on concrete: 'so ubiquitous on our planet that every square metre of its surface would contain about 1kg of it, if the building material were spread evenly… . The indelible imprint left by human beings on Earth has become so clear that it justifies naming a new geological epoch after mankind'. *The South China Morning Post* International Edition also publishes a few relevant items

Figure 6.1 Chinese novel *Anthropocene* (2016) © Zhao Defa.

including, in 2010, a review of the book *Harmony* by Prince Charles, 'Ours is, he says, the Age of Disconnection, an era marked as the Anthropocene – entirely dominated by man, that man defined by the Age of Reason'; a review of Gaia Vince's book *Adventures in the Anthropocene – Deadly Age of Man*; and a passing mention in 2015, 'Why Hong Kong Features in David Attenborough's New Planet Earth II Series'. In 2015, the Catholic weekly in Macao, *O Clarim,* publishes a thoughtful article by Robert A. Gahl, Jr (cleric and professor in Rome) on the widely acclaimed ecological encyclical by Pope Francis. Gahl declares: 'Many scientists and proponents of calling our age the Anthropocene claim that acting alone, or even as an entire nation or continent, would be ineffective to stave off the global effects of greenhouse gases. If the warnings of climate scientists are well-founded, then surely the Pope must forcefully proclaim the obligation for global coordinated effort to reduce the warming'.

As noted above, China's most important English-language publications are *People's Daily, China Daily*, and *Global Times*, all of which have articles on the Anthropocene. *China Daily* publishes internationally and has Anthropocene references in the US, Asia, and Africa editions, all rather neutralizing. *Global Times* appears to have responded to the geology conference in Cape Town in 2016, with three articles from that year. 'Anthropocene is Here, Say Scientists' is a long, thorough review of the AWG, without a hint of alarm; 'Don't Let the World be Wrapped in Clingfilm' (2015) is conventional, but with some local interest (a study by researchers at Shanghai's East China Normal University) showing the presence of up to 681 microparticles of plastic per kilogram of table salt sourced from the ocean; and 'The Golden Spike, or Start Date, for the Anthropocene, the Age of Man', on Will Steffen and the AWG. One notable exception to the neutralizing and reassuring narratives in most of the media coverage appears in *China Daily* in 2016 with 'The 11,700-year-old Holocene Epoch is Over. We're Now Living in the Anthropocene Epoch'. The article concludes: 'geologically at least, there will be no denying the influence mankind has had on the planet. And it's a hard one to put a positive spin upon, no matter how you frame it'. The research of Weigelin-Schwiedrzik (2017) provides an interesting coda to coverage of the Anthropocene in Chinese media, explaining that the Chinese government had established a national 15-year development plan of the earth sciences for the years 2005–2020, and this included an explanation of the Anthropocene. Her argument might suggest that Green GDP accounting (and ecological civilization), as practiced in China, has tended to sideline more alarming aspects of the Anthropocene debate, a conclusion at least partially borne out by our results. In his ethnographic study of ecological civilization and alternative food networks in southern China, Martindale (2019) enhances our understanding of how the dilemmas of reconciling economic growth, ecological civilization, and the potential risks of the Anthropocene are being worked out in theory and practice in China.[12]

Taiwan

The media ecology in Taiwan is generally considered to be relatively free and very competitive. Our searches produced around 35 results. All fall into the neutral or optimistic narrative frames, and concrete proposals for radical change are absent. In contrast with Mainland China, media articles from Taiwan exhibit more diverse positions on the Anthropocene.[13] The Anthropocene does not attract much media attention until 2013 when Taiwan's *China Times* reports that the Taipei Art Biennale in 2014 is to use 'The Great Acceleration' (an important idea in Anthropocene debates) as the theme to discuss art in the Anthropocene. Later, in 2014, *China Times* and *The Taiwan Times* publish several articles, introducing the concept of the Anthropocene together with the news about the Taipei Biennale. In *Apple Daily* 'Ocean's Seventy-One' (2015) highlights the serious stress to marine ecosystems caused by human activities.

In 2015, the digital platform *New Lens* reports the eco-pragmatist Steward Brand's view that 'it is a mistake to present climate change as a problem to be solved', and, in 2017, it addresses serious challenges in 'Ecological Anxiety in Anthropocene: What kind of Nature Do We Desire?' However, most articles tend to neutralize the risks. *The Taiwan Times* proclaims 'Human-Driven Climate Change Has Put the Next Ice Age Off By About 50,000 Years' (2016), as evidence of the ever-more-dominant influence of our species on the Earth's natural cycles. Two articles in the popular *Apple Daily* report the work of foreign authors. In the first (no date), Paul Robbins (from the University of Wisconsin-Madison) questions the simple image that human activity conflicts with biodiversity, and further argues that 'government policymakers should not only focus on conservation activities and ecosystem maintenance in specific areas, but should fully test various biogeographic characteristics, local labour, and livelihoods'. *Apple Daily* reviews Thomas Friedman's book *Thank You for Being Late*, in 'The Optimistic Prescription For the Anthropocene' (2017). He argues that human activities are no longer just about influencing but reshaping the earth and even new moral values. That's why the new geological era was named the Anthropocene. Although human beings are facing great challenges, he believes 'it will be OK ... for nature, biodiversity is the way to maintain resilience, while, for humans, it can be maintained through continuous learning'. In 2015, *The United Daily News*, with 'Take Off the Golden Branch and Master the Earth?', introduces the concept of the Anthropocene, and, in 'Geologists Proclaim That the Geological Era of the Earth is Moving From Holocene to Anthropocene', highlights the challenges for animal protection and urges human beings to get rid of their arrogant attitudes towards other species and to accept that other species share the Earth with us.[14]

South Korea

The media in South Korea produced 30 results from seven sources, plus two sources that had no results (Anthropocene is translated as *Ilyuse* in Korean). Most of these items send rather vague warnings about ecological damage that can be dealt with by human intervention. We start with four quality papers, roughly equivalent to *The Times, The Guardian* and *The Independent* in the UK. *Chosun Ilbo* (the largest-circulation newspaper) publishes its first Anthropocene-related article in 2014, with the headline 'Humans Have to Control Everything in the Globe if They Destroy Biodiversity', followed by 'Humans Pulling the Trigger of Extinction' (2015), one of the numerous responses around the world to the much-discussed Worldwide Fund for Nature (WWF) report, and a basic 'Anthropocene' account in 2016. *Korea Joongang Ilbo* (the second-largest circulation newspaper) offers several items in 2015 that send mixed messages. For example: 'Happiness is a Quiet Thing that Each Person can Enjoy', 'Human Evolution From Prey to Predator', and 'Extinction of Species Due to Humans ... Is it Now the Anthropocene?' An exception is 'The 6th Mass Extinction, More Dangerous for the Globe Than Nuke Is' (2015), highlighting the danger to flora and fauna, but not to humans.

Dong AIlbo (the third-largest circulation newspaper) published two items, the neutralizing 'I Met Swallows Again' (2015), and the optimistic, '2020 Climate Change Agreement is Just Around the Corner' (2016). *Hangeorae* (a popular progressive left-wing newspaper) offered a varied selection of Anthropocene items. In 2016, we find, 'The 6th Mass Extinction, How Human Race will Evolve?'; 'AI and the World Government'; and 'Human-Made Objects on Earth Amount to 30 Trillion Tons, What Have We Done?' All these are informative but mostly neutralize the risks of the Anthropocene. In 2017, the tone changes with 'A Lesson From the Apocalypse of Contamination', and 'Anthropocene, the 6th Mass Extinction is More Serious Than Expected'. Despite the emphasis on extinction, none of these quality media make any concrete radical proposals for change or suggest that humans might become extinct. Not one mention of the Anthropocene was found in three of the most-read tabloid newspapers *New Daily*, *Dispatch*, and *Sports Chosun*.

There are several English-language media in South Korea, generally targeted at foreigners (some of them are English-language editions of the Korean press). They all have some coverage of the Anthropocene, usually sourced from foreign agencies. *The Korea Herald* (the largest English-language source) offers a series of long and well-researched articles, starting in 2011 with 'Earth's Population of 7 Billion Faces 6th Mass Extinction?', reporting the findings from the UN Environment Programme, that two-thirds of the world's population will suffer from water shortages within 14 years, that primeval forests will be halved, and that oceans also face extinction of aquatic species. 2014 brings two more articles: 'With Their Mark on Earth, Humans May Also Name Era...the Time We Live In'. This looks like a press release about the AWG, informative but drawing no serious conclusions. One of the longest articles in the whole sample follows, 'The Worst Empires are Those of the Mind' (2014) by Andrew Sheng, a distinguished fellow of the Fung Global Institute in Hong Kong (now affiliated with Hong Kong University). This is a wide-ranging historical reflection on Barack Obama's presidential visit to Asia, citing a 'remarkable book' by Indian writer Pankaj Mishra, and mentioning in passing 'the common threat facing all mankind, the Age of the Anthropocene, in which we are all destroying the natural environment around us'. In *The Korea Times*, we find two syndicated 'Welcome to the Anthropocene' articles by Gwynne Dyer in 2014 and 2016, both well informed, both indicating serious risks from the Anthropocene, but neither advocating any concrete radical change. 2017 brings one of the few articles by a local contributor, 'Be Prepared for Trump' by Kim Joo-young, a student at the Graduate School of Environmental studies, Seoul National University, who accuses Trump 'of security and environmental issues in the middle of the Anthropocene, throwing away the US global leadership that his predecessors had worked so hard to accumulate'. The English-language *Joongang Daily* is associated with *The New York Times*, and its high-quality coverage is heavily dependent on outsourced articles. For example, in 'Risking it All in the Anthropocene Era' (2013), the eminent British astronomer Lord Martin Rees, argues here (and elsewhere) that 'The "Anthropocene" era, in which the main global threats come from humans rather than from nature, became especially risky with the mass deployment of thermonuclear weapons'.

It is not surprising that mainstream media often focus on the idea of sustainable development in their reporting on the Anthropocene. It can be seen as a ray of hope in what otherwise seems a distressingly gloomy story.[15] *Joongang Daily* provides several good examples. In 'Writing the Future' (2013), Jeffrey Sachs, the American economist and professor of Sustainable Development at Columbia University in New York (also a much-syndicated author), explains the risks of the Anthropocene. He argues: 'The simple fact is that humanity faces a stark choice. If the world economy's current growth patterns continue, we face ecological disaster. … I call today's growth pattern the business-as-usual option; the smart-technology growth pattern, by contrast, represents the sustainable-development option. Business-as-usual can work for a while, but it will end in tears, whilethe sustainable-development path can lead to long-term prosperity. So, what will it take to write the happy ending? First, we must recognize that we, as a global society, have a choice to make. Business-as-usual is comfortable. We think we understand it … on our current trajectory, short-term prosperity is coming at the cost of too many future crises'. The clarity of the message (that there can be a 'good' Anthropocene), and the fact that it appears in media all over the world make this a notable contribution.

A different angle on the issues Sachs raises appears in 2015, in 'Leaving Our Children Nothing' by Johan Rockström, the distinguished environmental scientist, former director of the Stockholm Resilience Centre, and leading researcher on planetary boundaries. The optimistic twist in the title is: 'we could be the first in human history to leave our children nothing: no greenhouse gas emissions, no poverty, and no biodiversity loss'. Rockström explains the difference between the UN Millennium Development Goals (MDGs), which focused almost exclusively on developing countries, and the more recent Sustainable Development Goals (SDGs), intended to be universal and to apply to all countries equally. He drives home an important political point, that the adoption of the SDGs 'indicates widespread acceptance of the fact that all countries share responsibility for the long-term stability of Earth's natural cycles, on which the planet's ability to support us depends', though this blurs the all-too-obvious truth that the ecological damage caused by some groups in all societies is much greater than that of others. Along the same lines, David Griggs, director of the Monash Sustainability Institute in Australia, seeks to 'Redefine Sustainable Development' (2013). Griggs declares: 'Today, humanity faces such a life-threatening problem …But where does one start? For almost three decades, sustainable development has been defined as development that meets the needs of the present, without compromising future generations' ability to meet their needs. Related policies have reflected the view that sustainable development rests on three equal pillars: the economy, society, and the environment. But this view is no longer tenable'. Griggs, like Rockström, concludes in endorsing the SDGs, but both are rather vague about how to achieve them. The articles by Sachs and Griggs, like several others, were sourced from the non-profit Project Syndicate, which provides high-quality news stories to media in developing countries at discounted rates.

Also in *Joongang Daily*, 'What Will the World Look Like in 2036. We Asked A Futurist' (2016), referencing the Anthropocene, the futurist James Dator, Director of the Hawaii Research Center for Future Studies, warns: 'Continued economic growth with more jobs and more products should no longer be the main goal of Korea, or any country'. To the first question, 'What will the world look like in ten to 20 years?' he answers: 'No one, certainly not I, can predict what the future will be, [but] substantial environmental change seems highly likely to me. It should be viewed as a new opportunity. If not, it will indeed become a catastrophe'. This item concludes with a poignant little poem by Kim Jee-Hee, as follows 'If we act as if it matters'/and it doesn't matter/ Then, no matter/but if we act as if it doesn't matter/and it matters/then I matter'. *Joongang Daily*, like other media in Korea, gives notice of Anthropocene-related arts events, for example 'The Huge and Minute Universe Displayed in Gwangju'[16] (2017) by the Argentine artist Tomas Saraceno. The theme of the exhibition is that 'in the Anthropocene, we must be free from the anthropocentric point of view and consider other species on the planet, even beyond the planet Earth'.

Japan

Searching the media in Japan for items about the Anthropocene was particularly difficult, not least because there appear to be at least four different versions of the term in Japanese: (アントロポセン、人類の時代、人新世、人類世).[17] As in many other searches, the English term also appears. Research on media coverage of environmental issues in Japan in English is sparse. However, one study stands out; Sampei and Midori (2009) usefully relate media coverage, public awareness of climate-change issues, and Japan's national campaign to reduce greenhouse gas emissions. Also interesting is Wakabayashi (2013) on voluntary business activities to mitigate climate change.

Mainichi Shinbun, a major national paper, gave just one result: 'Has Our Generation Entered a New Geologic Period – the "Anthropocene"?' (2016) reports on an international Anthropocene symposium at the National Museum of Nature and Science in Tokyo, advertised by a vivid poster (see Figure 6.2) and Kamei (2016).

Referencing microplastics, fallout from nuclear testing, Paul Crutzen, and debates around Anthropocene starting dates, Nobumichi Ariga, a researcher at the Museum, urges scientists to 'reconsider the long-lasting effects of human activities on the Earth in the last one million years and to begin planning for

Figure 6.2 Museums in the Anthropocene (Age of Man), conference poster (2016) © National Museum of Science Tokyo.

appropriate measures for the future'.[18] The national daily *Sankei News* discusses the WWF Living Planet report with 'In the Last 40 Years, Wild Animals Reduced By 58%' (2016), quoting Marco Lambertini (Director General of the WWF): 'the role of human activity in pushing the Earth and its natural systems toward peril is clear'. However, a note of caution is introduced by ecosystem conservation expert Professor Stuart Pimm from Duke University, who tells the BBC that the report's numbers are vague and mostly from Western Europe. The business-oriented *Nikkei Asian Review* offers several informative items, for example, 'Booming Discussions on the "Anthropocene": Six Actions To Avoid Water Risks' (2013) reports in some detail on an important conference in Germany, 'Water Problems of the Anthropocene', adding the significant fact that the UN plans to add 'water security' to its Sustainable Development Goals in 2015. This is followed by a well-researched review of climate change politics in the context of 'Crutzen's Anthropocene': 'Murky Politics Hobble Progress on Climate Change' (2015) by Brahma Chellaney, a global strategist and specialist in Asia and water issues (he is also a Project Syndicate author). Chellaney declares: 'Make no mistake: the future of human civilization hinges on sustainable development ... If we are to preserve our planet for future generations, we must move from the Anthropocene epoch to the 'Sustainocene' age'.[19] This is interesting insofar as it supports the argument that the idea (perhaps the ideology) of sustainable development binds international environmental organizations (notably United Nations Environment Programme, UNEP), big business, and some sections of academia and the media together, as we can see in the case of South Korea and elsewhere. We found only one item in the influential national daily *Asahi Shimbun* 'Where Did We Come From and Where are We Heading? (2017), which discusses the transition from the Holocene to the Anthropocene, 'a topic of recent debate amongst scholars'. The article concludes on the familiar note of connecting a 'boom in human populations' to 'environmental destruction'. *AFPBB News* (this appears to be Agence France-Press in Japan) prints two generally accurate accounts of how the Anthropocene is being promoted by scientists. The first, in 2011, 'Did Humans Change the World? The New Age of the Anthropocene' discusses debates on the name, citing Crutzen and Ellis in general support, but warning that it might take 20 years for it to be accepted by the scientific authorities, whereas, in 2016 'International Geological Congress Discusses Adopting the "Anthropocene" as a New Geologic Period', concludes: 'There are sceptics, however, as it is too early to decide the end of the Holocene, given the short time that humans have inhabited the earth'.

In March 2011, the Japanese edition of *National Geographic* published a series: 'The Age of Mankind, 7 Billion People's Earth: Anthropocene hanging the Earth'. This long article brings together the views of members of the AWG to explain geological history, predicting that, in the long term, nothing will remain of cities, deforestation may leave some sediment, species will become extinct, with many other technical details. The article poses the question: 'When did the Anthropocene (*Ninsei*) begin?', engaging with the debate between William Ruddiman's 'early Anthropocene' hypothesis and later dates proposed by Crutzen and the AWG. Crutzen is quoted: 'The word Anthropocene should be a warning to the world. I hope so'. Another article from the Japanese edition of *National Geographic* (also in 2011, inauspicious timing)[20] is apparently a translation from Kolbert's *Sixth Extinction*. The Japanese version of this article notes that, at the time, there is no defined term with which to translate the idea of the Anthropocene, but it uses '人新世 (jinshinsei) and mentions Crutzen and Zalasiewicz. Several Anthropocene-related items from disparate sources were found in 2016, for example a conference sponsored by Fujitsu discusses education in the new context of the Anthropocene, here translated as 人類世 (jinruisei)). *Nikkei Science* produces a special edition on thinking about the Anthropocene in the context of social inequality; and a translated article from the Japanese version of *Wired.com*, 'All the Things Produced in the History of Humanity Estimated at 30 Trillion Tons' on the much-reported article by Zalasiewicz in the *Anthropocene Review*, linking this with the advent of 'civilization' (also the topic of a Japanese translation of an article on the *Gizmodo* site in 2017). The *10 plus 1* website brings us a remarkable essay, 'What is a Human Being in the Anthropocene?' (2017) by Hiromitsu Yoshikawa. This piece is too long and complex to summarize here, but it certainly provides a splendid introduction to the science and philosophy of the Anthropocene with some quite original twists.[21] Also in 2017, the business website *social impact.com* introduces the concept as part of a new approach by the academic Norichika Kanie, connecting the Anthropocene with the SDGs. The article also advertises seminars aimed at businesses on how we can follow the UN's SDG strategy 'to change the world'. Another business website, the Canon Institute for Global Studies (*canon-igs.org*) hosted a meeting of the Breakthrough Institute in 2017 on 'Democracy in the Anthropocene', with presentations by Ted Nordhaus from the Breakthrough Institute and Sugiyama Taishi, a climate change mitigation specialist for Canon.[22] Other websites from Japan reported a variety of Anthropocene stories at this time, for example, on new types of minerals created as a result of human activities, sustainable development goals in the Anthropocene (the same article we found in Korea by David Griggs), and a translated article from the magazine of the UN University in Tokyo on the emerging 'urban Anthropocene' in 2014, also signing up to the UN goals (of particular interest in Japan, where about 90% of the population are said to live in cities).

The most successful Japanese searches were in *The Japan Times*, the most widely read English-language publication in the country. It yields around 30 items between 2009 and 2017 and here the significant presence of local contributors indicated above starts to decline. For example, 'Welcome to the Anthropocene: How to Survive a 'Fearful Age' (2009) by Rowan Hooper, a British science journalist, who graduated from *The Japan Times* to *New Scientist*, is a long and very well-researched reflection on the film of *The Road*, Cormac McCarthy's post-apocalyptic 2006 novel. The film, Hooper tells us, 'set me thinking about how humans might evolve in a climate-changed future. And a clue can be found from an ecological disaster in the distant past. Whether we change our ways or not, most scientists agree that we're living in what has been called the "anthropocene"—an era of Earth's history in which humans are having a massive impact on the environment and ecosystems. It is also a time of sweeping extinctions'.

'Beyond Copenhagen, There's More Than Just Cutting CO_2' (2009) by Stephen Hesse, a law professor at Chuo University in Tokyo, discusses the transition from the Holocene to the Anthropocene. Citing Rockström on planetary boundaries, he concludes: 'The danger, of course, is that in our eagerness to grow our economies and exploit Earth's bounty, we will push our planet beyond these thresholds – compromising the life quality of our children and grandchildren'. Martin Rees provides another version of the nuclear weapons article we found in Korea, in 'Blithe As Can Be About the Risks to Everything' (2013). 'Scary as Hell Ocean Research Storm Breaks', another contribution by the prolific Stephen Hesse (2013), offers a wide-ranging discussion of the breakdown of 'long-held barriers between nature and culture'. Paul Crutzen, Christian Schwägerl (from Yale Environment 360), and Bill McKibben's controversial book *The End of Nature* are all cited as an introduction to a 'perfect storm' of ocean research results. Hesse concludes: 'the conclusions are truly alarming. Or, in the vernacular, they're as scary as hell'. Food security is also identified as a major anthropogenic problem. This is certainly one of the best surveys of Anthropocene issues in any mainstream newspaper. The curiously headlined 'Anthropocentric Bent of "Alien" Fish' (2014), by Hiroaki Sato (a translator and essayist in New York), sends a serious message about the consequences of pro-active conservation: 'In the Anthropocene, humans have aided and abetted the process', for example, the introduction of the kudzu vine into the United States in the late 19th century to stop soil erosion, Asian carp into the United States in the 1970s to control weed and parasite growth in aquatic farms, and black bass into Japan from the United States in 1925. 'Experts Consider Calling the Time We Live In "Anthropocene", or Age of Humans', sourced from Reuters in 2014, is a positive review of Gaia Vince's award-winning book *Adventures in the Anthropocene*. In 2016, we find conventional reports on how the AWG promoted the idea of the Anthropocene at the geological conference in Cape Town, such as 'Anthropocene Epoch Mulled Over Earth's Human Imprint', sourced from Reuters in Oslo. Somewhat less conventional, 'Latour: Rethinking Ecological Crisis From the Ground Down' reports on a lecture series by the globe-trotting French polymath, Bruno Latour (see Chapter 12 of this book). 'Towards a Global Treaty on Plastic Waste' (2017), by the Berlin-based environmental campaigners Nils Simon and Lili Fuhr, is illustrated by a striking photograph of plastic waste on a remote Pacific island, one among many similar images in the global media. The caption to the photo reads: 'When researchers travelled to tiny, uninhabited Henderson Island in the middle of the Pacific Ocean in 2015, they were astonished to find an estimated 38 million pieces of mostly plastic trash washed up on the beaches' (from AP).

A small number of Anthropocene-related arts events were found in our Japanese searches,[23] for example the following from *The Japan Times*. 'Darren Almond: All Things Pass' (2014) reports a video exhibition, part of which was labelled 'Anthropocene', with short films, 'Sometimes Still' and 'All Things Pass', described as 'precisely choreographed multi-channel installations at an architectural scale. They are at once figurative and abstract'; 'Curator Okwui Enwezor Tackles Grim Realities at Venice Biennale' (2015) by Cameron Allan

McKean, formerly cultural editor of *The Japan Times,* latterly a PhD student in Anthropology at Deakin University in Australia, begins: 'Ugly, joyless, aggressive, didactic, morose, self-righteous, unpleasant; these are just some of the words used in the press to describe the recently opened 56th Venice Biennale in Italy'. This is connected with 'an attempt at articulating the horror and sense of dislocation we have with conceptualizing the scale of our new geological age, which many are calling the Anthropocene'; and a *Newsweek Japan* article by Ozaki Tetsuya, editor of a cultural magazine, discusses art and aesthetics in the Anthropocene in 2016. In the same year 'Black illumination: the Unhuman World of Junji Ito' describes in graphic detail (no pun intended) the artist's manga 'Long Dream' (now a TV film), referencing H.P. Lovecraft (who also influenced Donna Haraway), other masters of Japanese horror manga, even hints of Hieronymus Bosch or the much older tradition of the Buddhist jigoku-zoshi (hell scrolls). Tetsuya connects this with 'the anthropocene as our human-influenced epoch has been labelled', arguing that 'we seem to be embedded in the world and estranged from it at the same time. And the most difficult thought is that the world, the cosmos, even our own bodies, all are indifferent to our daily wants and desires. It is this indifference, this stark and shadowy blankness, that is both alluring and terrifying. And this is the terrain of Ito's manga, where, for a brief moment, we glimpse something of the order of deep time and the scale of the unhuman'. This is one of the very few articles that succeeds in connecting the art form with a meaningful Anthropocene imaginary.[24]

Finally, we searched six media sources from Afghanistan (*Afghan Online Press*, *Afghanistan Times Daily*, *Bakhtar News*, *Khaamess*, *Pajhwok Afghan News*, and *ToloTV*), finding some coverage of climate change, but nothing at all on the Anthropocene. Similarly, three online searches in Mongolia (*Montsame*, *UB Post*, and *Ddish*) gave some coverage of climate change but nothing on the Anthropocene.

Conclusion

The conclusions that we draw from coverage of the Anthropocene in the media of North Asia are tentative. While media in China are state controlled to a greater extent than in the other countries, we are somewhat more likely to find pessimistic narratives in Chinese media than elsewhere.[25] However, these narratives also tend to offer philosophical and technical answers rather than genuine strategies of radical socio-economic change. On the basis of our results, we can say that the media in North Asia, in general, provide their public with a good deal of accurate information about the Anthropocene, though not so much about the potential risks that might be faced in their or their children's lifetimes.

Notes

1 Young Jeong Kim, Rahul Patel, Emile St.-Pierre, and Li Weixiao also made contributions to this chapter.

2 It appears that the term was first used in China by a Chinese geologist, Liu Dongsheng, in 2004 (see Weigelin-Schwiedrzik 2017: 19).
3 Lian Ma (2015) is a study of how air pollution in Beijing was reported (also Lei Xie 2015, and Li 2019). President Xi (2017: 428-31), acknowledges serious problems of air, water, and soil pollution in China. For internal sources, see *Environmental Protection Web* (http://hbw.chinaenvironment.com/) on pollution and other issues, and *Chinese Environmental Newspaper (*https://www.cenews.com.cn/*)* for the rapid growth of local environmental courts (and fines). These sources provide legible English translations, but neither mentions the Anthropocene. Shytov (2017) explores problems of regulating environmental crime in China.
4 Chen (2018) offers a good survey of how the media in China cover climate change. Dan Lin (2017) presents a thorough analysis of how ecological conservation has developed in Shenzhen. For Shenzhen media, see Lee (2006), and Lee et al. (2007) unravels the complex story of the politics of the 'timid' media in Shanghai.
5 For critiques of the widespread uncritical deployment of sustainable development see, for example, Luke (2005), Lautensach and Lautensach (2013), and Sklair (2019a). Zeng, Y. *et al.* (2020) address the issue with reference to the UN goals.
6 This is reported all over the world. Zalasiewicz et al. (2019: Section 1.4.3.2) provides a pre-Coronavirus overview of health in the Anthropocene.
7 Tong (2014) suggests that some journalists in China criticize economic modernization on ecological grounds.
8 While these articles were tagged 'Anthropocene', it is not certain that the word appears in all of them.
9 Elvin (1998) provides a long-term perspective on the environment in China. See also Pomeranz (2017) and, on emissions, Wang Tao (2018). Cheng Li and Yanjun Liu (2017) explores two Chinese 'drought' novels through the lens of material ecocriticism.
10 The 'green' sincerity of the Chinese government is often challenged, notably in the context of the potential greenhouse gas emissions from the massive Belt and Road Initiative (BRI). There is no serious treatment, so far, of the topic in the *China Belt and Road Initiative Journal*. However, see Ning Zhang et al. (2017), Xiang Cai et al. (2018), Hilton (2019), and, most dramatically, Saha (2019). For an unconvincingly optimistic assessment of green financing, see 'How the "Belt and Road Initiative" Can be China's Path to Green Leadership' by an HSBC executive (*South China Morning Post*, 21 January 2018).
11 We are grateful to Zhao Defa for his interest in the project and to Peina Zhuang for her translation of the cover text. Zhuang (2019) presents a rather different perspective on the novel than we find in the media.
12 For an analysis of renewable energy in China, which is never discussed in our results, see Standaert (2019).
13 Yin-Yueh Lo and Peters (2015) present an interesting comparison of the 'medialization' of scientists in Taiwan and Germany. On environmental movements in Taiwan and South Korea, Hwa-Jen (2015).
14 In 2018-19 a series of meetings on the 'Pan-Pacific Anthropocene' was sponsored by the Research Centre for Future Earth at the National Taiwan University as preparation for the founding of a scholarly association (APPA).
15 Riemenschnitter and Imbach (2018) provide an overview of debates around sustainability in the 'Sinosphere'.
16 The city of Gwangju was the site of a massacre of civilians protesting against the Korean military dictatorship in 1980. A moving feature film was made of these events.
17 In addition, some articles were cut short due to restricted access.
18 The poster says, 'Toward Human History in the Biosphere and the Technological Sphere – Biosphere Technosphere'. For the full transcript see: file:///JAPAN%20MUSEUMS %20IN%20ANTH.pdf

19 On Sustainocene, see Furnass (2012) and Faunce (2012). So far, the term has not caught on.

20 2011 was the year of the Fukushima Daiichi nuclear disaster.

21 It may be found, with an English translation, at http://10plus1.jp/monthly/2017/01/issue-09.php

22 See Breakthrough Institute (2015) for the *Ecomodernist Manifesto*, generally associated with business-friendly approaches to the Anthropocene.

23 Surprisingly, in light of rising sea levels and flooding in Japan, we found no media coverage of the attention scholars have been paying to Hokusai's Great Wave in the Anthropocene (see Helmreich 2016).

24 Johnson (2014) provides an exhilarating introduction to Manga in the Anthropocene.

25 Also notable is media silence, especially in China, on urban and gender questions. Zhong (2018), based on survey datasets from Chinese city-dwellers, confirms the prominence of women and the young in green issues. Chinese women in the creative arts have spoken out on the Anthropocene (Jaguścik 2018). Huan (2014) provides a useful 'Red/Green' analysis of the growth of environmentalism in China.

References

Breakthrough Institute (2015) *An Ecomodernist Manifesto*. https://static1.squarespace.com/static/5515d9f9e4b04d5c3198b7bb/t/552d37bbe4b07a7dd69fcdbb/1429026747046/An+Ecomodernist+Manifesto.pdf

Brevini, B. & J. Lewis eds. (2018) *Climate Change and the Media*, Volume 2. New York: Peter Lang.

Cai, Xiang et al. (2018) Will developing countries become pollution havens for developed countries? An empirical investigation in the belt and road. *Journal of Cleaner Production* 198: 624–632.

Chen, S. (2018) Environmental journalism in China: Challenges and prospects. In Brevini, B. & J. Lewis eds. *Climate Change and the Media*, Volume 2. New York: Peter Lang, chap. 6.

De Burgh, H. & Z. Rong (2012) *China's Environment and China's Environment Journalists: A Study*. Bristol, UK: Intellect Books.

Elvin, M. (1998) The environmental legacy of imperial China. *The China Quarterly* 156: 733–756 (Special Issue: China's Environment).

Faunce, T. (2012) Towards a global solar fuels project-artificial photosynthesis and the transition from Anthropocene to sustainocene. *Procedia Engineering* 49: 348–356.

Furnass, B. (2012) *From Anthropocene to Sustainocene: Challenges and Opportunities*. Public Lecture. Canberra: Australian National University (21 March).

Helmreich, S. (2016) Hokusai's great wave enters the Anthropocene. *Environmental Humanities* 7(1): 203–217.

Hilton, E. (2019) How China's big overseas initiative threatens global climate progress. *Yale E360* (January).

Huan, Q.Z. (2014) Development of the red–green environmental movement in China: A preliminary analysis. *Capitalism Nature Socialism* 25(3): 45–60.

Jaguścik, J. (2018) Feminist responses to the Anthropocene: Voices from China. *International Communication of Chinese Culture* 5(1): 83–100.

Xi Jinping, (2017) *The Governance of China*, 2 Volumes. Beijing: Foreign Languages Press.

Johnson, K.L. (2014) Manga in the Anthropocene: Notes toward a cyberpunk ecology 2. *Southeast Review of Asian Studies* (*SERAS*) 36: 112–123.

Kamei, A. et al. eds. (2016) *Museums in the Anthropocene: Toward the History of Humankind Within Biosphere & Technosphere*. Tokyo: National Museum of Nature and Science.

Lautensach, A. & S. Lautensach (2013) Why 'Sustainable Development' is often neither: A constructive critique. *Challenges in Sustainability* 1(1): 3–15.

Lee, C.-C. (2006) Chinese party publicity Inc.: Conglomerated: The case of the Shenzhen press group. *Media, Culture & Society* 28(4): 581–602.

Lee, C.-C. et al. (2007) Party-market corporatism, clientelism, and media in Shanghai. *Harvard International Journal of Press/Politics* 12(3): 21–42.

Li, Cheng & Yanjun Liu (2017) Reading climate change in the Anthropocene: Material ecocriticism and Chinese environmental literature. *Green Letters* 21(2): 138–151.

Li, H.C. (2019) Smog and air pollution: Journalistic criticisms and environmental accountability in China. *Journal of Rural Studies*. https://doi.org/10.1016/j.jrurstud.2019.10.024

Lin, Dan (2017) Urban ecological conservation policy in Shenzhen: The production of a 'Biophysical Fix'. *Urban, Planning and Transport Research* 5(1): 38–58.

Liu, Hwa-Jen (2015) *Leverage of the Weak: Labor and Environmental Movements in Taiwan and South Korea*. Minneapolis, MN: University of Minnesota Press.

Lo, Yin-Yueh & H.P. Peters (2015) Taiwanese life scientists less 'medialized' than their Western colleagues. *Public Understanding of Science* 24(1): 6–22.

Luke, T. (2005) Neither sustainable nor development: Reconsidering sustainability in development. *Sustainable Development* 13: 228–238.

Ma, Lian (2015) The Beijing smog: Between media frames and public perceptions. *China Media Research* 11(4): 6–15.

Marinelli, M. (2018) How to build a 'Beautiful China' in the Anthropocene. The political discourse and the intellectual debate on ecological civilization. *Journal of Chinese Political Science*. https://doi.org/10.1007/s11366-018-9538-7

Marks, R.B. (1996) Commercialization without capitalism: Processes of environmental change in south China, 1550–1850. *Environmental History* 1(1): 56–82.

Martindale, L. (2019) *Tasting the Cosmological Rift: Alternative Food Networks in China's Ecological Civilization*. PhD dissertation, Lancaster University.

Neuweg, I. & N. Stern (2019) China's 14th plan, sustainable development and the new era, 20th China forum. London School of Economics. www.lse.ac.uk/GranthamInstitute/wp-content/uploads/2019/05/Chinas-14th-plan-sustainable-development-and-the-new-era.pdf

Pomeranz, K. (2017) Water, energy and politics: Chinese industrial revolutions in global environmental perspective. In Austin, G. ed. *Economic Development and Environmental History in the Anthropocene: Perspectives on Asia and Africa*. London: Bloomsbury, chap. 13.

Riemenschnitter, A. & J. Imbach (2018) Special issue Anthropocene matters: Envisioning sustainability in the Sinosphere. *International Communication of Chinese Culture* 5(1): 121–144.

Saha, S. (2019) China's belt and road plan is destroying the world. *The National Interest* (18 August).

Sampei, Y. & A.-U. Midori (2009) Mass-media coverage, its influence on public awareness of climate-change issues, and implications for Japan's national campaign to reduce greenhouse gas emissions. *Global Environmental Change* 19: 203–212.

Shytov, A. (2017) Environmental crime and communication to the public in China. *Journal of Chinese Political Science* 22: 57–75.

Sklair, L. (2019a) The corporate capture of sustainable development and its transformation into a 'good Anthropocene' historical bloc. *Civitas* 19(2): 296–314.

Standaert, M. (2019) Why China's renewable energy transition is losing momentum. *Yale E360* (September).

Tolan, S. (2007/8) *Coverage of Climate Change in Chinese Media*. Human Development Report Office (UNDP Occasional Paper).

Tong, J. (2014) Environmental risks in newspaper coverage: A framing analysis of investigative reports on environmental problems in 10 Chinese newspapers. *Environmental Communication* 8(3): 345–367.

Wakabayashi, M. (2013) Voluntary business activities to mitigate climate change: Case studies in Japan. *Energy Policy* 63: 1086–1090.

Wang, Tao (2018) Land use change and carbon emissions in china from 1990 to 2015. *IOP Conference Series Earth and Environmental Science* 186: 012077.

Weigelin-Schwiedrzik, S. (2017) Doing things with numbers: Chinese approaches to the Anthropocene. *International Communication of Chinese Culture* 5(1): 17–37.

Xie, Lei (2015) The story of two big chimneys: A frame analysis of climate change in US and Chinese newspapers. *Journal of Intercultural Communication Research* 44(2): 151–177.

Zalasiewicz, J. et al. eds. (2019) The Anthropocene as a Geological Time Unit: *A Guide To The Scientific Evidence And Current Debate*. Cambridge: CUP.

Zeng, Y. et al. (2020) Environmental destruction not avoided with the Sustainable Development Goals. *Nat Sustain* (2020). https://doi.org/10.1038/s41893-020-0555-0.

Zhang, Ning et al. (2017) Letters: Carbon footprint of China's belt and road. *Science* 357(6356): 1107. (15 September).

Zhong, Y. (2018) Environmental views of the Chinese urbanites. In Zhong, Yang ed. *Political Culture and Participation in Urban China, New Perspectives on Chinese Politics and Society*. Singapore: Springer Nature, chap. 6.

Zhuang, P. (2019) Desire and the body in Zhao Defa's the Anthropocene. *Neohelicon*. https://doi.org/10.1007/s11059-020-00515-4

7 South Asia

The 'provincializing' dilemma

Leslie Sklair, Jahnnabi Das, and Sunitha Kuppuswamy[1]

Introduction

Our searches of the media in 13 countries of South Asia brought up around 150 items from 188 sources, with only two countries (Laos and the Maldives)[2] producing no mentions of the Anthropocene. Almost half of the sample fell into the neutral frame, with most of the others being in the explicitly 'good' Anthropocene frame. Media in India, Pakistan, Bangladesh, and the Philippines made up over 80 per cent of the total, providing many high-quality items, many of which were home grown. Results from some smaller countries were impressive, notably Bhutan, where the single article 'How Will the Age of Anthropocene Fare for Bhutan? (2016), in the national newspaper *Kuensal*, compares the Anthropocene with Kali Yuga (the Hindu 'Dark Age') in a most informative essay. We found very few items on the Anthropocene outside the English-language media. This chapter will attempt to explain these patterns with reference to postcolonial arguments evident from some of the articles. The media in the subcontinent engage with the work of three scholars, Dipesh Chakrabarty, Amitav Ghosh, and Kathleen Morrison on issues around the provincializing of the Anthropocene (see Chapter 12 of this book).

We are fortunate in having several very useful scholarly accounts of the Anthropocene in Southeast Asia, which provide context in which to understand our media results from the region. For example, Corlett (2013: 1) explains: 'When I visited the Southeast Asian tropics as an undergraduate in 1973, it seemed as different from my European home as anywhere on Earth could be. The forest was already in retreat, but large contiguous tracts remained virtually intact. Fast-forward 40 years, however, and more than half of the original forest cover of Southeast Asia has gone, more than half of what remains has been logged or otherwise degraded, and the majority, logged or unlogged, has lost most or all of its large vertebrate fauna to hunting'. Corlett goes on to document the rise of plantation monocultures and urban areas and their associated infrastructure, and air pollution. 'Despite these similarities, Europe is considered a relative success story in conservation terms, while Southeast Asia is widely seen as an on-going biodiversity crisis' (ibid). Corlett suggests three main reasons for this. Southeast Asia supports 15–25% of global terrestrial species, while Europe has fewer species (5% of the global total) in a similar total land area, species in tropical forests are likely to be more vulnerable to extinction, and the ecological transformation of Southeast Asia has been extremely rapid. Hudson complements this in a broader Asian context: 'the concept of the Anthropocene can be said to work against the regional and bounded ideas of Asia and Asian studies. One of the great strengths of area studies lies in its local contextualizations, yet the cumulative effects of global human activities over at least the past 200 years have resulted in changes to the basic biological, chemical, and climatic processes of the whole Earth, changes that ultimately affect all humans' (2014: 943) but, it might be added, not all equally. Here, Hudson poses a question that might be asked of all the regions in our study: 'Does this mean that the Anthropocene is simply another process by which Asia becomes more like "the West"?... it is fair to conclude that most observers probably view the Anthropocene in such Euro-centred terms. For their part, there are various reasons why Asianists may have been slow in thinking about the Anthropocene' (ibid: 944). Many of our results suggest that at least some of the media in South Asia may have been ahead of 'Asianists' in this respect.[3]

India

Studies by Billett (2010) and Boykoff (2010) provide a good background for how media in India respond to the challenges of reporting climate change/global warming (CC/GW), and Nayak (2015) is a useful guide to environmental movements, this research anticipates the emergence of the Anthropocene in the Indian media. The first item we found appeared in the English-language *Deccan Herald*, a regional paper with seven different city editions. 'Developmental Activity Should Not Degrade the Environment But Has to Sustain the Earth' (2006) by the eminent space scientist, the late U.R. Rao, is an eloquent statement of issues raised by the Anthropocene. It was seven years before the next, 'Overpopulation a Problem For Our Finite Planet? Not Really' (2013), reprinted from *The New York Times*. This was by Erle Ellis, soon to become an influential figure in Anthropocene studies.

The national daily *The Indian Express* begins its coverage with an editorial 'Age of Man: A New Human-centric Geological Epoch could be Apocalyptic – or Bring New Hope' (2016), a summary of causes rather than risks, and this sets the tone for many of its articles. 'Earth's "Technosphere" Weighs 30 Trillion Tonnes' (2016) explains the concept, quoting a member of the University of Leicester research team. 'The Industrialized Human Race is on the Verge of Entering the Geological Record: What Would That Mean For the World? (2017) discusses terraforming and science fiction. *The Hindustan Times,* a daily founded in 1924 by Gandhi, provides two informative articles, both in 2016: 'The Anthropocene, or "New Age of Man," Would Start From the Mid-20th Century If the Recommendations are Agreed' cites the Great Acceleration, the golden spike, Paul Crutzen and Stanley Finney, and 'Earth Heading Towards 'Age of Plastic', Says New Study', borrows from the journal *Anthropocene,* another reference to research from the University of Leicester, noting: 'Plastics make excellent stratigraphic markers'.

The Hindu, a daily with national circulation, prints 'First Nuclear Bomb Set Off the Anthropocene' (2015), beginning the transition from the Holocene to the Anthropocene 'an epoch of catastrophic human-led change', referencing climate change, loss of biosphere integrity, land-system change, and altered biogeochemical cycles (phosphorus and nitrogen). These, we are told, are likely to cause serious threats to human well-being. Similarly, 'Anthropocene: Why This Naming Has Risen in the Earth's History' (2016) outlines the risks, but other items simply provide geological facts, as in 'The Anthropocene is Here, Propose Scientists' (2016), citing Jan Zalasiewicz on micro-plastics, sourced from AFP in Paris. *The Tribune* is both a leading daily in parts of India and a newspaper with worldwide circulation. 'Natural Progression' (2015) reviews Diane Ackerman's popular science book, *The Human Age*, on the Anthropocene. Despite the risks, we are reassured that 'We are Earth-movers. We can Become Earth-restorers and Earth-guardians'. Less optimistic is 'Living in the Anthropocene' (2016), explaining how scientists have launched the global hunt for a 'line in the rock' marking the 'scary' new man-made epoch' (2016). This message is reinforced by a quote from Clive Hamilton, that we should not be 'welcoming the Anthropocene'.

The Economic Times is a daily, headquartered in Mumbai, said to rival *The Wall Street Journal* in circulation. Its two items, in 2016, are the bland 'Human Activity Leading Earth into New Geological Epoch? and 'World May Lose Two-Thirds Of Its wildlife by 2020, Sixth Extinction on Cards', though the Worldwide Fund for Nature (WWF) is quoted: 'There is still considerable room for optimism'.

Asian Age, an English-language daily with editions in Delhi, Mumbai, and Kolkata, has several items of interest. For example, '2015 To Go Down In Annals of History as "Tipping Point" for Climate Change' addresses the politics of CC/GW, quoting the influential environmentalist Bill McKibben: 'The most compelling thing you can say about Paris [climate change agreement] is not that it saved the planet, but that it saved the chance of saving the planet'; 'Mankind's CO_2 Emissions Delay the Next Ice Age: Study' explains: 'the most recent part of the Holocene, following the Industrial Revolution, has become known as the Anthropocene'; 'Sea Swept: Celebrating "Earth Day" Has Become an Urban Fad' reports on a photography exhibition of the work of Delhi-based photographer and green activist, Ravi Agarwal, in the context of the local fishing industry. Finally, 'Under Water, Under Siege' (2017), by Professor Amita Bhide from the Tata Institute of Social Sciences, is a long, scholarly analysis of the problem of recurrent flooding in many Indian cities. Bhide argues: 'The poor, who are marginalized from more valuable lands, are pushed into occupying more contentious, environmentally risky terrains – riverfronts, nalas [drains], hill slopes, and edges of lakes … this segmented nature of the urban Anthropocene in India needs to be recognised because, often, decisions meant to deal with sites of environmental risk turn out to be a double whammy for the poor – they are the worst victims of disasters like floods; they are also victims of governmental actions to prevent future floods'.[4] English- language newspaper coverage of the Anthropocene is completed by a few of the more familiar stories in the *Deccan Chronicle*, *The Financial Express* and *The Telegraph*.[5] The Hindi language newspaper *Dainik Bhaskar* is considered to be the highest-circulation paper in India. In 2016, it published an informative article on the Cape Town Geology Conference, but nothing has been found since.

Two magazines in India have significant Anthropocene coverage. The fortnightly *India Today* provides about 14 articles on the Anthropocene, the earliest published in 2013. Most are unedited feeds from the Press Trust of India wire, including studies from scientific journals. These cover a wide variety of topics; for example 'Mayacene' (2014) provides a deep historical perspective. 2017 brings several items, for example, the report of a speech by the renowned Indian agriculturalist, M. S. Swaminathan, who says: 'the Anthropocene era is just catching on where humans are going to be the major determinants of the future'; the much-reported story from the *American Mineralogist* on the discovery of new 'Anthropocene minerals'; and 'Against the Backdrop of Dramatic Ecological Shifts That Has Come to Define the Anthropocene' reports on an Anthropocene-themed art exhibition in New Delhi. *India Today*'s coverage of current Anthropocene research continues with reports of the 'Anthropocene equation' from the Australian National University, on the technosphere and the 'age of plastics' from the University of Leicester, and the WWF report on species extinction. The magazine *Outlook* has eight items.In the first in 2015, Louie Psihoyos is interviewed about his documentary film, 'Racing Extinction' referencing the "so-called" Anthropocene; the remainder all appear in 2016. In 'Quo Vadis Earth?' the journalist Pranay Sharma gives a clear account of how the term 'Anthropocene' emerged, focusing on the work of the Anthropocene Working Group (AWG) and Zalasiewicz. Sharma concludes: 'From the future of a proposal to the future of a planet, no one's taking bets'. Other local commentators, Soity Banerjee, Ashwin Rammohan, Kajal Basu, and Dipsikha Thakur (for whom the Anthropocene is 'nothing but imperialism of normality') connect the Anthropocene in passing with a variety of topics, mostly neutralizing.

The term Anthropocene is also mentioned directly (in English) or indirectly (transcribing) in indigenous language magazines, for example *Desh-Videsh*, *Dainik Jagran*, *Vidyafeed*, *Hari Bhoomi*, *Nava Bharat*, and *Punjab Kesari* in Hindi, and *Anandbazaar Patrika* and *Aajkal* in Bengali. The Anthropocene also features in Tamil language media. For example, in 2015, *Thinakaran* (Sri Lanka), read by Tamil-speaking educated elites from India and Tamilnadu, offers a discussion on the adoption of modern cropping systems yielding greater harvests more quickly, but with negative effects on human survival, soil quality, and agriculture in general (see Roy 2017).[6] In 2016 *Kunkumam*, a popular Tamil weekly magazine (published in Chennai), has an unusually alarming story on the Anthropocene, linking it with plastic usage, increase in CO_2 emissions, sea-level rise, loss of endangered species, and deforestation. According to this article, probably the Anthropocene will be the 'last epoch of man'. Two years later, the devastating Gaja cyclone wreaked havoc in the state of Tamilnadu, with significant loss of life and green cover (see Figure 7.1).

Figure 7.1 Loss of green cover after the Gaja cyclone, Thanjavur district of Tamil Nadu © Ms. S. Mahalakshmi (2019).

Sri Lanka

Coverage of the Anthropocene by media in Sri Lanka is sparse. The *Daily News* prints two items, both in 2016, the first: 'Buddhist Solution for Environmental Problems' offers a rarely articulated commentary on the Anthropocene, the second: 'High-Income Economic Status in Next Two Decades: PM' reports a speech by prime minister Wickremesinghe at a new Science Centre, making an unlikely connection between Paul Crutzen's Anthropocene and national prosperity. The single item in *The Daily Mirror*, 'Let Us Renew Our Commitment to Social Justice' (2017), also references the prime minister, this time celebrating 'World Day of Social Justice' and implying that the UN endorses the Anthropocene. Similarly, the government publication *News* (2016) considers local effects of the Anthropocene, admitting: 'We are beginning to feel the adverse impacts of neglecting to protect the land on hills, wildlife habitat in Sri Lanka, plains to prevent soil degradation, and water resources'. Coverage in Sri Lankan media concludes with *The Sunday Observer*, 'World On Track to Lose Two-Thirds Of Wild Animals by 2020' (2016), a report on the 'Living Planet Index' from the WWF, and *The Sunday Leader* with 'The Anthropocene Extinction' (2016), on the album from the deathgrind band Cattle Decapitation from San Diego, USA.

Pakistan

Media coverage of the Anthropocene in Pakistan is marked by a good deal of borrowing from foreign sources, intermingled with some strictly local stories. We successfully searched 21 media sources which produced 42 results, with 12 of these sources (including two in Urdu) giving no results. The influential *Express Tribune* newspaper (owned by the Lakson Group, one of Pakistan's most important and best-connected transnational conglomerates) began its coverage of the Anthropocene with three articles that are notable in their own ways. The first, 'Trees And Us' (2012), explains that Ancient Greece was heavily wooded but the Greeks cut down the trees and the soil got washed away and now trees cannot grow easily. This is a warning for the Anthropocene 'because man is the leading cause of change to the environment'. The journalist confesses to have read 'some really scary figures' about this. 'Heatwaves, Droughts, Floods are All Adverse Effects of Climate Change' (2014) certainly sounds like an alarm call, though the article is illustrated with a photo of two apparently cheerful men pouring water over themselves to cope with the heat. The article is an account of a meeting organised by the Friedrich Naumann Foundation, attached to the Free Democrats, the German Liberal Party. The workshop, held at the Council of Pakistan Newspaper Editors (CPNE) office, was organised by the Foundation and CPNE 'to create awareness about the pressing issue of climate change and capacity building of journalists'. Two local experts briefed the workshop. We are told: 'the journalists were apprised of the role they could play while discussing the issues that they face on the ground when covering topics related to climate change and environment'. The Anthropocene is mentioned in passing in this rare case of the media discussing itself.

Also in 2014, *The Express Tribune* reports a talk given by the eminent New York-based postcolonial scholar, Professor Gayatri Chakravorty Spivak, in a conference in Karachi on postcolonial higher education: 'Do Not Let Education Join the Anthropocene – Human Activities Hastening the Extinction of the Earth', with no further details. From the mid-2010s, the reporting becomes less local; for example, in 2015, 'Health Impact: Global Report Outlines Health Challenges', on the launch of a Rockefeller Foundation–Lancet Commission report on planetary health 'Safeguarding Human Health in the Anthropocene Epoch'; and in 2016, 'Wildlife Populations Plunge Almost 60 Percent Since 1970', quoting an official of WWF, 'We are entering a new era in Earth's history: the Anthropocene'. In 2017, we find the local and the global in 'Pakistan's Endangered Species More Vulnerable Than Ever', concluding: 'Under the current trajectory, the future of many living organisms in the Anthropocene (current geological age) is uncertain'.

Pakistan Today reprints an item from the BBC, 'What Will be the Next Global Transformers?' (2013), explaining planetary transformation: 'humans have utterly transformed the planet on such a scale that many believe we are entering a new geological era, the Anthropocene', with a basic summary of what that means. The conclusion: 'Taxes and death notwithstanding, nothing is certain about the future. But it can be fun predicting what might shake things up'. It is uncertain whether the 'fun' comes from *Pakistan Today* or the BBC. In 2017, *The Pakistan & Gulf Economist* publishes an article by the influential Johan Rockström (who we encounter in most regions), 'We Can Still Rescue This Planet From Climate Change. Here's How', explaining research from the Stockholm Resilience Centre on tipping points in complex systems. He challenges those economists who argue that environmental destruction is a necessary evil on the way to prosperity: 'In the Anthropocene, this logic no longer applies. Even the wealth of the United States cannot protect against the levels of environmental destruction that we are unleashing', adding a fundamental truth about the Anthropocene, rarely clearly stated in the media: 'This story goes beyond climate. Other ways that industrialised societies are affecting the global commons include habitat destruction and the mass extinction of species, pollution of nitrogen and phosphorus from fertilizers leading to dead zones, ozone depletion, and ocean acidification'. Nevertheless, Rockström concludes on the optimistic note of his title, discussing the reassuring idea of 'safe operating spaces'. The story that 'Humans are Causing Climate To Change 170 Times Faster Than Nature' (2017) appears frequently. This is of interest because it references a technical research article by Owen Gaffney and Will Steffen, which is often illustrated with pictures of industrial chimneys belching out smoke. *Pakistan Today* and *The Nation* both reproduce versions of it, presumably from well-crafted press releases. *The Nation* also prints two more interesting items, 'Tipping Point for Climate Change' (2016), summarizing a little of the science, and quoting Saleemul Huq, director of the International Centre for Climate Change and Development in Dhaka, on the need for 'greater planetary consciousness' (an idea picked up by other media in South Asia); and in 2017 'Climate Change Hoax or Fact?', concluding that it is a fact, referencing Gaffney and Steffen (2017) on the 'Anthropocene Equation'. This item came to our attention, not through a discussion of the Anthropocene, but because the writer mentioned that the article appeared in the *Anthropocene Review*, a situation that occurred several times in our results.

The News International is the largest English-language newspaper in Pakistan, with an overseas version published in London. Most of its impressive Anthropocene coverage appears to be sourced from abroad. For example, 'Climate Apocalypse' in 2015 ranges far and wide and is from the US-based non-profit progressive news website *Common Dreams*. Two 2017 items, first published in the bi-monthly radical US magazine *Counterpunch*, follow. 'The Age of the Anthropocene' is excerpted from: 'What Does It Mean to Live in the Age of the Anthropocene?' It begins 'It's another day in the age of the Anthropocene where a global game of musical chairs continues to play out. As humans continue to plunder and pillage the Earth in a global economy that thrives' and continues in this challenging vein. The second item, 'Saving Species? Or Killing Them?', begins: 'Last year I read an Op-Ed in *The New York Times* entitled "Building an Ark for the Sociopocene." No, I lied. It was entitled "Building an Ark for the Anthropocene." But can't you imagine how the article might have read were it accurately titled to reflect the sociopathic nature of the world we have created?' This was excerpted from 'Endangered Species Don't Need an Ark – They

Need a Living Planet!' by the radical green activist and prolific author, Derrick Jensen. With 'A Warming Planet' in 2017, *News International* reviews Clive Hamilton's book *Defiant Earth*, concluding with Hamilton's suggestion: 'Earth's survival as a living organism may depend on humans transitioning from our role as Earth exploiters to a role as facilitators of Earth healing' but not how this might be achieved. *The Pakistan Observer* credits *The Guardian* for a 2016 article, 'Climate Change: Need For a New Financial System', making the unusual but interesting argument: 'reinventing our money system is crucial to our survival in the Anthropocene'; and, in the same year, borrowing from *The Washington Post,* an article by Engelke and McNeill (great acceleration researchers) argues, optimistically, 'While the Anthropocene Remains Young, and Our Options Remain Open, We Must Do Much More'.

Many sources have a single item. For example, *The Daily Times* (the organ of the Pakistan People's Party) reports on the Living Planet Index, which measures global biological diversity, based on trends in populations of vertebrate species from around the world (the Index is a joint venture of two influential bodies, the Zoological Society of London and the WWF). The weekly *Friday Times*, in 'Off the Grid But On The Ground' (2017), poses the important question 'what might the future [identified as the Anthropocene] spell if we continue with "business as usual"?' The paper gives no answers. *Dawn* is Pakistan's most important English-language newspaper, the country's journal of record. With about 14 results, its coverage of the Anthropocene is impressive. Beginning with 'Planet is In Critical State, Warns Science Declaration' (2012), it suggests that 'a scientific consensus that we inhabit a new epoch, the Anthropocene, in which humans are dominating planetary-scale processes' is emerging, then 'Entering the Anthropocene' (2013), by the award-winning environmental journalist Rina Saeed, who presents a well-researched critique of the unfulfilled promise of sustainable development and the 'triumphalism' of the Rio Earth summit of 1992 (this article was updated in 2019 with more acerbic criticism of the UN Rio+20 conference). However, Saeed expresses a little more optimism on the prospects that scientists can get to grips with development goals in the Anthropocene, citing the efforts of 'Future Earth' (identified in Chapter 1 of this book as a sponsor of the idea of the 'good' Anthropocene). In 2017, President Trump makes an appearance in effigy in the sardonic article 'Global Apartheid: On the Historical Superiority of the West' by the philosopher and public intellectual Nauman Naqvi. He begins: 'We are now well informed that the single most significant, the defining fact of our global historical situation today is that we live in the Anthropocene'. His conclusion is enigmatic. 'Personally, being modern and busy with my career, I find little time for lengthy prayer – so I pray simply: inshallah'. Finally, in 2015, *Dawn* publishes 'Climate Change a Bigger Threat Than Terrorism', quoting remarks by Dr Fawad Chaudhry, a former director-general of the Pakistan Met Office and vice-president of the World Meteorological Organisation, during a presentation given to members of the Parliamentary Task Force on Sustainable Development goals (SDGs). Chaudhry points out that Pakistan was 135th among carbon-emitting countries and contributed only 0.8% of global carbon emissions.

'Unfortunately, in our country, no government has given priority to the issue, even though climate change is considered a major challenge at the global level'. Citing the need to plant more trees, solve water problems, create green energy, and clean up industries, he declares: 'The era of the anthropocene is devastating planet Earth'.

Bangladesh

From 2009 to 2017, we found almost 50 articles on the Anthropocene in Bangladesh, in 15 English-language print and online media, while five other sources had none. As in most countries, the coverage peaked in 2016–17 (making up about two-thirds of the total). Most of these stories could be classified as neutralizing or reassuring, many drawing attention to potential and actual risks. The first and most articles (19) appear in *The Daily Star,* often considered the newspaper of record. Articles appeared predominantly in two forms – news and op-eds. While op-eds are mostly written by local expert columnists, Anthropocene news stories are usually sourced from international news agencies such as Reuters and AFP, with some local events reported by national newspapers. The news from international outlets focused on the debate surrounding the approval process of the Anthropocene (often labelled the 'new age of man') with reference to the Cape Town Geology Conference in 2016. In the same year, the word 'Anthropocene' attracted exceptional news attention in Dhaka, where an article in *The Daily Star* highlighted the effects of human activities, such as unrelenting urban development in Bangladesh, through an art exhibition 'Fragments Of the Anthropocene'. This profiled the work of two Bangladeshi artists that captured some geologically significant conditions emerging from loss of heritage, creativity, biodiversity, and the incessant appetite for aggressive urbanization in two prominent cities, Dhaka and Chittagong. The degrading impact of urbanization as a crux of human activities is the focal point of op-ed columns that appear in three other media outlets (*The Daily Sun, NTV, The Independent*). Based on a number of scientific reports on climate change and adaptation, these articles convey a sense of the Anthropocene to readers by highlighting the profound impacts of environmental degradation in the region.

Op-eds operate predominantly in three forms. First, some items give a dystopian view on climate change, warning readers to expect more catastrophic events, such as cyclones and flooding, which would become more extreme and common over time in the Anthropocene. 'The Extinction Factor', by the prolific local academic Dr Binoy Barman, in *The Daily Star* (2010) reports the views of the late Frank Fenner, an eminent Australian biologist, that the human race itself is at risk of extinction. Barman comments: 'Fenner's observation is important, albeit alarming. He has issued the last word, in the form of a warning, to mankind. A serious warning! *Sadhu Sabdhan*! Be careful or face doom! Humans will be destroyed and they themselves will be responsible for their destruction'. This type of representation not only articulates the severity of the state of the planet but also invokes what the environmental sociologists Lidskog and Waterton (2016:

395) identify as 'moral tension and urgency'. The sombre message is reinforced in *The Daily Star* with 'The Deep Impact' (2011), which specifically connects the Anthropocene and Bangladesh, as the location of the biggest delta in the world. 'Anthropocene is Here – AFP Paris' (2016) announces (prematurely) a ruling by the Cape Town Geology Congress on the AWG's efforts to make the Anthropocene official. 'Earth Becoming Inhospitable' (2015), sourced from *The Independent* in the UK, reports on Rockström's research on planetary boundaries, breached due to the Anthropocene, and leading to further dramatic weather events. 'Floods In South Asia and US: Is Climate Change the Link?' (2017), by the climate scientist Saleemul Huq, a regular contributor to these debates, reinforces this story. 'Hope In the Age Of Man' (2011) borrows from an article in *The New York Times* defending the use of the term 'anthropocene', not as a failure of environmentalism, but a recognition of the impacts which humans have had environmentally and geologically, and in the prospect of improving biological diversity and natural earth processes, an example of the 'good' Anthropocene narrative. Similarly, *The Daily Sun* provides its readers with a variety of informative articles. 'A Tipping Point For Climate Change. When Future Generations Write the History Of Humanity's Faltering Quest To Repair Earth's Climate System, 2015 Will Have Its Own Chapter' discusses at length two notable events: '[2015] almost certainly the hottest on record for the planet as a whole, and a rare moment of unity when 195 states [in Paris] pledged to curb the carbon pollution that drives global warming'. This article is unusual for bringing environmental movements and activists into the discussion.

With 'Anthropocene: This Term Refers To the Current Geological Era, Characterized By the Impact That Humans Have Had on the Environment' (2016), on the occasion of Earth Day, *The Sun* (in English and Bengali) proclaims: 'let us seize this green opportunity to familiarize ourselves with the trendy eco-lingo. Among ten words which have become a prominent part of Ecospeak, we find "Anthropocene"'. Paul Crutzen is cited and the American artist, Jonathon Keats is quoted: '[the Anthropocene] was meant to serve as a call to action, not just a physical observation'. Dr Saleemul Huq (a prominent advocate of sustainable development) again provides an optimistic conclusion, arguing that a 'greater planetary consciousness' on climate has shown up in business and politics as well. In his bid to negate existing alarmism in climate change narratives, he claims: 'I would say we are learning a lot about how to adapt to climate impacts, and, paradoxically, the poorer countries are leading the way on generating new knowledge through the practice of adaptation'. Also covered in *The Sun* are stories on 'Reproductive Health in the Era of Sustainable Development' on the importance of SDGs; 'What the Earth Will be Like in 10,000 Years', quotes various optimistic scientists and 'Carbon Emissions "Postpone Ice Age"' (all from 2016). The daily *Asia Age* (Bangladesh) provides impressive coverage, with long and well-researched reporting of several key issues by Anthropocene notables. Prime examples are 'We're Doomed. Now What?' (2015) by Roy Scranton; 'How Land Use Affects Climate Change' (2017) by Sujatha Byravan, a research scientist in Bangladesh; 'Stranger Than Fiction' (2017) on *The Great Derangement* (see Chapter 12 of this book); and 'Threats To the Survival of the Human Species' (2017), a reprint of a very long article by Noam Chomsky.

Several media offer only a few items. *BdNews24.com* is a relatively new online compendium with a good search facility. The first (of three), 'Humans Hastening Sixth Mass Biological Extinction' (2014) reports research by Professor Rodolfo Dirzo and colleagues, published in *Science* on 'defaunation extinction', often due to urbanization. The potential risk to humans is not highlighted here.[7] 'Wildlife Populations Plunge Almost 60 Percent Since 1970' (2016) picks up on the WWF–Reuters report on loss of biodiversity, but, by 2018, we find the typically neutralizing 'World Atlas Traces Human Scar' ... an impact so marked that many earth scientists now use the term Anthropocene to describe the present geological epoch'. *The Dhaka Tribune,* 'Is the Anthropocene World Different From the Stable Holocene Epoch of the Last 11,700 Years That Allowed Human Civilization to Develop?' (2016), is evidence of the exceptionally good media outreach of the AWG by Zalasiewicz and his colleagues, as is 'Plastic Now Pollutes Every Corner of Earth', based on research published in the journal *Anthropocene.* 'What COP23 Achieved' (2017) begins with the risks of the Anthropocene as expressed at the conference, and concludes: 'The biggest success of this conference is that more than 20,000 people around the globe have directly participated in the conference in order to express their opinions on making the planet safer and greener by reducing greenhouse gas emissions'. The writer is Saiful Islam, a Professor of Water and Flood Management in Bangladesh.

The Daily Observer problematizes extinction and the rights of animals in 'Upholding Animal Stewardship in the Anthropocene' (2016) by the Bangladeshi environmental sociologist Choyon Kumar Saha, a message reinforced again in 2016 in 'Extinction: The Decline is Yet Another Sign That People Have Become the Driving Force for Change on Earth, Ushering in the Epoch of the Anthropocene'. Here, Deon Nel, WWF global conservation director, is quoted: 'I don't speak at all about doom and gloom – we do see a lot of positive signs', one of which is the Paris talks. Again, the optimism is vague, and the issue of humans is absent. *The Independent* raises the connection between the Anthropocene and global health in 'Earth's Degradation Threatens Major Health Gains (2015), based on *The Lancet* report that we encounter in media all over the world. 'Urbanization Signal Detected in Evolution, Study Shows' reports the research of Professor Marina Alberti from the University of Washington, who argues: 'The reason these changes are important is because they change ecosystem function, therefore they have implications for human well-being'. *The Financial Express* prints 'In Search of Sustainable Development Model' (2017), proclaiming: 'the Anthropocene is a challenge which needs to be understood in order for Bangladesh to achieve sustainable development'.

Despite the fact that several of these Bangladeshi sources were bilingual, the only searchable Bengali newspaper we found was the prominent *Prothom Alo* (over six million readers per day). The first of three informative articles, in 2015, explains how Tim Beach and his team of researchers from the University of Texas found evidence for the 'Mayacene' in Central America – 'a microcosm of the early Anthropocene'. In 2016, 'Past Shifts in the Biosphere – the Realm of the Living – Show Up in Sediment and Rock, Especially Mass Extinctions Marking Boundaries When up to 95 Percent of all Life Forms Disappeared Within the Geological Blink of an Eye' provides a detailed introduction, referencing Crutzen, Big Oil, and the sceptical Stanley Finney; and in the same year 'The Anthropocene, Which is Argued to Start in the Mid-20th Century, is Marked by the Spread of Materials Such as Aluminium, Concrete, Plastic' tells a familiar story.

Overall, media representation of the Anthropocene in Bangladeshi media provides two main narratives, the neutralizing and the alarmist/reassuring. Neutralizing narratives are descriptive, drawing few conclusions. Alarmist/reassuring narratives identify humans as responsible agents but argue that humans are also finding adaptive solutions in tackling anthropogenic risks in Bangladesh. These outcomes are broadly consistent with research on the reporting of climate change in Bangladesh (see Das 2019, chap. 6), but we found no radical concrete proposals for change. 'New Year Revelry and Our Declining Chivalry' by the late Mohammad Badrul Ahsan in *The Daily Star* (2016) provides an eloquent coda: 'More wealth leading to more consumption has ushered in the Anthropocene … Revelry has displaced chivalry; every celebration substantiates that point. This New Year is another reminder of that irony in progress. Every loud shriek of pleasure will rise from a muted gasp of pain.'

Further South Asia

Of the remaining countries in South Asia, whose media permitted meaningful searches, the most results came from the Philippines, about 30 articles from ten sources, plus two sources with no results.[8] *The Philippine Daily Inquirer* (the largest circulation paper) began its coverage of the Anthropocene in 2011 with 'The Climate Pope' by Johan Rockström, who argues: 'Pope Francis' encyclical on the environment sends a powerful message not just to the world's 1.2 billion Catholics but to the rest of the global population as well. Firmly rooted in science ... [it] recognizes the need for urgent action, as the world confronts potentially catastrophic climate change'. Rockström also provides an expert summary of the Anthropocene debate. This was followed in 2016 by an almost unique article 'Climate Change to be Part of K+12 Curriculum', by the environmental journalist Marian Codilla: 'learning about climate change should not be limited to scientists, policy-makers, and advocates, it should also be discussed in the classrooms'. She quotes Climate Change Commissioner Lucille Sering to the effect that education is essential in addressing climate change impacts, reporting that the commission is currently discussing how to integrate climate change causes and impacts in the K–12 basic education curriculum. While the article itself does not mention the Anthropocene, it is illustrated with a photo from NASA, whose caption says: 'the way humans and their industries are altering the planet, especially its climate, have caused an increasing number of scientists to use the word Anthropocene to better describe when and where we are'. 2015 brings two more articles from Rockström.

'Planetary Boundaries and Human Prosperity' is an update of previous media and scholarly articles, focusing on safeguarding our planetary life-support systems and his research on planetary boundaries; and the syndicated 'Leaving Our Children Nothing', a call for action to ensure no greenhouse-gas emissions, no poverty, and no biodiversity loss, tied to the SDGs. *The Philippine Star*, a popular daily, with 'Foreign Comment: A World Without Coral Reefs' repeats a much-reported story as 'Coral reefs will be the first, but certainly not the last, major ecosystem to succumb to the Anthropocene – the new geological epoch' (2012). A rather convoluted article, sourced from Chengdu in China, proclaims, sensibly, 'Civilized Countries are Not Ones Where the Poor Have Cars, But are Ones Where the Rich Take Public Transit', a quote from Guillermo Peñalosa, Commissioner of Parks, Sport, and Recreation for the city of Bogota in Colombia. Speaking at a TED City 2.0 Salon in Chengdu, Peñalosa reconsiders mobility in the context of development. The Anthropocene appears via an article in the *Smithsonian* Magazine, and a call to make cities liveable again. Published in Cebu, *The Sun Star* is the only national paper in the country not based in Manila. It has one item, also on education, 'This Will Blow Your Mind' (2017) by the environmentalist, Dave Albao. He recalls how he studied a subject called 'Earth Sciences' at Negros Occidental Science High School but it is only now he is appreciating where humans fit into deep time. He says: 'If we are causing a mass extinction in this age of the humans – Anthropocene, as more scientists are now calling it – we are not just killing off a hundred species every year. We are threatening our life support systems. We are threatening our own existence'. The regional tabloid *The Daily Guardian* has one Anthropocene item, 'Waking Up Countries to the State of the Planet' (2012) by Ping Galang, a veteran radical journalist, showing that he continues to speak truth to power. Starting off with anthropogenic damage to vegetable growers in the large municipality of Candaba, he reports the research of scientists from all around the world: 'Consensus is growing that we have driven the planet into a new epoch, the Anthropocene, where many planetary-scale processes are dominated by human activities … society must not delay taking urgent and large-scale action'. Paying particular attention to the economic and social consequences of the Anthropocene, he reproduces results from the Eco2 Index, the ranking of economic and ecological health of 150 countries.[9] These results represent an almost unique opportunity for readers to discover the facts; for example, 'East Timor emerged as the top performer in the Eco2 Index rankings, with its lowest level of ecological deficit … other countries with "good" Eco2 Index ratings were Gabon, Bolivia, Angola, Central African Republic, Namibia, Paraguay, and Argentina … the "worst" performer in the index was Singapore, number 150, with a very high ecological deficit, despite its economic prosperity'. Other industrially advanced economies (including Japan, the Netherlands, Italy, the UK, Germany, and the US) had ecological deficits near the bottom of the rankings. Despite a positive record on the economic side, the Philippines had the poorest rating among the Southeast Asian countries, with the exception of Singapore. Ping concludes: 'Controversial as they may be, the results of the Eco2 Index study could be worth looking into, if only to evaluate the sustainability of current economic growth trends on the state of the nation's environment' (sourced from GMA News, a free-to-air television network).

The daily *Manila Times* showed several items with 'anthropogenic' but few with 'Anthropocene'. Reporting 'Science Film Festival 2017: The Past, Present and Future of Humanity' in the context of the education system, the article argues that STEM is given a boost by the annual international film festival, an initiative of the Goethe Institute from Germany (also reported in Vietnam) – the theme in 2017 is 'Anthropocene: Welcome to the Human Age'. More interesting is what comes next in the article. 'To quote further from the description of the theme provided us: "Industrialization, in particular, has contributed to the unmistakable and often irreversible fingerprint that we are making upon the Earth"', leaving it open to the reader to decide if this is true or not. On this interpretation, the Anthropocene could be seen as a post-colonialist strategy to combat economic competition from Asia.

Less controversial is 'The Golden Spike, or Start Date, for the Anthropocene' (2016) echoing familiar ideas on the Great Acceleration and the vested interests of fossil fuel producers. 'Legal Planet – A Global "Green" Pact?' (2017) profiles Dominique Bourg, philosopher and environmentalist from the University of Lausanne: 'We continue to proclaim the same principles as 20 or 30 years ago, in other words from another time, which was totally different from the reality of the world today. These objectives apply to a state of the planet that no longer exists … Anthropocene era, i.e. a planet whose functioning has been deeply modified by human activities', reminding us that, in 2016, Swiss voters rejected a law to reduce the Swiss ecological footprint. *The Manila Bulletin,* whose slogan is 'The Nation's Leading Newspaper', has one item, in 2017, 'A Bad Habit Badly Needing Change' which addresses the connections between plastic waste and the 'Anthropocene' era, based on an article in the *Smithsonian Magazine* from the United States. *Business Mirror,* a leading business daily, reprints 'Five Signs Earth is in a Man-Made Epoch' from *Bloomberg News* in 2016; and, in 2017, declares: 'No Room For Complacency, Scientists Warned' by Cecilio Arillo, a prominent political writer, who reports that the planet is approaching tipping points, and that this has consequences for national security. Finally, the country's leading tabloid in the Filipino language, *Pilipino Star Ngayon,* prints an unusual item, 'Maison & Objet Paris: It's Time To Re-Wild the Maison' (2016) by Ricky Toledo and Chito Vijandre (described as 'purveyors of bohemian chic') who declare 'Paris is at war … cyber zombies who can no longer distinguish between the virtual and the real. On a more physical level, the bleak future of an Anthropocene has to be neutralized, if not reversed'. Displaying a deft touch in relating science and the arts, they draw our attention to 'Maison & Objet' a professional trade fair in Paris dedicated to lifestyle, decoration, and design. Once again, media reporting of the Anthropocene leads us into strange places.

The media in Malaysia produced 17 results from four sources, whereas three other sources produced no results. We found articles on the Anthropocene in both English and Malay media (Anthropocene is translated as 'zaman manusia', 'age of man'). *Utusan Melayu,* a mainstream Malay newspaper, began its

Anthropocene coverage in 2014 with an article by Azizah Baharuddin, Deputy Chair of the Institute of Islamic Understanding, Malaysia. Discussing the need to save planet earth from the disruptive effects of human development, she argues: 'If humans don't do something and the Earth's temperature continues to rise, say, a total of four degrees more, humans and many other species will not be able to survive. The Anthropocene era will be the end'. This is an unusually uncompromising message. 2015 brings 'Prepare Yourself to Face Disaster' a rather technical question-and-answer session on the effects of the Anthropocene with Dr. Mazlin Mokhtar, an environmental chemist, largely about recent devastating floods in Malaysia, sending messages about religion, the market, and nature in general, concluding: 'Although the Anthropocene has not become a "buzzword" in our society, we don't have a reason not to take up what we can contribute to save the Earth from the Human Age'. *The Sun Daily,* Malaysia's first free tabloid, gives two items 'Earth's Degradation Threatens Major Health Gains'(2015) is sourced from Paris[10] and based on 'a sweeping scientific review', the Rockefeller Foundation–Lancet Commission on planetary health (see Whitmee et al. 2015). The second story, 'Mass Extinction of Animals' (2016), sourced from Oslo, adds to coverage of the WWF report on loss of biodiversity, concluding: 'The decline is yet another sign that people have become the driving force for change on Earth, ushering in the epoch of the Anthropocene'. These two articles together could certainly suggest to the attentive reader that human and nonhuman health are both affected by the Anthropocene. *The Malay Mail* (which appears to be connected with *The Independent* in the UK) publishes three Anthropocene-related articles in 2014. The first, a telephone interview with Elizabeth Kolbert on her new book, *The Sixth Extinction,* 'Is Humanity Behind Biggest Extinction Event Since Dinosaurs?' is sourced from Johannesburg. One exchange is of particular interest. Asked, 'Your book has a scary theme. What frightens you most about the issue?' Kolbert answers, 'I think what is most scary, as a parent of other human beings, is what will happen from the societal disruption that could potentially result from ecological disruption and climate change. But on a "future of the world" kind of level, what I found most sobering and frightening is what is happening to the oceans. I think ocean acidification is a kind of under-appreciated problem. And we seem to be changing ocean chemistry very dramatically and very quickly'. 'Navigating Towards the Future' by the Indian journalist Gaurav Daga (graduate of University of Cambridge and the Paris Institute of Political Studies ('Sciences Po')) considers the 'progressive fate of science in India', a wide-ranging reflection, concluding: 'All these debates reflect the idea of humans living in the anthropocene – where we risk heading into a hostile condition from which we cannot return'. Sourced from Rome (not surprisingly), 'Pope Francis Gets the Climate Change Agenda' (another case of 'climate change' as a metonym for the Anthropocene) reports a story that was picked up by media all over the world. Coverage of the Anthropocene in *The Malay Mail* appears to stop in 2015, with three items. Sourced again from Paris, we find: 'Earth's Degradation Threatens to Wipe Out Major Health Gains, Say Experts' and '2015 – a "Tipping Point" for Climate Change, Experts Warn', a well-researched explanation of 'the extent to which human activity, mainly burning fossil fuels, has played havoc with the planet's thermostat', connecting fossil fuels directly with the dangers of the Anthropocene. 'The End of Western Europe?' by Mark Beeson is reprinted from the online magazine, *The Conversation*. Beeson, a professor of International Politics in Australia, presents a rather philosophical perspective on refugees coming to Europe. He concludes: 'When seen from the perspective of the Anthropocene, perhaps this is just the way the world works.' Beeson, surprisingly, does not mention climate refugees.

New Straits Times[11] first addresses the Anthropocene in 2016 with 'Climate Change and Ethics' by Michelle Kwa from the Institute of Strategic and International Studies (ISIS), Malaysia. She begins: 'A revisit of 2015 reveals some of the most appealing and appalling climate change stories that were grabbing the global media headlines. It is no great exaggeration to say that climate change is real and deserves serious attention' (another rare example of journalist introspection). Kwa continues 'For the people who embrace change … the test of a first-rate intelligence [as F. Scott Fitzgerald stupendously wrote] is the ability to hold two opposed ideas in the mind at the same time, and still retain the ability to function'. This is an excellent summary of the ways in which governments and big business are dealing with the Anthropocene. 'Careless People Damaging Our Corals' (2017) by another local journalist, Rosli Zacaria, also from ISIS, also cites F. Scott Fitzgerald, but in the context of the anthropogenic damage inflicted on coral eco-systems, locally and globally. Finally, 'Neoliberalism is Wreaking Havoc on Our Planet' (2017) by Dr Azeem Fazwan Farouk (Universiti Sains Malaysia) follows Noam Chomsky by identifying three 'existential threats'. Two of these are man-made, human mastery of science and technology and 'on the environmental front, humanity is entering a new geological epoch called anthropocene'. The third threat is passivity. *The Borneo Post* is distinctive in having had a regular column, 'Nature Matters' by Alan Rogers, a local expat geographer-geologist teacher, who contributes two of our results in 2017. In the first, 'Is This a New Geological Epoch Made By Us?', Rogers tells us that he is rethinking what he teaches his students in the context of the Anthropocene; and in the second, 'The Plasticscene' he says: 'last week's column was about the proposed name for a new geological epoch, the Anthropocene to follow the Holocene and, before that, the Pleistocene. I would actually like to suggest the new epoch be termed the Plasticscene. Why?' Dr Rogers explains the name at length in 'A Re-Evaluation of Our Biosphere' (2017), an expert evaluation of invasive species: 'As a biogeographer, 'I am in a quandary or "caught between the devil and the deep blue sea." A huge part of my brain votes hard for the conservation and preservation of all species of flora and fauna, but another part envisages a more laissez-faire approach to allow nature to take its course'. This latter approach is adopted by many ecologists.

Borneo Bulletin has one article 'Earth Has Entered a New Era, New Smithsonian Book [*Living in the Anthropocene*] Says. That is Not a Good Thing' (2017), from *The Washington Post*. While conventional, this presents a useful summary of the issues of human overpopulation, urbanization, and the Holocene–Anthropocene transition. The article comments: 'Perhaps the Age of Mammon might be a better term'. One important idea in the book which would certainly add to public understanding is picked up, 'the Anthropocene goes beyond just climate and weather and marks a period when we have altered the landscape for our needs, from building highways, expanding cities, converting forest and savanna to agricultural use, constructing dams and flooding river valleys, and all the rest. There is an environmental price for all this progress, in the loss of habitats, species, and biodiversity'. The co-editor of the book, the Smithsonian scientist John Kress, is quoted on opportunities to move forward in the Anthropocene by conservation and habitat restoration, all predicated on a cooperative spirit between 'citizens, governments, social and religious institutions, the marketplace, and the private sector'. Also mentioned are Pope Francis's encyclical on the environment and the eminent biologist E.O. Wilson, who, in the book's afterword, argues optimistically 'we can save ourselves as a species only if we save other species, by large-scale habitat protection. If we get it right, 'if we pass through safely and bring most of the rest of life with us, human existence could be a paradise compared to today'.

Nepal's media gave 14 results from nine sources, plus five with no results. Displaying commendable eco-cosmopolitanism *The Nepali Times* begins to report the Anthropocene with 'The Greening of Greenland' (2011), combining a historical–geological perspective with the risks of the Anthropocene, followed by 'Nepal in the Anthropocene' (2017). Here, the editor, Kunda Dixit, profiles the Swiss photographer Fritz Berger, who chronicles the story of a Nepali village transformed by human development 41 years after his first visit, in 'Charikot is a Boomtown with High-Rise Hotels, Shopping Centres and Several FM Stations', advertising Berger's photobook of the process. *The Rising Nepal* (2018) 'Leaving Our Children Nothing' reprints a version of Johan Rockström's article on the urgent need for a new economic model but offers little indication of what this might be. *The Kathmandu Tribune* gives us 'Humanity Will Leave Earth by 2400, But Perhaps for Brave New World' (2018), taking an emerging approach to our plight, reporting research by astrobiologists from the National Academy of Sciences in the United States about 'life-compatible locations beyond our solar system'. This is an original approach, linking to several like-minded scientists, and taking the prospects of the collapse of our civilization seriously.[12] *The Himalayan* in 'Making a Difference' (2017), by a local writer, is another story about transformation, lengthy and moving, insofar as it is another of the few of our results to focus on educating children about the risks of the Anthropocene. To quote: 'Our education system prides itself on teaching our children about climate change, pollution, and biodiversity loss from a very young age. But it fails to show how the kids as individuals could play a more pragmatic role in fighting these looming problems today [creating] an atmosphere of apathy among the youths'.[13] The article (also published in *The Himalayan Times*) cites ecological problems that have emerged in recent decades in Kathmandu. In the peak coverage years of 2016 and 2017, *The Himalayan Times* printed versions of several popular Anthropocene stories, for example, 'New Minerals

Back Idea of Man-Made Epoch for Earth'; 'Wildlife Populations Plunge Almost 60 Percent Since 1970'; 'Age of Man'; and 'Global Warming Likely to Delay Next Ice Age'. These stories suggest that the publication was keeping up with breaking Anthropocene news. *The Kathmandu Post* also published several articles on familiar themes, for example, 'Earth in the Balance' (2013) and, again, Rockström's 'Leaving Our Children Nothing' (2015), both products of Project Syndicate. In 'Living in the Age of the Anthropocene' (2016), the radical Nepali journalist, Abhinawa Devkota, eloquently argues 'Nothing done by our ancestors can even remotely match the indelible marks we will be leaving behind for our descendants to see'. In 2017, we find (in the Entertainment section) 'Art and the Anthropocene' by the Kathmandu-based Indian artist Kurchi Dasgupta, who tells her readers: 'For the past decade or so, the word Anthropocene has been making an appearance with increasing regularity and across disciplines – so much so that no serious discourse on art can really afford its absence any longer', an indication that the Anthropocene was on the cultural and the scientific radar of Nepali media.[14]

Searches of media in Singapore brought 19 results from seven sources (six in English and one in Chinese), although three more sources gave no results. *Lianhe Zaobao*, the most-read Chinese newspaper in Singapore (though with somewhat limited access) prints an article in 2013 describing how the Earth has moved into the 'Anthropocene (Human Age) or the Sixth Extinction' and the efforts of the AWG to persuade the geological community to accept the name. 'Anthropocene' (2013) discusses the influential essay by Rockstrom and Raworth, 'Planetary Boundaries and Human Prosperity',[15] arguing that ecosystems also have the resilience to meet human needs. 'American-Style *Interstellar* and Religious Narrative' (2014), by the film critic Zhou Lei, is a review of the Anthony Nolan film (discussed in Chapter 13 of this book). *The Straits Times,* a popular daily, prints several original items. In 2014, 'Scientists Prepare for Change of Epoch to Reflect Humankind's Deep Impact on Planet' (sourced from Reuters in Berlin). John Palmesino, an architect who has worked with scientists to capture the impact of humans on the Earth on film, is quoted: 'You can no longer distinguish what is man-made from what is natural'. A meeting of the AWG at Berlin's House of Cultures of the World is discussed, as well as the exhibition of related artworks, videos, and seminars (see Chapter 13 of this book). In 2015, 'Humans Causing Extinction of Land and Sea Creatures' (from Reuters in Rome) repeats the usual story of loss of biodiversity, with an interesting rider: 'The biggest immediate losers will be large predators, like tigers and lions, while creatures domesticated by humans, including house cats, will thrive in the new world' quoting Jan Zalasiewicz (the leader of the AWG) from the *Anthropocene Review*. 'Scientists Warn of Major Threat to Human Health' (2015) tells of algae-filled water at a beach in Qingdao in eastern China as an example of potential threats, citing the alarming *Lancet* report. The article provides much technical detail (sourced from AFP). 2016 brings 'Dawn of New Geological Age Shaped By Mankind'?, explaining that human activity is leaving a pervasive and persistent signature on earth, citing a report in the journal *Science* by an international team led by Dr Colin Waters of the British Geological Survey, Will Steffen, and Erle Ellis, who is quoted to dramatic effect, that the Anthropocene is 'a challenge no smaller than a second Copernican revolution' (this well-researched article is sourced from Oslo). 'The Sounds of Nature' (2016), by the award-winning writer Yu-Mei Balasingamchow, is a long, contemplative, and political reflection on the pleasures of nature and degradation of the planet. She concludes: 'In Singapore, it is difficult to think about deep time. If we can stop, listen, think and admit that we, *Homo sapiens*, are not the centre of the universe, perhaps we can begin to reconsider our relationship with the natural world and what we are doing to it'.[16] 'The Age of Man' (2017) by Wesley Yang, a long and scholarly article, appeared originally in *The New York Times* magazine. The tone is set by a poignant picture of penguins in Antarctica, where human encroachment threatens the region's stability and an entire ecosystem. Yang poses the question: 'Scientists say the earth has entered the Anthropocene – the first geological epoch defined by human impact. But is the only remedy yet more interference?' He credits Crutzen for coining the term, explaining that it 'expressed his intuition that humanity had become tantamount to the great forces of nature and that our activities were shaping the state of the systems that regulate the conditions of life'. Summarizing the geological story, Yang references Diane Ackerman's optimistic book, *The Human Age*, Roy Scranton's pessimistic *Learning to Die in the Anthropocene*, Jedidiah Purdy's *After Nature: A Politics for theAnthropocene,* and Yuval Harari's *Homo Deus* (the latter, cited by Microsoft co-founder Bill Gates and Facebook chief Mark Zuckerberg, is a 'ruthless appropriation of the Anthropocene' in Yang's opinion). The online news site *Today* covers the Anthropocene mainly from a global perspective, beginning with 'One problem At a Time Just Won't Do' (2013) by Kevin Noone, Director of the Swedish Secretariat for Environmental Earth System Sciences. A Project Syndicate product, this is a long, thorough summary of the issues raised by the transition from the Holocene to the Anthropocene, and the necessity to start 'rethinking GDP …the top priority should be building resilience into all global systems'. Noone's conclusion, characteristic of what we might label the Stockholm School, is: 'Navigating the Anthropocene effectively and ethically is perhaps the most daunting challenge that modern humans have faced. Overcoming it will require a smarter approach to strategic decision making and a broader understanding of innovation. It is time for us to rise to the challenge'. In 'Scientists Ponder Name Change for Current Era To 'the Age of Humans' (2014), the emphasis is on debates around start dates for the Anthropocene. The president of the Geological Society of America, Hap McSween, is quoted: 'I actually think it's a great idea... It's a good way to point out the environmental havoc that humans are causing' (sourced from AP). 'Balancing Planetary Boundaries and Human Prosperity' (2015) is an extended version of an article by Rockström and Raworth (from Project Syndicate and also reported in *Lianhe Zaobao*). This version adds: 'We face an urgent need to define a safety zone that prevents us from pushing our planet out of the unusually benevolent Holocene state' with a discussion of the Planetary Boundaries framework, associated with Rockström, first published in

2009. 'Humans Top Predator' (2015) quotes Zalasiewicz 'Some human-made changes have genuinely not been seen before on planet Earth', from an event at the Thomson Reuters Foundation connected with the publication of the research in the *Anthropocene Review.* 'Learning from Boston' (2017) seems to be a rather vague message addressed to the political class in Singapore, calling for a 'new paradigm in governance – one that is Whole-of-Government, networked, innovative, exploratory, and resilient in the way it confronts the challenges of our time – challenges rooted in complexity and accelerating change of the Anthropocene'; similarly, 'Endless Possible Futures' (2017) looks to smart technology for answers, but alerts the reader to possible Orwellian abuse: 'To overcome these misconceptions, a mature discussion is needed, not a polemical one. The government has a central role to play in shaping this discussion … trust in a fast-changing and complex world – the world of the Anthropocene – is a vital asset to good governance'. Finally, a reprint of 'Tipping Point' (2016) repeats Dr. Saleemul Huq's call in Asian media for 'greater planetary consciousness'. *The Business Times* of Singapore steers clear of contentious issues, with two neutralizing items in 2016 and 2017. The first, 'Singapore Biennale's Full Line-Up' reports on the Sri Lankan artist Pala Pothupitiye's 'Other Map Series', and 'Desert Islands' by Hong Kong's MAP Office, 'imagining a world of islands in the Anthropocene, when the history of the 21st century will be written on water'. These are just two of 60 works by 63 artists and art collectives from 18 countries in Asia on display. In 2017, 'New Minerals Back Idea of Man-Made Epoch for Earth' revisits a story that toured the media world. 'Single-Minded Optimist' (2019) quotes Christiana Figueres, who led negotiations for the Paris Agreement, speaking at an event in Singapore: 'Scepticism about climate change should push us not into despair - but action' (the Anthropocene appears to be behind a paywall). Vietnam has a single item from one source (six other sources had zero items). *Nhân Dân* (the official paper of the Communist Party) offers 'Science Film Festival 2017 to be Held in Vietnam', the main theme of which is the Anthropocene. Explaining the term, the paper says: 'Today, we experience the Anthropocene through climate change and our natural world is facing a crisis' to be illustrated by the beauty and vulnerability of the ecosystems on our Earth. The festival is a pan-Asian initiative of the Goethe Institute.

Conclusion

Anthropocene reporting in the media in South Asia is often sourced from abroad, though the exceptions from local experts were generally of a high quality, particularly in India, Pakistan, and Bangladesh. Many Anthropocene-related arts events were covered. While there were few direct references to links between colonialism and the Anthropocene, such discussion was not far below the surface. A notable feature of our results was the prominence of rather alarming headlines, along with stories whose conclusions were more reassuring. A corollary of this was the almost total absence of radical concrete proposals for change to deal with the risks of the Anthropocene.

Notes

1 Geraldine Chan, Dina Delias, Jiri Krejcik and Brototi Roy also contributed to this chapter.
2 The Maldives often features in climate change reporting, particularly stories about an underwater meeting of the Cabinet to highlight rising sea levels. *The Guardian* (4 March 2017) reported a surprising change of direction by the new President: 'mass tourism and mega-developments rather than solar power and carbon neutrality would enable it to adapt itself to climate change and give its young population hope for the future'. See also Moore (2019), and Fletcher (2015, 2019) on neoliberal 'ecotourism', relevant all over the world.
3 See also Bloomgard (2017).
4 Mukherjee (2017) provides a vivid visual essay on 'Anthropocene Ruinations' in Mumbai and beyond. See Douglass and Miller (2018) for a survey on the rural Anthropocene in Asia.
5 *The Times of India*, a national daily, offers a limited drop-down menu on Anthropocene topics.
6 On the use of radio to communicate health risks in Tamilnadu, see Kuppuswamy and Rajarathnam (2009).
7 More recently Dirzo and his co-researchers warn of potential human extinction (see Ceballos et al. 2017).
8 See Visaya-Albanõ et al. (2016) for a *Climate Change Guidebook* for Philippine journalists, and Evans (2016).
9 There are several possible Eco2 indexes, and this may refer to Sumaila et al. (2015).
10 This and many other items appear to originate from Reuters.
11 Apparently, *The New Straits Times* is not for sale in Singapore, and *The Straits Times* is not for sale in Malaysia. Both media systems are strictly government controlled.
12 Malazita (2017) connects astrobiology with the search for an origin myth for the Anthropocene; see also Frank et al. (2017) and Haqq-Misra et al. (2017). On the closely related SETI (Search for Extra Terrestrial Intelligence) see Ćirković (2014). The understandably sceptical should note that these themes also appear in *The Washington Times* (2012), *The Houston Chronicle* (2012), *Physics Today* (2016), *Scientific American* (2016), and *MIT Technology Review* (2017). See also: https://www.nasa.gov/feature/what-is-astrobiology
13 Perhaps Greta Thunberg and the Extinction Rebellion movement make this a little dated.
14 Mathur (2017) offers an original Himalayan perspective on the Anthropocene.
15 See Rockström and Raworth (2015), behind a paywall at Project Syndicate.
16 'Deep time' is not often discussed in our media results. Fetz et al. (2018) explore 'cosmopolitan temporalities' in China and Brazil.

References

Austin, G. ed. (2017) *Economic Development and Environmental History in the Anthropocene: Perspectives on Asia and Africa*. London: Bloomsbury.

Billett, S. (2010) Dividing climate change: Global warming in the Indian mass media. *Climatic Change* 99: 1–16.

Bloomgard, P. (2017) The forests of southeast Asia, forest transition theory and the Anthropocene, 1500–2000. In Austin, G. ed. *Economic Development and Environmental History in the Anthropocene: Perspectives on Asia and Africa*. London: Bloomsbury, chap. 9.

Boykoff, M. (2010) Indian media representations of climate change in a threatened journalistic ecosystem. *Climatic Change* 99: 17–25.

Ceballos, G.P. et al. (2017) Biological annihilation via the ongoing sixth mass extinction signaled by vertebrate population losses and declines. *PNAS* 114(30): E6089–E6096.
Ćirković, M.M. (2014) Evolutionary contingency and SETI revisited. *Biology & Philosophy* 29: 539–557. doi:10.1007/s10539-013-9397-8
Corlett, R. (2013) Becoming Europe: Southeast Asia in the Anthropocene. *Elementa: Science of the Anthropocene* 1. doi:10.12952/journal.elementa.000016
Das, J. (2019) *Reporting Climate Change in the Global North and South: Journalism in Australia and Bangladesh*. London: Routledge.
Douglass, M. & M.A. Miller (2018) Disaster justice in Asia's urbanising Anthropocene. *Environment and Planning E: Nature and Space* 1(3): 271–287.
Evans, S. (2016) Journalistic norms, cultural values, and coverage of climate change in the Philippines. *Environmental Communication* 10(4): 492–507.
Fetz, M. et al. (2018) Cosmopolitan temporalities: A sociological analysis on climate imageries in Brazil and China. *Fudan Journal of the Humanities and Social Sciences* 11: 49–68.
Fletcher, R. (2015) Nature is a nice place to save but i wouldn't want to live there: Environmental education and the ecotourist gaze. *Environmental Education Research* 21(3): 338–350.
Fletcher, R. (2019) Ecotourism after nature: Anthropocene tourism as a new capitalist 'fix'. *Journal of Sustainable Tourism* 27(4): 522–535.
Frank, A. et al. (2017) Earth as a hybrid planet: The Anthropocene in an evolutionary astrobiological context. *Anthropocene* 19: 13–21.
Gaffney, O. & W. Steffen (2017) The Anthropocene Equation. *Anthropocene Review*. https://doi.org/10.1177/2053019616688022
Haqq-Misra, J. et al. (2017) *Astrobiology of the Anthropocene: A White Paper on 'Astrobiology Science Strategy for the Search for Life in the Universe' for the National Academy of Sciences*. Seattle, WA: Blue Marble Space Institute of Science.
Hudson, M. (2014) Placing Asia in the Anthropocene: Histories, vulnerabilities, responses. *Journal of Asian Studies* 3(4): 941–962.
Kuppuswamy, S. & S. Rajarathnam (2009) Disasters in Tamil Nadu, India: Use of media to create health epidemic awareness. *Estudos em Comunicacão* 6: 19–36.
Lidskog, R. & C. Waterton (2016) Anthropocene – A cautious welcome from environmental sociology? *Environmental Sociology* 2(4): 395–406.
Malazita, J. (2017) Astrobiology's cosmopolitics and the search for an origin myth for the Anthropocene. *Biological Theory* 20. doi:10.1007/s13752-017-0281-7
Mathur, N. (2017) The task of the climate translator. *Economic & Political Weekly* 52(31) (5 August).
Moore, A. (2019) Selling Anthropocene space: Situated adventures in sustainable tourism. *Journal of Sustainable Tourism* 27(4): 436–451.
Mukherjee, R. (2017) Anticipating ruinations: Ecologies of 'Make Do' and + 'Left With'. *Journal of Visual Culture* 16(3): 287–309.
Nayak, A.K. (2015) Environmental movements in India. *Journal of Developing Societies* 31(2): 249–280.
Rockström, J. & K. Raworth (2015) Planetary boundaries and human prosperity. https://www.project-syndicate.org/commentary/environment-boundaries-human-prosperity-by-johan-rockstr-m-and-kate-raworth-2015-04?barrier=accesspaylog
Roy, T. (2017) Land quality, carrying capacity, and sustainable agricultural change in twentieth century India. In Austin, G. ed. *Economic Development and Environmental*

History in the Anthropocene: Perspectives on Asia and Africa. London: Bloomsbury, chap. 8.

Sumaila, U.R. et al. (2015) Eco2: A simple index of economic–ecological deficit. *Inter-Research* 530: 271–279.

Visaya-Albanõ, I. et al. (2016) *KLIMA 101: A Climate Change Guidebook for Philippine Journalists*. Manila: Network for Environmental Journalists.

Whitmee, S. et al. (2015) Safeguarding human health in the Anthropocene epoch: Report of the rockefeller foundation–lancet commission on planetary health. *The Lancet* 386: 1973–2028.

8 Latecomers to capitalism, latecomers to the risks of the Anthropocene

Vladimir Vuletić and Eni Buljubašić[1]

Introduction

This chapter focuses on media coverage of the Anthropocene in 11 countries in Central and eastern Europe (CEE). Our goal is to determine the prevalence of articles on the Anthropocene and to discuss a sample of them in terms of the criteria explained in the introductory chapter (Chapter 1 in this book). This will allow for comparisons with other regions, as well as between countries in the CEE region, also showing whether and to what extent the Anthropocene, as a global discourse, has become standardized or not.

The CEE countries had a period of intensive industrial development in the second half of the 20th century, during the communist era, when they were (albeit briefly) part of the Soviet bloc. This was simultaneously the period of awakening of environmental consciousness and the emergence of the first environmental movements in CEE, which subsequently played a role in changing the political system and transitioning towards capitalism.[2]

We begin by outlining basic characteristics of the development of environmental issues in CEE countries. Then, we present the findings of the discourse analysis of Anthropocene media coverage in the region. In the conclusion, we discuss the scope of Anthropocene discourse as portrayed in the media, in the context of the specific circumstances arising from historical and territorial characteristics of the region.

The question of the research approach and methodology in the analysis of media content needs to be briefly addressed. Discourse analysis is a controversial term, with wide and varied uses (van Eeden 2017a). It is important to note that, for us, discourse denotes a system of statements which cohere around common meanings and values (Coates 2013), and, thus, Anthropocene discourse is taken as a discourse carrying particular ideological and moral issues. In line with Foucauldian analysis, we see the media communication of an object as a certain way of creating that object, i.e., the understanding of it. Thus, the media is a site where struggles over the creation of dominant discourses are performed, the final goal of which is normalization of the discourse through, in Foucault's terms, disciplining and/or creating an acceptable and self-understandable way of using certain terms. In short, discourse analysis should help observe consequences of using the term Anthropocene in specific ways and in specific contexts, showing what ideological responses and types of mobilization are being advocated by the media (see Hansen 2012). Moreover, the introduction of the term Anthropocene in the year 2000 coincides with the rise of the Internet, global media, and social networks. Despite that parallelism, however, almost two decades after it was introduced into scientific literature, the term is still not generally known to the public in CEE nor is it widely disseminated in the academic community. This could be explained by the lack of interest in questions provoked by the Anthropocene. In order to discern the reasons for lack of interest in the Anthropocene, as well as to examine alternative interpretations that exist in Central and eastern Europe, we begin by summarizing the history of environmental issues in the region.

Brief history of environmental issues in Central and eastern Europe

The interest in environmental issues in most European socialist countries appeared in the late 1960s and early 1970s, as was the case in Western European industrial countries. This is unsurprising since the imperative of industrial development was the shared dominant strategy in both East and West. The conclusions of the Club of Rome in their path-breaking book *The Limits of Growth*, published in 1972, influenced governments of socialist countries to introduce environmental protection policies. Even before this, however, Yugoslavia had formed a state commission to deal with critical environmental issues (Oštrić 1992). Similar developments occurred in Romania and other Eastern bloc countries. This was part of Cold War era competition in moral superiority, expressed through the adoption of large and ambitious packages of environmental legislation from the 1960s onwards (see van Eeden 2017b, on Poland and Romania). Environmental enthusiasm waned by the mid-1970s when environmental activities became bureaucratized or suspended. This was due to the fact that industrial and technological power centres, together with state administrations, wielded huge political influence to develop resource exploitation policies, irrespective of the effects on the natural environment (Vargha 1992). Within conservative political structures, an impression was left that any type of movement, even an environmental movement, could be a threat to the stability of the regime, so environmental issues were left on the margins of political interest (Oštrić 1992).[3]

It took almost a decade for environmental topics and movements to re-emerge, which coincided with the escalation of the economic crisis of the socialist system. In this regard, labour strikes, particularly those organized by Solidarnošć in Poland, and economic problems such as shortages of consumer goods, raised questions of the legitimacy of the political system in an increasingly obvious manner. In the mid-1980s, the loss of political legitimacy created space for the development of environmental movements. Vargha (1992) argues that the public linked the environmental crisis to the political system, which created the hope that a change of political system would automatically lead to the resolution of environmental problems. This provides an example of how the development of a powerful discourse can play a significant political role and contribute to the change of political system. As shown by Barbara Jancar-Webster (1993), experts from eastern Europe almost unanimously agreed that the planned economy and state socialism were the main reasons behind environmental problems in their countries, even suggesting that the socialist system and communist ideology by themselves were important polluting factors. The nuclear disaster at Chernobyl played an important role in this context. Although it encouraged or strengthened the anti-nuclear movements,[4] such movements were active in certain CEE countries even before the accident, e.g., in Hungary (Koszegi and Thompson 1982). In Yugoslavia, even before Chernobyl, the anti-nuclear movement was the strongest wing of the environmental movement.[5] The case of the Yugoslav anti-nuclear movement shows how the media helped spread anti-nuclear public opinion and how the poor media presentation of certain supporters of the nuclear programme turned the public against it, leading to a moratorium on construction of nuclear power plants (Oštrić 1992).

To some extent, we could agree with Vargha, who argues: 'one of the distinguishing features of the eastern European revolutions of 1989 is their strong environmental movement in the pre-revolution days, which served as a rallying point from which broader demands for political change emerged' (1992: 33).[6] So, why did the environmental movements in CEE not survive the political and economic changes to which they contributed, and which were expected to improve environmental protection? Transition-related problems (notably economic recession, poverty, uncertainty, and strengthening of nationalism), as well as organizational weaknesses, conflicts within the movements, and civil society leaders becoming involved in government, led to a sudden weakening of green parties and environmental movements during the first half of the 1990s (Oštrić 1992, Jancar-Webster 1993, Vukelić 2014). Jancar-Webster (1998: 69) aptly summarizes the situation:

> The environmental movement in CEE has undergone profound transformation; the movement has shifted from being a mobilizing agent for populist protest against the totality of the Communist regime and, in its place, has emerged pragmatic, goal-oriented professional organizations. Western aid agencies and environmental peer groups have had a strong influence on this transformation, which has brought advantages to local environmental NGOs. However, the transformation has also resulted in a loss of the local perspective, with its distinct modus operandi and bottom-up input, and this has impoverished political discourse in the transition states.

Paradoxically, after system change, i.e., after economic and political systems in eastern and western European countries became closer to each other, and with subsequent accessions to the European Union, the development of environmental awareness and the strength of environmental movements in these countries started going down different paths (Chaisty and Whitefield 2015). The question why these differences occurred and what caused them remains to be fully researched.

The Anthopocene in CEE media

Speaking in the spirit of media ecology theory, if media are seen as a dish or a pot, used to prepare cultural content, it could be said that, in Central and eastern Europe, the Anthropocene is still raw or insufficiently processed. This is suggested by the fact that, over almost two decades, from the time it entered into public use in 2000 until the end of 2017, just over 300 articles mentioning the Anthropocene were found in 11 CEE countries. From the data researched for this work, about two relevant articles a year were published on average in each country. However, there are significant differences between CEE countries in both coverage and timing. In the Czech Republic, Croatia, Slovakia, and Russia, the first Anthropocene articles appear as early as 2002–2004. In Serbia and Poland, the first articles were published in 2010 and 2011, respectively. In Romania, Estonia, Lithuania, Latvia, and Ukraine, the Anthropocene was first mentioned in the period 2014–2016. With 125 articles, the Czech Republic is far ahead of any other CEE country in terms of number of published articles. Russia with 50, Slovakia and Serbia, each with about 40 articles published, followed by Croatia and Poland (15 and 13 articles, respectively), Romania (nine), Latvia (six), Estonia (five), Ukraine (five) and Lithuania (four) all published articles. However, various methodological issues in the collection of data need to be taken into account.[7] While it is clear that media in CEE countries differ in terms of interest in the subject, there are similarities across the region, suggesting that the influence of historical paths taken is still important. This, we argue in the conclusion, results in a tendency in the CEE region towards the provincializing of Anthropocene.[8]

Czech Republic

The Czech Republic stands out in terms of both total number of Anthropocene articles published, and in the number of articles written by local authors.[9] The first item in CEE media on the Anthropocene (in 2002) was the work of the Czech scientist, Ivan Turec, in the most-visited website in the country, *inky.cz,* 'Co je antropocén?' (What is Anthropocene?). Turec raises issues (rarely addressed in subsequent media coverage) about ideological conflicts that can be linked to the Anthropocene, such as the conflict between environmentalists, extractive industries, and consumerism. As is the case in other CEE countries, most of the articles written on the Anthropocene in Czech daily newspapers and online portals are complete or summarized versions of articles taken from foreign media, mostly from English-speaking countries. The majority of these articles can be considered neutralizing, presenting the views of advocates of the term 'Anthropocene', rarely raising any alarm. The names that are mentioned most often are Paul Crutzen and Jan Zalasiewicz.

The situation differs somewhat in weekly magazines and in magazines that give greater prominence to articles by local authors. Figure 8.1 is the cover image of a special issue on the Anthropocene of the magazine *A2*, mostly written by Czech authors (the image itself is open to more than one interpretation).

Figure 8.1 Special issue on the Anthropocene in the Czech magazine *A2* (2016) © *A2*.

Some authors dilute the importance of the idea of the Anthropocene by referring to alternative theories, according to which global warming is not necessarily caused by human actions but may also be caused, for example, by solar activity. The consequences of this approach are important. If climate change is caused by humans, we might have a good chance to avoid the worst effects of warming, since it is only up to us. If climate change is caused by solar activity, we are powerless. The combination of these two explanations suggests that the acceptance of the idea of Anthropocene could lead to more conscious human action. In 2011, for example, *Respekt.cz* in 'Vek človeka' (Age of Man) states: 'Without human civilization and its intervention into global climate, the planetary ecosystem would end in a new Ice Age'.[10] The idea that the Anthropocene could be used to raise awareness about the ability of humans to influence the planet, and that such influence can also be positive and not only negative is widely represented in Czech media. In that sense, the Anthropocene leaves enough space for human freedom and is not only the result of unintended consequences of human activities, as is claimed by more pessimistic ecologists. 'Good' Anthropocene narratives bring the promise of a long future for humanity. Czech business magazines also deal with the subject, raising new questions and promoting the need to develop new technologies, such as electric cars, but also advocate that the Anthropocene should be seen as the epoch of the 'new humankind', which will be capable of encouraging collective political activities of planetary proportions. Anthropocene is also seen as a useful term, despite not adequately reflecting the unequal contribution of different groups of people to global climate change. This is the reason why some authors consider that the term Capitalocene would be more appropriate, since it would underline the disproportionate contribution of capitalism to the risks of climate change, as argued in *Advojka.cz* 'Dela človek historii' (Man-Made History) in 2016. An idea discussed only in Czech magazines (for example, in *Ekonom* and *Vesmir* and nowhere else in the CEE media) is the problematization of the 'Chinese Anthropocene', i.e., China's rapid development aimed at matching two centuries of European development in just a few years. Becoming a global economic player means that China also becomes a global climate factor. In short, three types of Anthropocene narratives are found in the Czech media. Most articles are predominantly neutral. They present the facts about the effects of human activities on the planet, some with and some without catastrophic scenarios. On the other hand, there are many articles, especially in magazines, that discuss the possibilities for humankind if we accept the idea that mankind can influence nature. Anthropocene understood in this way also opens possibilities for the expansion of human freedom. If approached seriously, the argument goes, the idea of Anthropocene may change consciousness.

Slovakia

Although Slovakia formed a state union with the Czech Republic during most of the 20th century, there are many differences in the way the Anthropocene is treated by the media in these two countries. First of all, the way in which Anthropocene is presented in the Slovak media is much more influenced by foreign sources, most prominently by German media. For example, 'Scientist Have Not Yet Sufficiently Explained Human Impact on Nature' which appeared in *Sme.sk* in 2002, was based on an article in *Frankfurter Allgemeine Zeitung*. *Sme.sk* followed this with 'Planetárne hranice pre ľudstvo' (Planetary Boundaries for Humanity) in 2009, referencing Johan Rockström's research from the Stockholm Resilience Centre (a story that was reported all Around the world). Harmful human impacts on global climate are definitely anticipated, but the author notes that man is only mentioned as a sub-factor. This article mentions Paul Crutzen as one of the 'fathers of the Anthropocene'. The article was re-published on the same portal seven years later (2016). It seems that more recent articles in the Slovak media tend to take the climatic problems and the risks of the Anthropocene more seriously. Renáta Zelná, the science and technology editor of *Sme Tech*, in 'Humans Have Changed the Earth. Scientists Want To Declare a Human Era' (2016), refers to the Anthropocene Working Group (AWG) at the International Geological Congress in Cape Town, according to which we are probably already living in the Anthropocene. However, she explains, the professional geological authorities have not yet accepted the Anthropocene as a new epoch. On the basis of our results, it can be seen that many Slovak media portals simply mention the term, others briefly define it, whereas still others try to point out the need for deeper reflection. The common feature of Slovak Anthropocene reporting is the rather frequent comments that the term, despite the relatively clear logical context, is still not firmly established and is still subject to discussion by experts. A significant proportion of articles also focus on the issue of start dates, others to congresses and conferences. Based on our current findings, the media coverage of Anthropocene within the Slovak Internet news portals is rather sporadic, and therefore we hold that the term Anthropocene is unlikely to be familiar to Slovaks who rely on news portals for information.

Serbia

In Serbia, it is interesting, but expected to a certain extent, that web portals are the most important for sharing information on the Anthropocene. Most Anthropocene articles are richly illustrated, for example, with photos of pollution (usually smoking industrial chimneys) or other symbols of anthropogenic risk. Serbian media usually explain the meaning of the term, stating that it is a new geological age during which human activities have influenced the planet. Paul Crutzen is usually mentioned as the person who named this age (although one article credits Eugene Stoermer, who used the term informally in the 1980s). In addition to Crutzen, we also find the names of Jan Zalasiewicz (leader of the Anthropocene Working Group, AWG), Nicole Boivin (Director of the Max Planck Institute), geologist Colin Waters (also of the AWG) and the anthropologist Felix Pharand, commonly cited in articles on the Anthropocene. The basics of debates around start dates are discussed in *National Geographic Srbija* 'When Did the Anthropocene Begin?' (2014) and 'Scientists Establish the Year of the Beginning of the Anthropocene' in *B92.net* (2015). An idea mentioned in many articles is the prediction that future geologists will recognize our age by remnants of plastic, aluminium, and concrete in sediment, as contemporary geologists currently identify remnants of meteors that have hit the Earth. This is often used to emphasize the catastrophic consequences of humanity's behaviour for the planet. A special emphasis among these consequences is placed on the extinction of flora and fauna, and the finding that almost two-thirds of species have become or are heading towards extinction due to human activity, for example in 'Entering Anthropocene – The New Geological Epoch' (2017) in *Sensa.* 'What Scars Will Mankind Leave on the Planet' (2014) in *Novi Magazine* is the only article that attempts to explore the complexities of the term by arguing that it is more important to raise awareness about extractivism, boring into the Earth. It should be noted that the prevailing attitude of most of the articles is the one expressed in *Blic* (2016): 'the concept of Anthropocene is provocative since it suggests that our current way of life, especially in developed parts of the world, is not sustainable … Experts warn that our planet is so polluted that it has entered into a new geological period'. *Opanak.rs* (2017) expands on this with: 'We are Going into a New Geological Epoch', as does *Kurir* (2017), in 'Alarm Warning: Earth Enters into the Anthropocene, Here's What it Means for Nature and Humans'. However, despite these warnings, there are no concrete radical proposals for change. A public lecture on the Anthropocene by Milan M. Ćirković, the eminent Serbian astrophysicist and author of popular science books, was reported in the cultural portal *Art Anima* in 2015. Ćirković expresses the interesting view that the Anthropocene is the best way to stand against anthropocentrism, which caused the problem in the first place. With few exceptions, we found little extended discussion of the Anthropocene in the Serbian media. There are some sensationalist articles, taken from Western media, which are usually put in the entertainment section. Also noteworthy for its rarity is 'Anthropocene: the Critics, the Pseudoscientists, the Fraudsters' (2015) in the website *Mišljenje o mišljenju* (Opinion on the Opinion).

Romania

Media interest in the Anthropocene appears relatively late in the Romanian media. According to search results from Google in Romania, it appears that only three pages are in any way related to the Anthropocene, and only nine articles in the online media discussed the Anthropocene (those being in 2016 and 2017). These mostly list concrete evidence of anthropogenic impacts, such as global climate change, greenhouse gases, and melting glaciers. Readers are informed of the importance of the term 'Anthropocene', debates about its start date, and attempts to introduce it into geography textbooks. Several articles announce the Cape Town International Geological Congress which was expected to provide answers and formally establish the Anthropocene as a geological epoch. The Romanian magazine *Ziare.com.ro* in 'Farewell Holocene: We are in a New Geological Era' (2014) reported an event in Germany during which it was suggested that the Anthropocene would officially succeed the Holocene, a geological epoch that started over 10,000 years ago. *HotNews.ro* prints the story that testing of nuclear weapons marked the start of the new epoch in 'Are We Already in the Anthropocene?' (2016). We can conclude that the media in Romania generally accepts that the Anthropocene represents a conceptual and historical perspective dominated by the view that, if humanity continues to act uncontrolled, the sustainability of human civilization and even the survival of the human species could be jeopardized. The paradox of this new age, according to the latter article, is that, although we are able to create artificial lives at present, the activities that we have carried out have caused such a decline in terrestrial biodiversity that, at the scientific level, the phenomenon is described as being the sixth great extinction since the appearance of life on Earth. Finally, an item from 2018 is worth mentioning. In the environmental magazine, *greenly.ro.*, 'Developing Human Communities in the Anthropocene' declares: 'The Anthropocene of the Age of Man is Changing the Planet'.This is the only article we found in Romania that discusses the ideological background of the name Anthropocene itself, viewing it as a concept that criticizes the idea of progress, the values of the capitalist market, and high technology, which comprise the supreme neoliberal dogma.

Baltic countries

We found little coverage of the Anthropocene from media in the three Baltic countries, only four bland articles published in 2016–17. However, one thing that distinguishes Baltic countries from some of the other CEE countries is an interest in artistic approaches to the Anthropocene. For example, according to the magazine *Dienas Bizness*, during the Venice Biennale of Architecture in 2016, the Baltic countries created a joint exhibit: 'with the trendy concept Anthropocene, which marks the planetary geology of the new era, in which human activity has left an indelible impression on the climate and the planet's ecology'.

Croatia

In Croatia, the first Anthropocene item we found appeared in the newspaper *Večernji list* in 2002, summarizing an article from the *Frankfurter Allgemeine Zeitung*. The best-selling daily *24sata* had nothing (although, in 2019, it reported the celebrated British naturalist David Attenborough's speech at Davos, recognizing the Anthropocene). Other media had a few results each, mostly in 2016–2017. In 2016, *Croatian News Agency* reported on the meaning and etymology of the word 'Anthropocene'. Citing the journal *Science*, it gives useful details on the research of Zalasiewicz, Reinhold Leinfelder, Colin Waters, and Crutzen. The leftist portal *H-Alter.hr* ran an original article, 'Welcome to the Anthropocene', a week after an influential Anthropocene paper was published in *Science*, drawing from the *Science* article, and (reportedly) direct communication with its lead author, Waters. However, not all Anthropocene reporting in Croatian media is as straightforward. For example, the regional daily *Slobodna Dalmacija*, reporting on the Cape Town congress, also gave news of the *Australopithecus afarensis* hominoid, with no connections apparent, and the leftist portal *Telegram*, in 'Humanity's Devastating Effect on the Planet in 19 Photos' (2015) stated that scientists have 'ironically' suggested the term 'Anthropocene' (this may be a veiled reference to the Capitalocene). The views of Croatian journalists or public figures are rarely voiced. In that context, another article in *Slobodna Dalmacija* stands out: 'Scientists: After 12 Thousand Years, it's Time for a New Epoch' (2014). It discusses the AWG, and quotes Croatian geology professor, T. Marjanac, one of the few who denies anthropogenic climate change: 'We're misinterpreting nature and climate change [he says], which fuels arguments for political action against ourselves, such as the enforcement of taxes for global warming for which there is still no consensus in the scientific community'. The weekly political and cultural publications and portals share characteristics with daily newspapers. They mostly run articles translated from foreign media, including *The Independent, The Guardian, Huffington Post, BBC*, and *AFP*. The portal *Express.hr* in 'Man Has Brought Himself To the Verge of Nature, to the End of History' (2017), via *The Independent*, profiles the work of the radical Slovenian philosopher Slavoj Žižek: 'Hurricane Irma Will Happen Again – So We Need the Answers To Some Difficult Questions About Global Politics ... This is What We Call Anthropocene'. Lifestyle magazines gave no results, though *National Geographic Croatia* has a translation of a largely reassuring article by Elizabeth Kolbert, 'The Dawn of the Anthropocene: Age of Man'. In general, the number of Anthropocene articles in the Croatian media is relatively low, mostly featuring summaries from foreign (English, French, or German language) media, and rarely addressing Anthropocene issues in the local context.

Before 2015, the Anthropocene debate is presented in its own terms, while later it appears as background to current events (such as Intergovernmental Panel on Climate Change (IPCC) and Worldwide Fund for Nature (WWF) reports, the UN Paris meetings, and the Cape Town International Geological Congress. Simultaneously, debates on the onset, relevance, and validity of the Anthropocene still appear, along with some outright anthropogenic climate change denialism. Reporting on the Anthropocene usually offers the neutral discourse, accompanied by some sensationalism. Relatively few articles present the Anthropocene as an issue that can be technologically fixed, or as a civilizational shift opportunity.[11] Articles on climate change or local extreme weather phenomena (e.g., floods or droughts) are much more numerous than Anthropocene items, and contain certain elements of the Anthropocene narrative, without mentioning the term itself, being more examples of metonyms of the Anthropocene. As in the Baltic states, reporting on Anthropocene-related creative arts appears to have flourished recently in Croatia, and, though this occurs outside our data-collecting period (after the end of 2017), it seems significant. In 2018, the left-wing news portal *H-alter.org* profiles the exhibition in Zagreb of Slovenian artist Maja Smrekar in 'Gear For Surviving in the Anthropocene – Trailer', indicating that invasive species and biodiversity loss are motifs in this critique of neoliberal treatment of climate change. In 2019, the leftist *Novi list* reports that the artist Tin Dožić won a national young artist award for 'Poems for the Anthropocene', a series thematizing the Anthropocene in relation to technology. The niche arts portal, *Plesna scena*, covers several Anthropocene-related arts events in 2018; for example, an international art exhibition 'Nature, Society and Other Stories' which featured 'Art in the Anthropocene'. The ethnologist Suzana Marjanić explains that these are artists 'who detect the ecological pre-apocalypse'. The Croatian Association of Fine Artists also drew attention to the Anthropocene. Leila Topić, a prominent arts curator, however, warns that thematizing the Anthropocene may distract from global power and geopolitical issues. She is quoted: 'Most experts emphasize that the causes of global ecological issues are sociogenic, not anthropogenic. Therefore, Anthropocene as a concept cannot explain the ecological crisis or provide a recipe for overcoming it. ... The shift to a sustainable world, if possible, will require engagement, willingness for frugality, vision, imagination, and joint efforts'. Finally, in 2019, the Anthropocene is also reported in the context of Croatian art in the exhibition 'Land Art, Earth Art, Earthworks: Earth and the Anthropocene' at the Museum of Contemporary Art in Zagreb in the cultural portal *Vizkultura*.[12]

Poland

Similar to most media, the Polish media are rich with results on climate change, but sparse on the Anthropocene. Nine outlets combined had over 900 hits for 'climate change' between 2011 and 2017, but only a dozen articles on the Anthropocene in newspapers and portals combined. Several items in *Gazeta Wyborcza* pick up common themes.[13] The first in 2015, 'Dinosaur Extinction is a Trifle' (implying that what is next will be worse), and, in 2017, a report on a paper by Zalasiewicz et al. in *Anthropocene Review*, which argues that, today, each individual's material usage 'surpasses the number of things that were held by whole groups of hunter-gatherers living before the agricultural revolution'. 'Go Ahead, Foil My World. Anthropocene, Age of Humans, Plastics' (2017) presents the Anthropocene as the age of man, referencing a much-reported item from 2016, 'We Have Produced So Much Plastic That We Could Wrap the Globe With it', mentioning Zalasiewicz and the Polish researcher, A. Gałuszka (University of Jan Kochanowski), as members of the AWG. Items in the peak years of 2016–2017 mostly draw from the *Anthropocene Review* and news from the AWG, covering plastic pollution, the technosphere, and biodiversity loss. For example, the *Gazeta Wyborcza* journalist Tomasz Ulanowski voices his personal opinions, making a strong connection between the Anthropocene and the 'technosphere'.[14] Three articles in the leftist magazine *Krytyka Polityczna* in 2018 stand out as unusually radical commentaries on the Anthropocene. The first is an in-depth interview with the Polish sociologist Ewa Bińczyk, author of *Epoch of Man: Rhetoric and Apathy of the Anthropocene* (2018, in Polish).[15]. Bińczyk (2019, in English),discusses local–national issues in relation to Poland's post-WW2 history and, unusually, envisions the future in the Anthropocene in terms of de-growth vs. neoliberalism. This is one of the few hints in our results that growth-obsessed capitalist corporations may be blocking effective action to address the dangers of the Anthropocene. Figure 8.2 is a powerful expression of how the real politics of the Anthropocene work in Poland and elsewhere (see Kundzewicz 2017).

'Androcene' by journalist Kinga Dunin gives a feminist perspective on the Anthropocene: 'Capitalism, Capitalism, Capitalism – and Where is Patriarchy

Figure 8.2 Big polluters bankrolling COP 24, Poland (2018) © Corporate Accountability and Corporate Europe Observatory.

in All This? Maybe a Better Name for Anthropocene is Androcene'.[16] Dunin criticizes the lack of women at the Katowice climate summit, emphasizing that women are more frequently the victims of the Anthropocene's natural and social disasters. Finally, 'It's Time the Law Prosecutes Ecocide' (2017) is an interview with the artist and environmental scholar, Nabil Ahmed, who advocates criminal law prosecution of 'ecocide'.[17] He voices suspicion on the usefulness of the term Anthropocene and alleges (probably correctly) that it is more popular in the humanities than in climate science, concluding: 'The question asked by activists fighting the practical effects of climate change: is the term Anthropocene useful in this fight? … will it give us real tools to deal with the effects of climate change? I doubt that this term is really useful in that sense'.

Ukraine

Ukrainian media coverage of the Anthropocene began in 2016–17, with *Gazeta po Ukrainski* and *Express.ua* reporting the new 'Era of Man', introducing the Anthropocene with the usual neutralizing basic descriptions. *Expres.ua* describes it as an 'informal geochronological term [denoting] a geological era in which the level of human activity plays a significant role in the ecosystem of the Earth' and more Earth science information, mentions of 'official data' and human impacts at the geological scale through plastics, concrete, and deposits that remain after nuclear tests. The privately owned *Gazeta po Ukrainski*, on the other hand, discusses William Ruddiman's early Anthropocene theory, explaining: '[about 7000 years ago] to free land for crops and pastures, people burned the forest and, as a result, the amount of carbon dioxide and methane in the atmosphere has increased to such an extent that instead of cooling, there was a warming'. The news agency *Firtka* in 2016 writes on animal extinction, reporting that the estimate of a 60% decrease in animal numbers in the past 40 years is 'worrisome' and linked to 'the new era Anthropocene'. Most sampled articles in the Ukrainian media present a neutral discourse, though *Firtka* concludes with some judgments, for example: 'The worries about the decreasing population of animals have been justified. Scientists claim that the Earth is entering the new era, the Anthropocene'.

Russia

In the Russian media between 2004 and 2017, we found over 50 articles in which the word *Antropotsen* (Антропоцен) occurs. The first mention of the term appears in 2004, 'The Earth Entered a New Geological Epoch' in the online newspaper *Lenta.ru*, highlighting the impact of human activity on the environment; followed by 'Scientists: By 2050 We Will Need a Second Earth' (2006) in *Izvestia* and 'Anthropocene Era' (2007) in *Nezavisimaya Gazeta* (hereafter *Gazeta)*,[18] a relatively independent daily mostly read by the elite. These items depict a malleable Anthropocene, depending on human activities but there is no suggestion that radical change is necessary. The reporting picks up in 2008, with *Gazeta*, *Izvestia*, and *Lenta* all reporting Anthropocene stories. Most of these represent it neutrally, as a concept for a new geological epoch, quoting English language media, British scientists, and institutions involved in the formation of the Anthropocene, with Crutzen named specifically, as in 'Mankind in a New Era' in the venerable geographic magazine, *Vokrug sveta*. In contrast, two articles (both in the news site *Gazeta.ru* in 2008) critique the concept's Western origins. One mentions the eminent Russian scientist V.I. Vernadsky's notion of noösphere, while the other, 'Searching for an Era that is Driven By the Psyche', by geologist Rudolph Balandin, argues that Russian and Soviet scientists had conducted research on 'Anthropocene' much earlier than Western colleagues, under the terms 'psycho-sphere' and 'anthropogony'. These articles express frustration that Western academics ignore the early Russian contribution to what is now called Anthropocene research.[19] Between 2009 and 2014, the Anthropocene is sporadically discussed in the Russian media, with around a dozen mentions. Reports on the Stockholm memorandum in 2011, which endorsed the risks of the Anthropocene, however, contain elements of 'malleable Anthropocene' discourse, as in *Izvestia*, 'Nobel Laureates Proposed a Plan to Save Humanity' (2011), we find 'to save the planet for future generations, scientists offered to take a number of urgent actions' and the daily tabloid *Komsomolskaya Pravda* reported (rather optimistically): 'Nobel Laureates Will Save the World in 2012. Scientists Have Prepared a Plan for the Salvation of Mankind'. Following the global trend, a spike in coverage appears in 2015–16. Neutralizing items reappear throughout this period, as well as some new topics. For example, reports that 2014 was the hottest year on record, and imagining the world without people in *Gazeta.ru*, biodiversity loss in *Lenta, Trud, Gazeta, Moskovskiy komsomolets*, and the former state news agency *RIA*. Debates on start dates for the Anthropocene (including nuclear tests from the late 1940s and the discovery of the Americas) appear in *Lenta, Moskovskiy Komsomolets, RIA*, and *Vokrug sveta*, tinged with discourse on 'fake news' and the role power plays in academia. More frustration for the perceived exclusion of Russian science from the European/Western science model is repeated in an article in *Gazeta.ru* (2015) discussing a paper from the Russian journal *Diversity and Distributions,* in which scientists developed a model of the planet without humans. In the daily (and often sensationalist) *Moskovskiy Komsomoletz* (2015), a Russian scientist criticizes 1610 as the onset of the Anthropocene concept, and dismisses the Anthropocene as a 'brand' invented by 'ecoalarmists'.[20] In *Gazeta*, we find: 'It Seems That the Entire Planet Becomes the Disaster Zone' (2015)', criticizing the notion of the Anthropocene as something that overestimates human capabilities. The science historian Elena Zheltova discusses the connection between Russian literature (particularly Tolstoy) and Bruno Latour's Actor Network theory, citing Latour's article 'Agency at the Time of the Anthropocene'. *Gazeta* also published Anthropocene-related stories by local scientists on nutrition, modern agriculture, and loss of biodiversity. In *Lenta*, we find reports on local events, such as a lecture by Dominique Boer (University of Lausanne) at the Moscow French Institute, providing a very detailed overview of the geological characteristics of the Anthropocene, possible geopolitical changes, the rise of elites, and transhumanism. This article comprises imaginaries of all three Anthropocene narratives (neutralizing, 'good' Anthropocene, and radical change) as possibilities, arguing: 'Anthropocene will be a period of changes in all spheres, including economic and political. It will also be the era of the emergence of new types of freedoms. Human life will open new facets, and humanity, in addition to obvious problems, will also have a lot of new opportunities'. Two items in *Gazeta.ru* in 2015 add discussions of risk to the neutral discourse, 'A Total Anthropocene Came to Earth. In 250 years, Mankind Changed the Earth in Unrecognizable Way' and again 'Climate isn't Afraid of Sanctions. The Year 2014 Has Become the Hottest in the History of Meteorological Observations'. This is followed by an interview with Dr. Vladimir Pitulko, 'Who Slayed Mammoth Zhenya?' (a local reference). Pitulko strongly opposes the idea of the Anthropocene as a distinct geological period. He is quoted: 'Now mankind has the ability to destroy itself, but it cannot affect the planet on a geological scale. If the development of human technologies reaches the level at which it will be possible to influence global processes, then it will be possible to think about defining the Anthropocene as a separate epoch'. The government-owned daily *Rossiyskaya Gazeta.ru* also provides some coverage of the Anthropocene, for example, 'Chicken [bones] is a Marker of Our Time: We Live in a New Geological Era' (2016), deflated by a comment from Dr. Yuri Lavrushin: 'the announcement of a new era in our history has no serious scientific justification and is clearly premature', while 'Got a Story? Man Opens a New Geological Era' (2017) is less dismissive, presenting various possible causes of the Anthropocene. *Vokrug sveta* rounds off coverage of the Anthropocene in Russia with five articles in 2016–17 on the Cape Town International Geological Congress, the technosphere, and start dates.

To sum up, the number of articles in the Russian media reporting on the Anthropocene (mostly daily newspapers and portals) is, though the most numerous of the CEE countries, commensurate with the country's size and population. In that respect, except for 2015–2016, Anthropocene reporting from Russia is relatively weak. However, several themes stand out: (1) frustration around Eurocentrism, i.e., neglect of Russian research in the mainstream 'Western' Anthropocene narrative; (2) scepticism, which follows from that, and (3) simultaneously, acceptance of the mainstream narrative and frequent borrowing from English language media outlets and research journals. All three types of discourse are present, led by neutralizing narratives (over 30), followed by a number of articles containing elements of 'good' Anthropocene', and very few exemplifying the discourse of the necessity for radical societal change.[21] It is also worth noting that, despite its ongoing impacts, Chernobyl does not appear in any of our results (see Brown 2019).

Conclusion

There are notable differences, as well as similarities, between the CEE countries analyzed in this chapter. They differ in size, dominant religion, position in the system of states, landlocked or coastal. Some of them share the 'crossroads', the national metaphor of being the bulwark 'in-between' the West and the East (historically, marked by the border with the former Ottoman Empire, or even between Catholic and Orthodox Christianity). The key shared characteristic of the CEE countries in our context is the heritage of socialist industrialization and the earlier described emergence of environmental movements. Two questions arise. First, why is there a relatively weak interest in the Anthropocene and ecological subjects in the media in CEE countries (especially compared with Western Europe and North America)? And second, why do CEE media outlets predominantly rely on English language sources?

CEE countries appear to be lagging behind and following ideas developed in Western Europe and the USA. In this light, Vargha's lucid comment in 1992 that, as CEE countries during the socialist period oriented towards Moscow, so they nowadays orient towards Western Europe (Vargha 1992: 38), seems prophetic. Western Europe's influence in CEE countries is three-fold, manifested in knowledge transfer from the West, institutional support (most importantly in the work by NGOs), and EU regulations and project funding. Our research results tend to agree with van Eeden (2017a) that, in CEE countries, the dominant Anthropocene discourse is being standardized. This willingness to standardize, whether or not the standardization is consistent with societal practices in CEE countries, is, it might be said, truly provincial. In general, then, adherence to the neutralizing (often neoliberal) discourse in Anthropocene reporting may in this light be perceived as a part of the ongoing 'Westernizing' discourse. That would mean that ecological issues take second place to the perceived priority of socio-economic 'catching up' with the West or moving, as it were, 'from Europe's East to Europe Proper'. This sentiment was explicitly expressed in the Polish articles on the Anthropocene, while resisting Western discursive hegemony was, not unexpectedly, voiced in Russian media.

With the possible exceptions of the Baltic countries and Croatia, the Anthropocene concept is not much represented in the creative arts in most of the CEE. It is explored in multiple ways: as a predominantly philosophical provocation, a scientific, technological concept (Russia, Baltic countries), and as both a socio-cultural and a natural issue. Unlike in mainstream media reporting, the need for change is regularly mentioned, though with little attempt to localize the Anthropocene or offer specific strategies for change.

Notes

1 Grigor Asmolov, Nathalie Goshylyk, Lukas Likavčan, Juraj Skačan, Richard Stahel, Dragos Obreja, and Viola Egikova also contributed to this chapter.
2 See Carmin and Fagan (2010), especially its extensive bibliography.
3 Professor Supek, a passionate advocate of the ideas presented by the Club of Rome, noted that, although seemingly politically 'innocent', environmental movements raise questions that may cause chain reactions; for example, installing filters on factory chimneys can lead to questions about what those factories are producing, why they are producing it and what are the working conditions like in them (Supek 1973).
4 For example, a strong anti-nuclear movement emerged in Russia in 1988, preventing the construction of new nuclear power plants for the next five years.
5 In addition to an anti-nuclear wing, environmental movements expressed themselves in many spontaneous local gatherings related to various incidents, as well as through forming youth organizations dealing with environmental issues.
6 The case of Slovenia shows, without a doubt, the significance of environmental movement in mobilization of the population towards system changes.

7 For example, weekly magazines in Slovakia do not have online editions, so we could not establish the exact number of articles. In the Czech Republic and Romania, only magazines and portals that published at least three articles on Anthropocene were taken into account.
8 Like most of the regions in this book Central and eastern Europe is heterogeneous by many criteria. It includes some very large and some relatively small countries.
9 As was the case in most other countries, two-thirds of the articles were published in 2015 and 2016.
10 As portrayed in the film 'The Day after Tomorrow'.
11 This increases to at least five articles if the years 2018–2019 are considered.
12 For more on Anthropocene-related creative arts, see Chapter 13 in this book.
13 *Gazeta Wyborcza* was founded as the mouthpiece of the anti-communist Solidarity movement and is now best described as a liberal-conservative newspaper.
14 A press release from the University of Leicester (30 November 2016), in the name of Professor Mark Williams, reads: 'The technosphere is a major new phenomenon of this planet – and one that is evolving extraordinarily rapidly'. An international team, led by University of Leicester geologists, has made the first estimate of the sheer size of the physical structure of the planet's technosphere – suggesting that its mass approximates to an enormous 30 trillion tons.' This was reported all over the world. See Haff (2014).
15 The book was a publishing sensation in Poland, garnering 40 interviews in the general media – TVN 24, national radio stations like Trojka Radio and TOK FM – and leading newspapers and periodicals, including *Gazeta Wyborcza*, *Polityka*, *Tygodnik Powszechny*, *Dziennik Gazeta Prawna*, *Polska*, *The Times*, and *Krytyka Polityczna*).
16 Raworth (2014) famously suggested the term 'Manthropocene'. It is striking that media generally almost totally ignore the large feminist literature on the Anthropocene (for example, Gibson-Graham 2011, Caputi 2016, Grusin 2017). Donna Haraway is the exception (see Chapter 12 in this book).
17 See Crook and Short (2014) on genocide and ecocide. The term '*Ekobójstwo*' (Ecocide) is not yet common in Polish.
18 The coverage of this *Gazeta* and *Gazeta.ru.* is similar. It seems likely that all media in Russia is controlled, to some extent, by those with links to the government.
19 In fact, Vernadsky is frequently mentioned as a forerunner of the Anthropocene by scholars outside Russia (see, for example, Oldfield and Shaw 2006, Guillaume 2014). Vernadsky (1945) demonstrates that he was such a forerunner.
20 In this connection, see Lee (2019).
21 See Poberezhskaya (2015) on media coverage of climate change in Russia.

References

Bińczyk, E. (2019) The most unique discussion of the 21st century? The debate on the Anthropocene pictured in seven points. *The Anthropocene Review* 6(1–2): 3–18.

Brown, K. (2019) Learning to read the great chernobyl acceleration: Literacy in the more-than-human landscapes. *Current Anthropology* 60(Suppl. 20): S198–208.

Caputi, J. (2016) Mother earth meets the Anthropocene: An intersectional feminist analysis. In Godfrey, P. & D. Torres eds. *Systemic Crises of Global Climate Change: Intersections of Race, Class and Gender*. London and New York: Routledge, chap. 3.

Carmin, J.A. & A. Fagan (2010) Environmental mobilisation and organisations in post-socialist Europe and the former Soviet Union. *Environmental Politics* 19(5): 689–707.

Chaisty, P. & S. Whitefield (2015) Attitudes towards the environment: Are post-Communist societies (still) different? *Environmental Politics* 24(4): 598–616.

Coates, J. (2013) *Woman, Man and Language*, 3rd edition. London: Routledge.

Crook, M. & D. Short (2014) Marx, Lemkin and the genocide–ecocide nexus. *The International Journal of Human Rights* 18(3): 298–319.

Gibson-Graham, J.K. (2011) A feminist project of belonging for the Anthropocene. *Gender, Place and Culture* 18: 1–21.

Grusin, R. ed. (2017) *Anthropocene Feminism*. Minneapolis, MN: University of Minnesota Press.

Guillaume, B. (2014) Vernadsky's philosophical legacy: A perspective from the Anthropocene. *The Anthropocene Review* 1(2): 137–146.

Haff, P. (2014) Humans and technology in the Anthropocene: Six rules. *Anthropocene Review* 1(2): 126–136.

Hansen, M.B.N. (2012) Foucault and media: A missed encounter? *The South Atlantic Quarterly* 111(3): 497–528.

Jancar-Webster, B. (1993) Introduction. In Jancar-Webster, B. ed. *Environmental Action in Eastern Europe: Responses to Crisis*. New York: Routledge.

Jancar-Webster, B. (1998) Environmental movement and social change in the transition countries. *Environmental Politics* 7(1): 69–90.

Koszegi, F. & E.P. Thompson (1982) *The New Hungarian Peace Movement*. London: Merlin Press. http://digitalarchive.wilsoncenter.org/document/113738

Kundzewicz, Z.W. et al. (2017) Climate Change in the Media: Poland's Exceptionalism. *Environmental Communication.* https://doi.org/10.1080/17524032.2017.1394890

Lee, J. (2019) Why Vladimir Putin suddenly believes in global warming. *Bloomberg Opinion* (29 September).

Oldfield, J.D. & D. Shaw. (2006) V.I. Vernadsky and the noosphere concept: Russian understandings of society–nature interaction. *Geoforum* 37: 145–154.

Oštrić, Z. (1992) Ekološki pokreti u Jugoslaviji 1971–1991 (Ecological movements in Yugoslavia 1971–1991). *Socijalna Ekologija* 1(1): 83–104.

Poberezhskaya, M. (2015) Media coverage of climate change in Russia: Governmental bias and climate silence. *Public Understanding of Science* 24(1): 96–111.

Raworth, K. (2014) Must the Anthropocene be a manthropocene? *Guardian* (20 October).

Supek, R. (1973) *Ova jedna zemlja (This only Earth)*. Zagreb: Naprijed.

van Eeden, P. (2017a) Materializinig discourse analysis with James, Schmitt and Latour. *Palgrave Communications* 3. doi:101057-palcomms.2017.39

van Eeden, P. (2017b) Politization and depolitization of ecology: Polish and Romanian perspectives. *Interface: A Journal for and About Social Movements* 9(2): 83–113.

Vargha, J. (1992) Environmental perspectives in central and Eastern Europe. In Verhoeve, B. & G. Bennett eds. *The Netherlands and the Environment in Central and Eastern Europe*. Zwolle, The Netherlands: Dutch National Council for Environment Policy. 31–38. https://doi.org/10.1177/2053019616688022

Vernadsky, V.I. (1945) The biosphere and the noösphere. *American Scientist* 33: xxii, 1–12.

Vukelić, J. (2014) *Potentials of Emergence and Development of Environmental Movement in Serbia in the Post-Socialist Context*. Ph.D. thesis, University of Belgrade (in Serbian). http://nardus.mpn.gov.rs/bitstream/handle/123456789/4846/Disertacija445.pdf?sequence=1&isAllowed=y)%5b%5bIS. (English summary).

9 Western Europe

Planetary Eurocentrism

Boris Holzer and Leslie Sklair[1]

Introduction

The media in France, UK, Germany, and Italy supplied about two-thirds of the total Anthropocene results in Western Europe. Quality national papers in most countries dominate the coverage (notably, in France, *Le Monde* and *Libération*; in the UK, *The Guardian, The Telegraph,* and *The Independent*; in Germany, *Frankfurter Allgemeine Zeitung*, in Spain, *El Pais* and *El Mundo,* in Italy, *Corriere Della Sera* and *La Repubblica*, and *La Libre Belgique* in Belgium and *Le Temps* in Switzerland). There is also significant coverage of the Anthropocene in mass circulation tabloids such as *The Daily Mail* in the UK, *Bild* and *Bild am Sonntag* in Germany, *La Stampa* in Italy, *ABC* in Spain; and *Aftenposten* in Norway. We also find coverage in the business press, for example *The Economist* and *The Financial Times* in the UK, *Wirtschaftswoche* in Germany, and *L'Echo* in Belgium, and in magazines such as *New Statesman* and *The Spectator* in the UK, *L'Obs* in France, *Der Spiegel* and *Focus* in Germany, *Grande Reportagem* in Portugal, *L'Espresso* in Italy, and even in free tabloid newspapers (*Metro* in Belgium and the UK, and *The Evening Standard* in the UK). Coverage of Anthropocene-related events and ideas from the social science, humanities, and creative arts are found in all types of media.

The overall results from European media indicate that what we conceptualize as neutralizing and optimistic reassurance narratives account for most of the items found. While many articles communicated some pessimism and called for change, the nature of the required change was usually left vague. This finding demonstrates that, although there appears to be a relatively large number of informative articles over a wide variety of media in western Europe, almost none of them present concrete proposals for radical change to cope with the potential threats to human survival in the Anthropocene.[2] The chapter concludes that, despite coverage of the sometimes alarming writings of scientists, notably Paul Crutzen (Nobel prize-winner) and Jan Zalasiewicz (chair of the Anthropocene Working Group, AWG), the risks of the Anthropocene are rarely presented as critical to human survival. This is mainly due to reporting which focuses on disputes among scientists rather than on the conflicting political implications of the Anthropocene.

Three foundational articles for quality media coverage of the Anthropocene: France, the UK, and Germany[3]

The media in France, the UK (mainly England), and Germany produced over half of the results found in western Europe. One article from each of these three countries can be considered as 'foundational' in the sense that they establish a sort of template for how the quality media approach the difficult and sensitive idea of the Anthropocene. We first analyze these foundational articles in the context of the results from each of these three countries, and then discuss the rest of the media coverage of the Anthropocene in western Europe.

The term 'Anthropocène' was adopted early in France. In 2004, *Le Monde* published 'Earth: Sick of Man', a wide-ranging, and well-informed essay, referencing the Intergovernmental Panel on Climate Change (IPCC) and the release of Roland Emmerich's disaster film, 'The Day After Tomorrow'.[3] *Le Monde* argues: 'For a long time, the Earth has only experienced natural changes. Until the arrival of *Homo economicus* … a very recent and very brief period in view of the long history of the Earth, which Paul Crutzen (Nobel Prize laureate for Chemistry in 1995) called the Anthropocene. With its rampant demography, its activity and growing mastery of nature, the human species is changing the climate and its environment'. Acting seriously on greenhouse gases 'forces us to call into question an economic model based on the massive use of fossil fuels and on growth in perpetuity for the industrialized countries or those in the process of being'. The article concludes by drawing attention to the perils of a 'situation driven by climate change, rising sea levels, deforestation, and increasing populations living in flood plains. It is the poor countries, especially in Asia, that are likely to pay the heaviest tribute for these water overflows. The challenge is immense, and it is vital to take it up'. We call this article 'foundational' because it highlights the need to reduce dependence on fossil fuels and the difficulties of doing so but stops short of concrete radical measures that would challenge the status quo. This is the first occasion in which an article in our global searches provides links to other relevant sources, though not all the links are still active. This practice is now commonplace, plus video links, TED talks, author promotions, and so on. The article calls into question 'growth in perpetuity', the central tenet of neoliberal capitalism but offers no alternative; finally, it introduces the idea that those in poor countries who contribute least to eco-system damage are liable to suffer more than those in rich countries, who are responsible for most of the greenhouse gas emissions. This may also be the first time that a newspaper has commissioned artwork to illustrate its views on the Anthropocene (see Figure 9.1).

Coverage of the Anthropocene in *Le Monde*, while always informative and often lengthy, appears to have been mostly neutralized thereafter, with typical headlines such as 'The Earth has Entered a New Geological Era: the Anthropocene' (2008); 'Fukushima Or the End of the Anthropocene: Emergency Exit from the Vanity of Our Mode of Growth'. The solution to the problem offered is: 'Companies need to pull themselves together to invent systems that are human-sized, resilient, and cooperative' (2011); 'Man Brought the Earth into a New Geological

Figure 9.1 *Le Monde* man (2016) © Nini La Caille.

Epoch: The Impact of Human Activities on the Terrestrial System has Become Predominant' (2015), referencing the 'great acceleration'; 'Are We Going to Enter the Anthropocene in 2016? Researchers Must Agree on the Beginning of This New Geological Era, Shaped by Man' (2016), and 'Anthropocene: Geological or Societal Subject?' (2016). Similarly, the French morning daily, *Le Figaro*, focuses more on the emergence of the Anthropocene idea than its potential risks, as in 'A Geological Era of Which Man Would Be the Hero: The Idea Launched in 2002 By the Nobel Prize-Winning Chemist Paul Crutzen to Adapt the Geological Classification of Time to Consider the Impact of Man on his Environment is Slowly Making its Way' (2011). The somewhat more alarming 'Man is As Devastating as the Asteroid That Put an End to the Dinosaurs' (2014), cites two French Anthropocene scholars (Bonneuil and Fressoz; their book was published in 2017); and two bland reports, one on the AWG in the Cape Town International Geology Congress (2016), the other on 'The Planet Enters a New Era: Man Brought the Earth into a New Geological Era, the Anthropocene' (2016).

The leftist daily, *Libération*, also has significant coverage of the Anthropocene, for example, discussing the rather sceptical opinions of the French glaciologist Claude Lorius, 'A Conditioned Era: The Glaciologist Demonstrates That Man has Become a Climatic "Geo-Engineer" as Powerful as the Geological Forces, and Announces the Anthropocene, the Era of Man' (2011); 'Anthropocene: There is a Storm in the Era', highlighting William Ruddiman's research on the 'early' Anthropocene thesis (2015); and 'Anthropocene Calls for New Rights for the

Earth: Man has Become Stronger Than the Planet. Unless We Quickly Emerge from This Posture of Domination of Nature, We Can Destroy Nature' (2015). This article is one of the very few to raise the issue of 'ecocide', calling for the International Criminal Court to be equipped 'with means that are up to the challenges of the Anthropocene to stop the ecocide in progress'.[4] The article cites the work of Latour, Haraway, and Viveiros de Castro (see Chapter 12 of this book for further discussion of these theorists). In the weekly news magazine, *L'Obs*, we find mostly conventional neutralizing stories, for example, 'In 1610, the Human Species Took Control of the Planet' (2015), one of many discussions of the Columbian exchange thesis of Lewis and Maslin (see Chapter 13 of this book); and another, much reproduced story, 'Anthropocene: Plastic, Radiation, Gas ... These Indelible Traces That Man Leaves on Earth' (2016). Most of these articles do indicate, directly or indirectly, that the Anthropocene comes with risks of various types, but the tone is almost always neutral or reassuring.

An extensive search of small-town and regional media in France revealed very few results, but one source does stand out. Between 2015 and 2017, *La Montagne,* a small, progressive newspaper based in Clermont-Ferrand, published several Anthropocene-related articles. The first (in 2015) reports on a meeting on global warming at a local pub organized by Attac (a leftist social movement), where the activist Anne Bonnérat 'animated the debate and began the evening by recalling the change of geological era (now called Anthropocene by many scientists)'. But, even here, despite the radical critique of the economic and political system, the message appears to be one of hope for the future if we can transform existing institutions, but not how to do this.

It is worth noting that many French intellectuals were obviously interested in the Anthropocene. In November 2013, a conference, 'Thinking the Anthropocene', was organized in Paris by several prestigious universities and research bodies (the Australian ethicist, Clive Hamilton, was also involved). The focus was on environmental humanities and the invitation circular, all welcome, free, in English, proclaimed (controversially): 'in an epoch in which "Gaia" has been reawakened, the "social-only" conceptions of autonomy, agency, freedom, and reflexivity that define modernity, and the idea of the human on which these disciplines have been constructed, must be rethought'. A selection of the papers from the conference was published (Hamilton et al. 2015). A few years later, the venerable journal *Les Annales* published its first statement on 'L'Anthropocène' (Editors 2017).

The second 'foundational' article, from the UK, appears in *The Financial Times* of London, also in 2004 (published about two months after the article in *Le Monde*). Written by its science editor, Clive Cookson, it reports: 'Scientists Warn of New Anthropocene Age. Should We Recognize New Epoch of Human Influence?' This item is notable for several features. First, it is written by a science editor; second, it is illustrated by a photo captioned: 'Nobel-winning atmospheric chemist Paul Crutzen, who has championed the idea of designating a new geologic age, reviews NASA climate data', thus adding a prestigious face to the concept; and third, it begins to provide some institutional context for the process of legitimizing the Anthropocene for the scientific community. Cookson cites the Euro

Science forum in Stockholm in 2004, identifying Will Steffen (chief scientist for the International Geosphere–Biosphere Programme), John Schellnhuber (director of the Tyndall Centre for Climate Change at the University of East Anglia), and Katherine Richardson (Professor of Oceanography at Aarhus University, Denmark) who reminds us of the serious impacts of carbon in the oceans. These scientists contribute arguments about the global significance of the Anthropocene and its threats to ecosystems all over the world.

Reports of scientific meetings and academic publications are regular components of Anthropocene reporting in the mass media. Coverage in *The Financial Times* appears to settle into a variety of reassurance narratives, spiced with occasional slight alarm and suggestions that scientists can't agree, as in 'Scarred Earth to Enter the Anthropocene Epoch' (2008). 'Muddle Over the Moment Mankind Made its Mark on Earth', in 2015, argues: 'The whole Anthropocene discussion deserves a wider hearing, not least for its power to inspire… It also poses a question worthy of public debate: do we want our arrival in the planetary logbook to be heralded by the ingenuity of the industrial revolution, or our indifference to plastic rubbish?' More blandly, we have the typical headline, 'Earth Entered New Epoch in 1945, Say Scientists' (2015); 'The Anthropocene. Dawn of a New Epoch?' (2016) indicates various potential geological markers. Similarly, in *The Telegraph*, we find: 'Earth Entered New Anthropocene Epoch in 1950, Scientists Say' (2016), a long article on James Lovelock in 2014, explaining his big idea, Gaia (the Earth as a self-regulating system that maintains favourable conditions for terrestrial life), and his new book *A Rough Ride to the Future* where 'Lovelock points out that, three hundred years ago, when Thomas Newcomen invented the steam engine, he also unknowingly launched the Anthropocene epoch'. Also in 2014, *The Telegraph* correspondent declares: 'Until a Few Years Ago the Word Anthropocene Would Bring a Puzzled Look to Most People's Faces, Rather As If One Had Alluded To the Akond of Swat. Not Anymore'.[5]

In the UK, 'Letters to the Editor', as well as the articles themselves, are of interest. For example, in *The Independent*, 'Capitalism is Heating Up the Planet' (2013), we find: 'Your leading article's statement "ultimately, the solution [to climate change] lies with the market" is astonishing since climate change is caused by none other than the market itself'; 'Capitalism – a System That is Incessantly Expansive and Inherently Wasteful – is the Precise Opposite of What is Required to Combat Climate Change' (2013), from Fawzi Ibrahim, author of the book *Capitalism versus Planet Earth: An Irreconcilable Conflict*. Are these irate correspondents justified in their criticisms that this progressive liberal paper is not taking the Anthropocene seriously enough? Headlines over the years suggest they are; these examples from *The Independent* are typical: 'Climate Change Causes New "Epoch"'(2008), arguing that the start of the industrial revolution shows that we are now living in a new epoch called the Anthropocene; 'How We Must Adjust Our Lifestyles to Nature: Welcome to the Anthropocene' (2015), citing the 'good' Anthropocene advocate Christian Schwägerl; and 'Humans are Causing Climate to Change 170 times Faster Than Natural Forces, Referencing the "Anthropocene Equation"' (2017). All of these are valuable to readers who wish to understand the basic facts of the Anthropocene, but not so valuable for understanding potential risks.

The Economist also has extensive coverage of the Anthropocene, mostly in open access comments from readers, rather than articles. The much-referenced 'Welcome to the Anthropocene' leader and commentary of May 28, 2011, jostled for attention with features on Australia's Promise, Austerity in southern Europe and the IMF, and Hope in Nigeria. Despite these distractions, we find many critical responses from readers, for example in 2011: 'Welcome to the Anthropocene, says *The Economist* article in May. Humans have changed the way the world works. Now they have to change the way they think about it, too. And that means changing economic planning and reversing population growth, quickly, very quickly'. It appears that a journalist from *The Economist* attended what might have been the first public meeting on the Anthropocene, organized by the AWG and hosted by the Geological Society of London on May 11, 2011 (see Figure 9.2).[6]

The tone of coverage in *The Economist* is also typically expressed in 2013, with: 'Beware the Age of Loneliness. Man Must Do More to Preserve the Rest of Life on Earth, Warns Edward O. Wilson'. The eminent biologist poses the dilemma in his new book: 'will we stop the destruction for the sake of future generations, or go on changing the planet to our immediate needs? If the latter, planet Earth will enter a new era of its history, cheerfully called by some the Anthropocene, a time for and all about our one species alone. I prefer to call it the Eremocene, the Age of Loneliness'.[7]

It is surprising to find that *The Guardian* (along with *The New York Times*, the most-quoted and borrowed-from source in all our results) rarely bucks the reassurance trend. Four articles from 2016 (all picked up by media elsewhere) provide evidence for this view. The first, 'Generation Anthropocene' by the celebrated natural history writer Robert Macfarlane is discussed in Chapter 12 of this book. The second, 'How the Domestic Chicken Rose to Define the Anthropocene' by the environment editor Damian Carrington, is not as frivolous as it sounds, arguing cogently that the bird could become the 'type fossil of the Anthropocene'; the third, by the eminent scientist Sir Martin Rees, 'The Anthropocene Epoch Could Inaugurate Even More Marvelous Eras of Evolution' is self-evident; as is the fourth: 'Books to Give You Hope: *Adventures in the Anthropocene*', an enthusiastic review of Gaia Vince's 'dazzling work of global reportage, which won the Royal Society Winton prize in 2015'. This review is exceptional in that it actually engages with the contents of the book, whereas most reporting focused on the prize and the gender of the recipient. However, the reviewer, Claire Armistead, literary editor of *The Guardian* reveals: 'I love this book for what it says about the resourcefulness of humanity, the capacity of ordinary people to make a difference, and for those small differences to add up to something that might even turn the planet away from its apparently unstoppable hurtle towards environmental destruction. Vince doesn't shirk the hard stuff – the whats, whys and whethers of conservation'. This takes us into the difficult territory of journalist (and scholarly) 'hope versus despair' (see Chapter 14 of this book). It may also be noted that, though important, conservation is not the central issue of the

Figure 9.2 First public international conference on the Anthropocene (2011) © Geological Society of London.

Anthropocene. In 'Geologists Press for Recognition of Earth-Changing Human Epoch' (2017), *The Guardian* does buck the reassurance trend, printing an excerpt from *Defiant Earth* by the Australian realist/alarmist Clive Hamilton, exposing 'The Great Climate Silence: We are on the Edge of the Abyss But We Ignore it'.

This is immediately balanced out by 'After 200,000 Years of Modern Humans on a 4.5 Billion-Year-Old Earth, We Have Arrived at New Point in History: the Anthropocene' (2017), reporting the 'good' Anthropocene narrative by the ecological geographer and Anthropocene notable, Erle Ellis.

It may come as a surprise to followers of the British press that the coverage of the Anthropocene in the tabloid *Daily Mail* places it in the global top twenty in our results (see Table 2.2 in chapter 2 in this book-). Here, we find a mixture of reassurance and sensationalism. For example, 'How the World Would Thrive Without Mankind', is a review of the book, *The World Without Us*, by Alan Weisman. We are alarmed, then reassured: 'there is no doubt that we are wreaking terrible damage. So much so that scientists talk about the "Anthropocene" – the destructive Era of Man. We are making a terrible mess in places – but, without us, the world would surely be a duller place' (2007); 'Vatican's Scientists Back "Dodgy" UN Climate Change Report that Even its Authors Admitted was WRONG', a Controversy That 'Has Sparked Anger Among Climate-Change Sceptics' (2011),[8] and the banal, 'Impact of Humans has Created New Geological Age' (2014). More surprises await us. In *The Sun* (the tabloid most despised by cultural elites, though read by some) we find: 'BRAVE NUKE WORLD [sic]: Here's What Planet Earth Will Be Like in the "Plutocene" Era, Following an Apocalyptic Nuclear War' by Andrew Glikson, climate scientist at the Australian National University. This piece was originally written for *The Conversation*, a well-respected online publication, supported by universities in Africa and elsewhere. Though he is pessimistic, Glikson concludes: 'Humans Will Survive in Relatively Cold High Latitudes and Altitudes. A New Cycle Would Begin' (2017). While the significance of nuclear fallout as a marker of the Anthropocene is widely reported, the Plutocene is not. It is difficult to know what to make of this item in *The Sun*, though we can speculate on the Australian connection. In *The Daily Express,* a conservative tabloid in decline, we find several items, for example: 'Anthropocene: Will Scientists Herald a New "Age of Humans" and Name New Epoch? (2014)', accurately, if minimally, reporting debates about start dates, the observations of an astronaut, the great volume of scientific studies on the Anthropocene, and the contrarian views of Stanley Finney. *The Evening Standard*, a free London paper owned by a Russian businessman, has a curious mixture of Anthropocene items. It is one of several media sources to review the controversial book by Mark Lynas, *The God Species*, charting the end of the Holocene and the start of the Anthropocene. Lynas himself is controversial, having abandoned socialist environmentalism in the light of the Anthropocene.[9] *The Evening Standard* concludes: 'he is robust in his dismissal of climate change "deniers"... a clear-eyed, hard-headed assessment of the ecological challenges facing us – and all the more bracing for it' (2011). This article is followed by 'We're All Doomed – Yet Again', continuing 'such is humankind's impact that some scientists now suggest this brief era, the Holocene, be characterized primarily as the time of the planet's sixth great extinction … known as the Anthropocene' (2014); 'The Anthropocene Era Has Now produced [the BBC TV series] Blue Planet II, and That's a Pretty Wondrous Thing' (2017). This is strange, as the series hardly mentions the Anthropocene directly, though it does raise questions of anthropogenic ecosystem damage. Another free paper, *Metro,* now with Europe-wide distribution in various guises, offers, 'The Planet is Probably in a New Geological Age - the Anthropocene Epoch' (2016). The article states, rather neutrally,'Us Humans Have Left Our Mark on the World' and goes on to explain the work of geologists, how humans have changed the atmosphere, oceans, and wildlife, dumped plastic rubbish in oceans, and used high-carbon-emission concrete in buildings. The eminent Anthropocene scholar, Colin Waters (of the AWG), is quoted: 'the changes to the world are as big as what happened at the end of the last ice age'. Jumping the gun, like many other media, *Metro* declares: 'The International Commission on Stratigraphy Will Decide to Accept the "Anthropocene Epoch" as an Additional Unit in the Time Scheme Used to Describe the Planet's 4.6 Billion Years of History' (2016). Despite some lapses, this article might have set commuters thinking, perhaps even encouraging some to join the Extinction Rebellion protests that closed down parts of central London in 2019. Maybe one of them was the subject of 'OK Commuter: It's Hot and We're Out of Our Minds … the Hottest Day of the Anthropocene and a Bloke Slouching Off the Bus at Forest Hill is All in Black – Jeans, Heavy Trainers, and a Hoodie Pulled Up Over His Baseball Cap', in *Metro* (2017). Yet another free paper, *City AM*, reports 'Plastiglomerates, Rocks Made of Melted Plastic, Found in Hawaii … Because Plastic Doesn't Degrade, These New Rock Forms Will Most Likely Stay On the Record, Becoming Part of the So-Called Anthropocene Era –Where Human Activity on the Earth Leaves Permanent Evidence' (2014).

Outside the national dailies and quality magazines, there is little coverage of the Anthropocene in the UK. However, in 2006, *The Oxford Mail* offers: 'Time for Heated Words on Global Warming', explaining that the eminent scientist Lord May did not want to be a 'messenger of doom' in a lecture in Oxford. Lord May suggests we might call the post-1780 period, the Anthropocene, that is, seeing humans as the most important entity. We must not be stunned into inaction because of the enormity of the problem, he maintains…The problems may be global, but the answers have to be national and, ultimately, local and down to the individual'. *The Oxford Mail* also prints a few items on a local band with Anthropocene interests (see Chapter 13 of this book).

A third foundational article was published in Germany in *Frankfurter Allgemeine Zeitung* under the title 'At the End of the Holocene' (2000). Written by Jürgen Kaube (who joined the newspaper's board of publishers in 2014), it reports preliminary results from the Third Report of the IPCC, focusing on Europe. The article starts with a quote from Max Frisch's novel *Man in the Holocene*: 'Nobody is reckoning on another flood'. This serves to illustrate how humans underestimate and misjudge the forces of nature and how little significance their existence has against the backdrop of a whole geological era. Kaube contrasts Frisch's disaster scenario (a summer downpour) with the predictions of environmental science: due to global warming, the future might not just bring a lot of rain during summer but no more snow during winter. He responds to allegations of 'climate alarmism' against such scenarios with the argument that evidence of climate change has accumulated and that preventive action will be taken regardless of such objections, because sources of European economic growth (agriculture, manufacturing, tourism) are at risk. The article then refers to scientists Paul Crutzen and Eugene Stoermer and their coining of the term 'Anthropocene'. Kaube interprets the term to imply that earth history now depends on human intervention – and expresses the expectation that humanity will 'take action against its own effects', thus ending on a seemingly positive note, that sounds like a more fatalistic undertone when he specifies that genetic engineering may soon be able to make man 'heat-resistant'. Several features of this article re-appear in subsequent contributions. First, regarding the timing of articles, international events such as scientific or political conferences, and accompanying reports about climate change (used as a metonym) often provide opportunities to engage with the Anthropocene. Second, regarding actors in the Anthropocene story, Crutzen is usually mentioned, often quoted, sometimes interviewed, when the Anthropocene is discussed. Third, regarding the 'framing' of the topic, the problem of potential alarmism is frequently brought up, not only by critics of the Crutzen hypothesis or climate change, more generally, but also authors sympathetic to the cause see the risk of exaggerating human impact. This is against the backdrop of experience; for example, a number of environmental topics – such as the *Ozonloch* (ozone depletion) and *Waldsterben* (forest dieback) – attracted much attention, particularly in Germany, often in an almost apocalyptic manner, but were eventually deemed to be manageable by political and technological intervention. Some observers use that experience as an argument against the Anthropocene discourse, others simply do not want to be accused of scaremongering.[10]

The coverage of the Anthropocene in the German news media took off slowly but has been quite extensive, with over 200 articles between 2000 and 2017. Most articles appeared in the national dailies *Frankfurter Allgemeine Zeitung* (FAZ) and *Süddeutsche Zeitung* (SZ), but the online outlets of three weekly news magazines (*Der Spiegel, Focus, Stern*) almost matched the level of reporting of the national dailies. Even in the *FAZ*, however, it took several years after the foundational article until the next. This discussed Crutzen's proposal to regulate global climate change by technological means, namely geoengineering. He is quoted: 'I hope this experiment never needs to be conducted' (2006), followed, about a year later, by 'Man, the Climate Perpetrator' (2007), discussing the work of palaeoclimatologist William Ruddiman, who argues that man's impact on global climate dates back some 8,000 years. Echoing concerns about alarmism, Ruddiman is depicted as a 'climate realist'. His research shows that previous phases of global warming due to anthropogenic greenhouse gases may have prevented another ice age, but he argues that this has little relevance for future scenarios of global warming.

The scientific debate about the beginning (and therefore the *past* history) of the Anthropocene plays a prominent role in the German language press. *FAZ* prints a flurry of well-informed Anthropocene stories, especially in the peak years of 2015 and 2016. 'On the Plastic Cliffs' (2015) discusses the question of new forms of rock sediment, including plastics, associated with the Anthropocene. The AWG receives further coverage in the following year in 'Is the Anthropocene Coming? Old Certainties Fade in the New Age' (2016). The future dangers and political consequences of the Anthropocene are most clearly presented in interviews with natural and social scientists, notably Crutzen, quoted again: 'It Frightens Me How Vulnerable the Atmosphere is' (2013). The sociologists Ulrich Beck and Bruno Latour are referenced in 'The Apocalypse Does Not Respect Constraints' (2014). Book reviews and lectures also give opportunities to address the potentially fatal consequences of human activities on the environment. For example, Elizabeth Kolbert's *The Sixth Extinction* is reviewed and discussed with reference to the Anthropocene in 'Beware If *Homo sapiens* Establishes Himself' (2015) and, once again, Bruno Latour provokes discussion with his book, *Facing Gaia* in 'The Age of Gaiagraphy' (*FAZ* 2017).

The second German quality daily in our sample, *Süddeutsche Zeitung* (*SZ*), has also published regularly on Anthropocene issues, starting in 2010 with an article in the science section, explaining the term, with references to Crutzen and Zalasiewicz, 'The Epoch Man'. In subsequent years, there are relatively more articles that problematize the Anthropocene and link it to cultural events and political decisions. Commenting on the UN summit in Rio de Janeiro, 'Who Participates, Gains' (2012) discusses the problem of global environmental cooperation and expresses the hope that technological innovation will contribute to climate protection (more geoengineering). In 2013 and 2014, several articles mention the Anthropocene in the context of a two-year project at the *Haus der Kulturen der Welt* (House of World Cultures, HKW) that reflects upon the impact of humanity on the environment, 'At the Beginning of a New Geological Era' (2013), and 'Out of Joint' (2014). A similar exhibition at the *Deutsches Museum* in Munich focused on technological aspects, 'A New Chapter in the History of the Earth' (2015).[11] Both events served to disseminate the notion of the Anthropocene more widely and motivated discussions about the 'Risk of Progress' (2015), in a rather risk-averse way, still taking 'progress' for granted. The Anthropocene project at HKW was also taken up by Germany's most widely circulated tabloid, *Bild*. An article mentioning the Anthropocene (out of only three in total) describes the upcoming event in a rather factual manner, 'House of World Cultures Plans Two-Year-Long Thinking Project' (2012). Although reporting in the tabloids is sparse, both the timing and the topics reflect the overall trend. *Bild*'s first Anthropocene article had appeared in 2008, discussing the new term and its implications from the perspective of geological science. In 2016, the newspaper reports the ongoing scientific discussion and hints at its symbolic meaning in neutral terms.

The online offers of the three most-widely read German weekly magazines, *Der Spiegel*, *Stern*, and *Focus*, are characterized by large differences in the volumes of their reporting about the Anthropocene; whereas *Der Spiegel* published over 20 articles, *Focus* and *Stern* contributed over 60 and only three items, respectively. The relatively few results for *Stern* could be due to the fact that the web archive only shows recent articles, because there are no articles dating from before 2016. The first discusses the geological debate about 'A New Epoch of Earth History' (2016) and adopts a reassurance narrative by expressing the hope that humans will be made aware of their own role in the earth's future. The second is more pessimistic, reporting warnings from the physicist Stephen Hawking about the possibility of self-extinction of mankind, mentioning the Anthropocene in this context.

The weekly *Der Spiegel* has an established track record of publishing on environmental topics. For instance, it carried influential articles on the forest dieback (*Waldsterben*) of the 1980s, including a cover story that proclaimed 'The Forest is Dying' in 1981, with a memorable image.[12] It therefore comes as no surprise that the magazine's online outlet *Spiegel Online (SPON)* has also published several articles on the Anthropocene. The reporting begins with some items that focus on the arguments of Crutzen and other Earth scientists. For example, 'The Well-Tempered Planet' (2007), 'Anthropocene: Geologists Argue That a New Geological Era Has Begun' (2008), and 'Sediment Cores: Hidden Climate Archive' (2009). Subsequent articles emphasize the various and sometimes conflicting accounts of the beginning of the Anthropocene and their consequences in terms of the responsibility of mankind, such as in an interview with Jan Zalasiewicz, 'We are the Predominant Predator on Earth' (2011). Emphasizing the responsibility of mankind for ecosystem degradation, several articles raise the topic of extinction and discuss evidence on the role of humans in the disappearance of species, for example in 'Palaeozoology: Murder in XXL' (2015), on the consequences of human evolution and expansion during the past 50,000 years. *Der Spiegel* also highlights controversies about the Anthropocene, including some items that oppose the concept and its usefulness. For example, in 'Plan For a New Geological Era: An Epochal Mistake' (2016), the geologist and science journalist Axel Bojanowski argues that there is no evidence that humans have significantly changed the planet, so the proclamation of the Anthropocene would be poor science. Bojanowski proposes that we regard the Anthropocene as a cultural movement. In our terms, this would mean absorbing the Anthropocene into the Anthropo-scene. Others also argue against exaggerating the role of humans in different ways. For example, rather than rejecting the Anthropocene, the reporter Georg Dietz accepts it dramatically: 'The End is Near, and That is Fine' (2014). Optimism and fatalism, it seems, can be the flip sides of the same coin.

The relatively large number of Anthropocene articles in the news magazine *Focus* does not mean greater variety in terms of topics covered and positions advocated. 'How Mankind Can Survive' (2009) presents the arguments for the Anthropocene and the idea that staying within critical limits may still be an option to survive its consequences. Technological fixes and their potential side-effects are discussed in 'Biodiversity: Human-Made Extinction' (2010), and, two years later, the focus is on the problem of global collective action highlighted by the Rio Conference and its aftermath: The chances to 'Just Quickly Save the World' (2012) appear to be slim but the new ecological 'grassroots movement' might create momentum. Further articles reiterate the geological discussion about finding clear markers of the Anthropocene (often referred to as 'golden spikes'). The combination of statements and open questions typical of this style of reporting is brought into sharp relief with the title 'Man Has Fundamentally Altered the Earth – But When Did His Era Begin?' (2015). The German business magazine,

Handelsblatt, provides only a handful of articles that mention the Anthropocene. Those primarily reflect upon geological discussion about start dates for the Anthropocene, neither challenging nor endorsing it. An interesting exception to this kind of matter of factual reporting is a guest article by the German philosopher Peter Sloterdijk: 'The Long Way to World Society' (2017). However, he only mentions the Anthropocene as one dimension in a broader account of globalization, focusing on problems of transnational migration.

Anthropocene reporting in Austrian print media follows a similar pattern to that in Germany. Many articles focus on the question of when the Anthropocene began, particularly highlighting Ruddiman's 'early Anthropocene' thesis. *Die Presse* in 'Climatology: The Burnt Ice Age' (2009) contrasts the 'conventional' view, that the Anthropocene is a by-product of the industrial revolution, with Ruddiman's view. Subsequently, 'Forward Into a New Age' (2014) and 'Anthropocene: The World is No Longer and Will Not be What it Used to Be' (2015) basically rehash this discussion, providing more details on the AWG and debates among geologists. Another newspaper that has published more or less continuously on the Anthropocene from 2015 onwards is the *Salzburger Nachrichten* and most of these articles stem from the regular column of the distinguished journalist Viktor Hermann. He discusses particular consequences of the Anthropocene, such as the loss of biodiversity in 'The Sixth Mass Extinction is Happening Now' (2015), but also suggests a reassurance narrative by interpreting the Anthropocene as 'Science Recognizing the Significance of Humanity' (2016). He argues that, whereas technology has been able to fix acid rain and the hole in the ozone layer, we are still waiting for a similar solution in the face of climate change.

In the German-language media in Switzerland, the two major dailies *Tages-Anzeiger* and *Neue Zürcher Zeitung (NZZ)* account for about half of the Anthropocene-related articles, with the weekly *WOZ* providing the remainder In all cases, the reporting starts rather late (2013) and predominantly opts for a moderately optimistic position that highlights the dangers of Anthropocene but also expresses the hope that humanity will rise to the challenge. *NZZ* dominates Swiss Anthropocene reporting with nine items, primarily awareness-raising articles that are motivated by new publications, scientific reports, or current exhibitions. The weekly *WOZ* follows a similar path, with a particular focus on cultural events in the year 2016. Interestingly (but hardly surprising from a Swiss perspective), the *NZZ* article 'Do We Disappear in the Anthropocene?' (2013) refers to the Swiss-born Max Frisch's novel *Man in the Holocene* that already provided the backdrop for our foundational article in the German *FAZ*. The author Uwe Justus Wenzel, humanities editor of *NZZ*, first points out how the Anthropocene is often a vehicle for human hubris and technological optimism and then raises the question as to whether man might actually disappear in 'his' age. Also notable are two excellent surveys in *NZZ*, the first by Marco Metzler, 'How Science Fiction Changes the World' (2009), followed by 'At Some Point, Nature Strikes Back' by Martin Zähringer, one of the first media articles to connect science fiction and the Anthropocene. Finally, the politically conservative weekly *Weltwoche* did not feature a single article on the Anthropocene.

The most impressive media coverage of the Anthropocene in Switzerland comes from *Le Temps*, the only nationwide Francophone newspaper, which informs its readers with an exceptionally high standard of articles (and with especially interesting images to accompany each one).[13] 'Sustainable Development Or Decay, the Environmental Dilemma' (2011) is a rare discussion of the ideas of Jacques Grinevald, eminent development theorist and disciple of the father of degrowth theory, Nicholas Georgescu-Roegen.[14] The article is in the form of a scholarly debate with Beat Bürgenmeier, former dean of the Faculty of Economics and Social Sciences at the University of Geneva, with the last words expressing Bürgenmeier's faith in technology. 2012 brings 'The Anthropocene, Or the Part of Man', a wide-ranging reflection on the geopolitics of the Anthropocene, adding some local interest with a quote from Philippe Gillet, vice-president of the Lausanne Polytechnic. 'The Paradox of a Zero-Risk society' (2013), by the British astronomer, Lord Martin Rees, warns of ecological and nuclear catastrophes that 'lie in wait'. This is another lengthy product of Project Syndicate. In 2015, we find a five-minute read, 'Collapse' (*Effondrement*), with an evocative graphic of a farmer carrying two buckets of water with the words 'Anthropocene', 'ecofeminisme', 'finitude', 'nimby', and 'collapse' etched into a cracked landscape.[15] These terms, we are told, all appear in the *Dictionnaire de la Pensée Écologique* edited by Sandrine Rui (University of Bordeaux). 2016 brings several informative articles, notably 'On the Origins of the Ecological Crisis', which reports the contribution of the historian Jean-Baptiste Fressoz at a conference on 'the political history of CO_2'. The article discusses the politics and the history in great detail and cites *The Shock of the Anthropocene* by Bonneuil and Fressoz (2017), one of the most influential popular science books on the topic. We also find in *Le Temps* 'Green Initiative: the Swiss Rendez-Vous With the Anthropocene' on the referendum;[16] and 'To Govern is Not To React, it is To Foresee'. The dailies *Tribune de Geneve* and *Le Matin* (now online during the week) both reprinted articles from *Le Temps.* Finally, the venerable daily *24 Heures* offers two items in 2016, the first on a prize-winning exhibition at the Museum of Nature in Valais, 'Objectif Terre: Vivre L'Anthropocène' in which the judges congratulated the artists 'for their courage to avoid sensationalism'; and the second, 'The Earth Enters a New Geological Epoch, the Anthropocene' on the AWG in the International Geological Congress in Cape Town. The regional *ArcInfo* prints a single Anthropocene-related article of local interest, 'One of the Discoverers of Global Warming Lives in Neuchâtel', citing an anecdote by the eminent glaciologist Claude Lorius, a frequent contributor to Anthropocene debates in France. As indicated, most of these articles explore their topics (some at length) in a serious manner, though with few indications of alarm.

With some exceptions, the increasing urgency of anthropogenic environmental problems is not a prominent topic in the French, British, German, and Swiss press. The geological debate about the timing of the Anthropocene is portrayed in great detail and with sustained interest. New arguments and statements by prominent scientists serve as criteria of newsworthiness, despite the fact that the *future* implications of the Anthropocene do not depend on the outcome of the scientific debate about its beginning in the more or less distant *past*. That stands in contrast to the smaller number of articles that address some of the more pressing and largely irreversible aspects of the Anthropocene, such as species extinction (possibly including the human species). In recent years, the number of alarming reports has increased but, taking into account the predominance of those emphasizing the need for new technological solutions, one might be inclined (to quote Max Frisch again, even if out of context) 'Man does not have that much time'(*so viel Zeit hat der Mensch nicht*).[17]

A Mediterranean (or southern Europe) Anthropocene?

A distinctive feature of coverage in the Italian media (over 100 items found) was the prominence of Anthropocene-related creative arts events, notably in *La Repubblica*, where 27 of the 46 items we found were creative arts events (discussed in Chapter 13 of this book). General coverage of the Anthropocene in *La Repubblica* was more conventional; for example, 'Here Comes the Era of the Anthropocene: This is How Men Manipulate the Earth' (2005) and Welcome To the Era of Anthropocene: The Trace of Man is Too Strong' (2016), which comments that not all forecasts are pessimistic, quoting the opinion of the British scientist Martin Rees, that the electronics revolution could lead us to conquer other planets, hoping that 'it is not the destruction of the old one that drives us to new worlds' borrowing from an article in *The Guardian*. 'Anthropocene is the Geological Era Characterized by the Presence of Man on Earth' (2016) discusses the first use of the term in 2000, pointing out that scholars are divided on its origins. Introducing a more worrying tone, 'Apocalyptic or Integrated?... The Age of Man, the So-Called Anthropocene, is Remembered For the Sixth Extinction' (2016), described as 'an apocalypse comparable to the one that put an end to the dinosaur sixty-five million years ago'. As usual, there is no indication that humans could also become extinct. *Corriere Della Sera* (the daily with the largest circulation in Italy) offers only five articles, beginning in 2010 with 'Humans Extinct Within 100 Years', describing the catastrophic prediction of the Australian biologist Frank Fenner, who implicates the demographic explosion and out-of-control consumption. In 2013, we find 'What Will Remain of Us? The Anthropocene's Technofossils', another well-worn theme, and, in 2015, 'Anthropocene: 1610, the Year in Which Man Changed the Planet', back to start dates. *La Stampa* prints two alarming articles in 2006. 'The Apocalypse is in 2050. We Will Need Another Earth' reporting: 'Worldwide Fund for Nature (WWF) launches the alarm: we are in ecological deficit... We have reached a point of no return. If we continue with "business as usual" we will be extinct by 2050', but no calls for action. By 2016, the tone has changed: 'Welcome to the Anthropocene: in the Layers of Rocks, Traces of the Era of Plastic' referencing the Cape Town AWG conference, and concluding blandly: 'If, this morning, you have thrown a plastic bottle in a bin, you have made history. You have contributed to the era of the Anthropocene'. Finally, in 2017, the tone becomes more gloomy, 'The Disasters of the Anthropocene and the Fever of the Planet', profiles the scientist and TV personality Luca Mercally, who is quoted at an event on the Anthropocene: 'Things are only getting worse and we are starting to lose hope', repeating the pessimistic message of a book he wrote in 2013, where he advised, with a little more optimism: 'Let's get ready to live in a world with less resources, less energy, less abundance ... and perhaps more happiness'. The weekly political and business magazine, *L'Espresso*, in 'Save the Planet, Let's Burn Ice' (2016) engages with geoengineering, quoting the Italian Nobel laureate in physics, Carlo Rubbia, to the effect that methane trapped in glaciers can become the future energy source (another contentious topic). Perhaps the most unconventional item we found in Italy appeared in both *Wall Street Italia* and *Corriere della Sera* (in 2012) under the eye-catching title 'Italy, We Need Bankruptcy'. Based on an interview with Serge Latouche, economist and originator of the term '*décroissance*' (degrowth),[18] this is not a message we would expect to find in either of these two media platforms. Though the degrowth movement is expanding all over the world, this key concept is almost never discussed in the media in the context of the Anthropocene.

Media coverage of the Anthropocene in Spain is notable for the naming of local journalists and/or local references. Three articles in the most important newspaper, *El Pais*, give a flavour of the coverage. Starting in 2014, the environmental economist Mari Carmen Gallastegi sets the tone with 'The Geological Stewards of the Earth: The Anthropocene: Human Domination of the Environment', asserting the responsibility of humanity for climate change (especially China), arguing that the damage may already be insurmountable and that geoengineering is tempting but dangerous. Her conclusion is that life in the Anthropocene brings with it great responsibility and challenges. In 2016, the environmental journalist, Javier Salas Quesada, strikes a slightly different note with 'Welcome to the Anthropocene. We've Changed the Natural Processes of the Earth', defining the idea, quoting scientists, discussing start dates, alluding to its social significance, and pointing out some local anthropogenic signs in Bilboa. He concludes by telling us that industrial pollution in the Ria de Bilbao since 1940 constitutes a monument to the Anthropocene. The article avoids taking any position as to how society or politicians should react. In 2017, the science journalist Nuno Dominguez writes 'The Deepest Ocean Trenches have "Extraordinary" Levels of Pollution', a detailed explanation of anthropogenic damage by pollutants in the Mariana Trench. Quotes from scientists support the conclusion that even the most remote of places on the planet is not free from damaging human impacts. In 2017, we find in *El Pais* 'Is the Gasoline Over?' (based on an article in the *Anthropocene Review*, but with no discussion of the Anthropocene). However, this is one of the very few serious discussions of renewable energy from any region, citing the director of the Global Carbon Project to the effect that 'transition towards the green electrification of the economy' is happening.[19]

Subsequent articles in *El Pais* discuss the Cape Town International Geology Congress, the work of Bruno Latour, and the Sixth Extinction, the latter leaving the fate of humans open. Coverage in *El Mundo*, the second 'newspaper of record' in Spain (mainly in 2016) begins with two well-researched articles by the environmental journalist Carlos Fresneda, the first, 'Anthropocene: the Geological Impact of Man', explains why scientists search for the golden spike, the ultimate test of the new geological age marked by human activity. Jan Zalasiewicz is pictured in his office and the focus is on the question of official ratification of the Anthropocene. The second, 'Caught in the Anthropocene', also featuring Zalasiewicz and the AWG, likewise gives little indication of the potential risks or what to do about them. Maria Perez Avila's, 'The Anthropocene Begins: the Earth Transformed by Human Beings' quotes a recent study in *Science* and tells the same story as the previous items. Similarly, several articles in *La Vanguardia* in 2016, for example, 'Our footprint Leads the Earth to a New Geological Age: the Anthropocene', by the nature journalist Joaquim Elcacho, raises no alarms. However, an article in this paper, also in 2016 and attributed to EFE (the major international Spanish news agency), concludes that the formal recognition of the Anthropocene would support those who recognize excessive human impacts on the earth and the need to modify our behaviour as a species. This is as near to appreciating the potential risks of the Anthropocene as we find in the Spanish media. *ABC*, the third 'newspaper of record' in Spain, offers only two items (in 2014 and 2016), both on the technosphere, both borrowing from the *Anthropocene Review*, both neutralizing. The free national newspaper *20 minutos* entered the field of Anthropocene reporting early, with an article in 2008 by the geographer and environmental journalist, Cesar Javier Palacios, 'A New Geological Age Begins, the Anthropocene'. His conclusion is refreshingly direct: 'Humans have had so many significant negative impacts on the earth, that even geologists are warning us not of future collapse, but of our current age of destruction'. The influential business daily *Expansión* gives us two articles that further reinforce these main trends. The first, in 2010, 'From Atapuerca to the Anthropocene' is an interesting reflection on nature and man's place in it. Atapuerca, near Burgos, is where the earliest remains of humans in western Europe were found, dating back about one million years. The author argues that, as the history of volcanoes shows: 'humans can't overcome large natural phenomenon, and might not even be the major cause of climate change – which is scientifically questionable, we shouldn't bend over backwards to reduce CO_2 emissions – cheaper and more efficient to manage our resources sustainably, but we shouldn't try to reach impossible climate objectives'. This is the clearest expression of climate change denial in our results from Spain (there are a few others in the western Europe media). The second item in *Expansión*, 'Economists and Environmentalists Come Together To Work Out a Future Plan' (2016), follows a similar direction, but on a different path. The subtitle of this piece is revealing: 'Some of the best scientists in the world gathered in Oxford to expose the theoretical guidelines for the organization of future societies. They are called ecopragmatists or ecomodernists. And they represent the future'. The article speaks in passing of 'a positive Anthropocene that makes conservation a priority'.

Most of the local and regional papers searched in Spain contained only one or two Anthropocene-related articles, for example *Tiempo* (2017), *Telva* (2017), *La Razon* (2017), and *Diario De Cadiz* (2016). These covered the usual topics in the usual ways. *Diário de Notícias* extends a 'Welcome to Anthropocene' (2010, reprinted 2017). Rather more ambitiously, *El Periodico* de *Catalunya*, with 'We Live in the Anthropocene and We Don't Know It Yet. Or Maybe Capitalocene?' (2016) and cites the British anthropologist, Tim Ingold. *La Voz De Galicia* is an exception, with about ten items (between 2008 and 2017), mostly long analytical reflections with many references to Spain. None of these articles paid much attention to the risks of the Anthropocene and what we should be doing about them. Most of them offered basic facts with reassurance narratives.

In Portugal, the magazine *Grande Reportagem* began to cover the Anthropocene in 2010, with an article reporting the view of Zalasiewicz and Mark Williams, that the Anthropocene started with the industrial revolution, and then, in 2015, presented a different view: '1610, the Year That Man Changed the Planet', one of many reports from all over the world that the start of the Anthropocene coincided with the Columbian Exchange, a theory of particular interest in the Iberian peninsula. In 2017, the topic was 'Rethink the Anthropocene' defined as a term used by some scientists to describe the most recent period in the history of Planet Earth.[20] 'Earth in the Age of the Anthropocene' (2014) in the daily *Publico* raises the very important but rarely discussed issue of educating the young about the Anthropocene (see Chapter 14 of this book). Perhaps the most eye-catching item in the coverage of the media in Portugal appeared in *Visao* in 2013 under the enigmatic title 'It's Bliss, Asshole! The International Day of Happiness is Coming. What a Fright!' arguing that, to be happy in the Anthropocene, means ensuring that your human influence on the functioning of the planet is the least harmful possible. The point most strongly made is that, though we might never know our distant neighbours on the planet, we must strive to make sure all our 'backyards' are as clean as possible. Almost half the results from Portugal (25 in total) reference Anthropocene-related creative arts; for example, *Grande Reportagem* in 2017 reports 'Forum of the Future' in Porto, in which the curator Guilherme Blanc (cultural advisor to the Mayor) is quoted: 'Let us question the age of the Anthropocene, from the various perspectives, and also propose a critical reflection on our vision of nature'; in the municipality of Mesão Frio, Theatre of the Cold presents its creation 'Eco' in the park, coinciding with 'Natural Culture in the Anthropocene'; the exhibition 'Eco-Visionários: Art and Architecture after the Anthropocene' in MATT, one of Lisbon's major museums (with the support of the British Council); and finally a six-screen Anthropocene installation, based on the planetary implications of climate change, by radical British artist John Akomfrah, originally displayed at the Venice Biennale of 2015: 'Purple' is choreographed as six meditations on what the Anthropocene philosopher Jane Bennett calls 'The Adventures of Vibrant Matter'.

In Greece, no single newspaper or magazine delivered more than a few results, but a significant number delivered at least one. However, 36 magazines were searched without a single result. Overall, the balance of stories was in line with other European results. One of the most popular stories had a local connection, namely research, reported in a US-based scientific journal, suggesting that 4% of the Earth's minerals are anthropogenic and that some have origins in the ancient mines of Lavrio in Greece. This is reported in *Eleftheros Typos* in 2016 and 2017, in *Eleftheria* and *Avgi* in 2017, in *To Vima*, and *Thema* in 2017 (as well as in other sources in many countries). The Greek results appear to be somewhat dependent on foreign sources. There are many examples. *Popaganda* in 2016 reviews an article from the journal *Science* in which Colin Waters, together with the AWG, explain that we have left the Holocene and entered the Anthropocene, a geological epoch characterized by intensity of human activity; *Kathimerini* 'World Population of Mammals, Birds, Amphibians, and Reptiles Have Declined by 60% Since 1970', quotes Marco Lambertini of WWF, who mentions the 'new era of Earth history, the Anthropocene', which the article defines as 'a term that scientists have introduced to emphasize humanity's sovereignty over the planet'. 'A New Geological Epoch is Being Investigated' (2016) summarizes a paper published in *Science Review*, outlining the procedure according to which geologists are attempting to achieve formal, geological recognition of the Anthropocene. In 2014, an opinion piece by Erle Ellis is reprinted in *Kathimerini* from *The New York Times*. Finally, in *Avgi*, we find 'Are We All Responsible?' an article that originally appeared in *Le Monde* (2015), reviewing various arguments, accepting and denying anthropogenic climate change, and concluding that current climatic changes are indeed anthropogenic. The magazine *Antinews* provides three informative items. The first, 'The Dawn of a New (Geological) Epoch has arrived' (2010), summarizes the geological argument for a new epoch, the Anthropocene, as a result of the recognition by scientists (such as Crutzen) of the magnitude of human interference in Earth System operations. It describes the different nominated start dates, from the advent of agriculture 10,000 years ago to fallout from nuclear weapons tests in the 1950s; the second, 'What the Future Holds' (2013), summarizes a talk by the much-quoted economist Jeffrey Sachs, who concludes: 'Admit that humans are now a geological force, that technology can help by increasing efficiency (for example, agricultural – to manage with growing populations in a manner that is economically and ecologically viable), and thirdly, to make a pact, something analogous to a new social contract, to take climate change seriously'. The third article in *Antinews*, 'The Anthropocene: Planet Earth Has Entered a New Epoch, Say Experts' (2016), is distinctive, in that it is the only Greek source to cite the Australian ethicist Clive Hamilton, reporting his opinion that geologists should not simply be discussing the Anthropocene, but should be trying to prevent it instead. *Avgi* also provides a useful service in 'How Much Time Do We Really Have Left?' (2017), a review of books concerning climate change and the Anthropocene, citing volumes by Naomi Klein, Vandana Shiva, Jean Jouzel, Susan George, and Timothy Morton. The results from Greece contain only one Anthropocene-related creative arts event, a listing in *Avgi* (2017) for an exhibition at the Museum of Contemporary Art in Crete entitled 'Anthropocoenos, 20 Degrees Celsius'.

The two Mediterranean islands in our sample both produced minimal Anthropocene results. In Malta, the most important papers, *The Times/Sunday Times of Malta* and *The Malta Independent* both had many thousands of stories on climate change, but very little on the Anthropocene, and what there was usually came from syndicated sources, such as *The Times of Malta*, 'Extreme Makeover: Are Humans Reshaping Earth? Anthropocene' by Marlowe Hood from AFP (2011), and 'Human Imprint Thrusts Earth into a New Geological Epoch' by Alister Doyle from Reuters (2016). One of the few in-house authored articles, by Anne Zammit in *The Sunday Times of Malta* introduced a note of sarcasm with 'Let's Send the Climate Change Denialists into Orbit So That They Form a Protective Shield to Keep the Earth Cool Enough for the Rest of Us to Live Down Here' (2017). The two items from Cyprus are both from *The Cyprus Mail online*, 'Welcome to the Anthropocene' by the much-syndicated Gwynne Dyer, and 'Wildlife Extinction' (based on the WWF report), both in 2016. Summing up, the evidence suggests that there is no specifically Mediterranean Anthropocene, although perhaps there are signs of an Iberian Anthropocene, whereas the creative arts Anthropo-scene appears to be flourishing.

A Nordic Anthropocene?

Between them, Denmark, Finland, Iceland, Norway, and Sweden produced around 200 results, an impressive total per capita. Denmark and Norway produced dozens of items each. The Danish daily *Information* contributed the most for a single source in Denmark, starting with 'Welcome to the Anthropocene' in 2011, a general piece in which Zalasiewicz, Crutzen, and Steffen are all cited. *Information* proved to be a particularly problematic source when first searched in 2017–18; however, when revisited in 2019, it became easier to negotiate, though many of the results from Anthropocene-tagged searches contained no mention of the term in the text – in some cases, it appeared in accessible comments. For example, 'It is Too Early to Exclude Man-Made Climate Change' (2017) led to an interesting link on 'Climate Change and Anthropocene Extinction 24' (at *bits of science.org*), apparently an exposure of climate denial sources. The necessity for this is underlined by a comment from Rolf Schuttenhelm (of *'bits of science.org'*) on the article 'Self-Proclaimed Oracles Make Us All Dumber' (2016), a riposte to the oracles of climate change denial. Also in 2016, *Information* published a long article by two Danish humanities scholars, Anna Cornelia Ploug and Søren Mau, 'The Anthropocene Ideology', in which they argue forcefully in favour of replacing what they call 'the Anthropocene hype' with the Capitalocene.[21] This view appears only once more in our West Europe results. Other items in 2016 reproduce reassurance narratives, for example, in an article on the important issue of the non-degradability of plastics, citing Jan Zalasiewicz from the journal *Anthropocene*, but raising no alarm. Coverage includes two creative arts events (both in 2016). 'Arctic Beauty and Doomsday Prophets – How to Make Art About the Climate?' reports the involvement of a Danish artist in an exhibition called NEOARCTIC, partially inspired by the Anthropocene Project at HKW in Berlin. The second poses a searching question in its title: 'How to Make Performing Arts About the Climate Crisis and the Anthropocene Age Without Falling into Trivial Images of Polar Bears?' reporting on how Kirsten Dehlholm tries to give one of the great and difficult topics of the time an 'artistic form that can appeal to emotions and counteract powerlessness'.

The Danish daily *Berlingske Tidende* had a single item, 'A New Geological Epoch With the Name Anthropocene [antropocæn], the Time of Humans' (2016) This article sends an alarming message in a strangely un-alarming fashion: 'When one day we will not be here anymore, our walk on this Earth will have been reduced to a line in the sand to an inconsiderable geological layer on the third planet from the sun. The only evidence of the shortest of all geological periods: the Anthropocene'. The popular daily *Kristeligt Dagblad* published two substantial articles in 2015, 'Human Beings Have Become a Devastating Force in Line with Volcanoes and Meteor Impacts'. Referencing an article (on Archaeology and the Anthropocene) in the journal *Anthropocene*, this gives a good summary of the issues around naming the new era, with a useful sidebar, but any potential risks are muted. The second article, 'We Have Taken Over the Role of God as a Creator', reports that, in Germany, the Ludwig-Maximilian University in Munich held an international conference entitled 'Religion in the Anthropocene', and, in Denmark, the University of Aarhus has established a research programme, AURA, on the Anthropocene.[22] The article concludes with a quote from Sigurd Bergmann, a theologian connected with the Rachel Carson Centre in Munich: 'basically, the Anthropocene also raises the question of what charity is like'. This was followed by two articles in 2017, 'Can We Stop Our Ruination of the Planet?' and 'Nature and the Antropocæne Man', announcing a major project by the Aros Museum of Fine Arts comprising a four-kilometre-long art zone along the Gulf of Aarhus to explore human relationships to nature.

In Sweden, the mass circulation daily *Aftonbladet* prints 'Geologists Say That the Earth Passed into the Age of Man, Anthropocene' (2011) and 'Did the Human Epoch Begin in 1610 or 1964?' (2016) –two of many neutralizing reports of debates around start dates, updating a previous article in 2015. More controversially, in 2017, a debate between two Swedish scholars (Sörlin and Hornborg) was reported in *Dagens Nyheter* (*DN*) and *Aftonbladet* (discussed in Chapter 12 of this book). *DN* stands out for its coverage of the potential risks of the Anthropocene in several other articles, notably, in 2015, 'Climate Change is a Terrorism That Seems To Be Slow', one of the few media references to Rob Nixon's idea of ecological 'slow violence', and, in 2017, 'Deep Opposition. A Global Minority Destroys the Planet', another reference to the damaging effects of capitalism. In 2015, *DN* sponsored a six-part series in collaboration with Sweden's leading technology university (KTH) to highlight research in environmental humanities ahead of the Paris climate summit. The messages projected by the articles in the series on risks are mixed. *Göteborgs-Posten* and *Dagens Industry* both published one item in 2015, on start dates.

The results from Norway were particularly problematic for the usual reasons, with many more tagged positives than accessible articles, and possible duplicates. Nevertheless, it is true to say that the media in Norway frequently printed stories on the Anthropocene (*antropocen* in local dialects, as well as English spelling). Seven out of twelve sources searched produced items on the Anthropocene.

Aftenposten, the largest newspaper in the country, began its coverage in 2010 with: 'All These Strangers' in which a Norwegian climate scientist argues that the alien species responsible for the greatest changes to the planet's ecosystems is us, 'changes so dramatic that geologists have given our current age its own name – the Anthropocene'. The article concludes that all is not lost. Also, in 2010, we find: 'A Common Disbelief', similar to the previous, but with the addition of climate change denial. This was followed in 2016, by 'Mankind Establishes a New Geological Epoch—the Anthropocene', a basic presentation of the facts. *Dagens Næringsliv* **(*DN*),** Norway's main business newspaper, offered only a few lines of text for several items in front of its paywall from 2015 to 2017. These included a jazz concert in Trondheim; 'We are in many ways as destructive as the asteroid was for the dinosaurs'; a photographer working on an Anthropocene theme; a band possibly named 'Anthropocene'; and 'The End is Near', reviewing the book by Gaia Vince referred to earlier in this chapter. The only completely accessible item is 'New Study: The Age of Man Started in 1945', reporting influential research published in *Anthropocene Review*, which concludes that everything about the Anthropocene seems controversial. *Dagsavisen* (formerly the paper of the Norwegian Labour Party) offers a single result, 'The Human Epoch' (2013), summarizing the complexity of the Anthropocene with the help of a Norwegian philosopher. The results from the left- wing daily *Klassekampen* (Class Struggle) are particularly difficult to evaluate, due to a paywall that changed during the search period. Many items appear to refer to cultural events, especially book reviews. The only fully accessible article, 'In the Shadow of the Hologram' (2017) refers to 'Anthropocene Zombies'; there is no shortage of 'zombies' online in the Anthropocene. *Morgenbladet* (Norway's equivalent to *The New Statesman* in the UK) has the same search problems as *Klassekampen*, many results but few fully verifiable. 'With a Silver Spoon in Your Mouth? Man is Not Alone in the Urge To Control and Adjust the Environment' (2010), by the evolutionary scientist Markus Lindholme, is tagged 'Anthropocene' but does not appear to mention the word. 2016 brings 'The World is an Anthropocene', an enthusiastic review of the Anthropocene exhibition in Rio's Museum of Tomorrow, by the Norwegian-American photographer John Erik Riley, and 'On the Way To a New Epoch', a report on the AWG. The rural daily, *Nationen*, prints two unusual items. 'We Continue to Produce Food, Folks!' (2015) is a guest-blog by a farmer, Lise Thorsø Mohr (from a famous Norwegian farm), who is generally happy but worries about 'the erosion of soil and the troubles of agriculture in the Anthropocene', and 'God and Market' (2017) by a student who connects religious agnosticism, the market, and the Anthropocene. Finally, the tabloid *Verdens Gang* (reportedly, the most-visited online news site in Norway) has a short Anthropocene-related video in 2016 'Five Signs that We Have Created a New Age'(2016) by Norwegian geologist, Henrik Svensen, who presents the Anthropocene as a geological, economic, and social phenomenon.

Reporting of the Anthropocene (antroposeeni) in Finland is, likewise, sparse and generally neutralizing.[23] The mass circulation newspaper *Helsingin Sanomat* picks up the Anthropocene story in 2004 with a bland account focusing on Crutzen, but it is a decade later before other Finnish media take notice, mostly addressing start dates. For example, the national broadcaster *YLE* reports, 'The Carpet of Hundreds of Millions of Abandoned Stone Tools Tells About Lost Vibrant Life in Sahara' (2015), explaining that the start of the Anthropocene might have been during the industrial revolution, or perhaps 10,000 years ago when humans began to clear forests for agriculture. The year 2016 brings 'New Epoch Named After Humans May Have Started – Greatest Transformation since the Last Ice Age', reporting that scientists believe that the new Anthropocene epoch started with nuclear testing, citing an influential article in the journal *Science* on the difference between the Holocene and the Anthropocene. On a somewhat critical note, *YLE* reports 'Professor Katherine Gibson: There are No Jobs for Everyone, a Rethinking of Economics is Needed' (2016). She is quoted: 'the root cause of the many problems of our contemporary age is the economic system ... At the same time, we have the knowledge and ability to change direction'.[24] *Uusi Suomi*, a publication with an uncertain history, introduces a note of alarm with 'Geoscientists Voted: "New Age of Humans", the Anthropocene, Has Started on Earth' (2016), reporting on the AWG in Cape Town, and concluding, 'The concept of the Anthropocene can spread the awareness of the critical situation on the planet'. The weekly family magazine, *Suomen Kuvalehti,* 'Humans Had Significant Impact on Environment Thousands of Years Ago' (2013), mentions Crutzen and Stoermer. *Kansan uutiset,* the newspaper of the Left Alliance party, reports the new epoch in 2016, 'It is Anthropocene Now' and, rather more critically, 'Wild Nature is No More' (2017), which quotes an historian who criticizes the 'official Anthropocene narrative' and questions the role of 'we or humankind' as a political subject in the Anthropocene, a topic energetically problematized in the environmental humanities.

The media in Iceland appears to start its modest Anthropocene coverage with an article in 2015 in the daily *Morgunblaðið* 'The Age of Exploration, the Beginning of the Anthropocene' (mannöld in Icelandic).This is based on an item from the BBC, arguing for 1610 as the starting point for the Anthropocene. *DV*, a bi-weekly newspaper with daily online news, invites us: 'To Learn to Think Like Sea-Weed', in an article about a transnational workshop 'Future Fictions' as 'One Way to Tackle the Anthropocene'. The monthly *Stundin* (also with daily online news) reports, sardonically 'on radio-active desert of the late anthropocene era', in which 'anthropocene' is written in English, a not uncommon practice in global media. The now defunct *Fréttatíminn,* prints 'Wild cows Awake From the Dead' (2016), a long well-researched article on extinction and cattle breeding, though the Anthropocene is mentioned only in passing. *Kjarninn* (daily online news site) declares 'Man Has Become a Geological Actor' (2016), one of many purely descriptive articles on the Holocene to Anthropocene transition. Finally, *RÚV* (Icelandic National Broadcasting Service) reports, in 2017, on the radio programme 'The Morning Shift', an interview with the distinguished Icelandic anthropologist Gísli Pálsson, an early exponent of the need to bring social sciences and environmental humanities into debates around the Anthropocene (see Pálsson et al. 2013). From all this evidence we can conclude that while Anthropocene coverage in the Nordic countries is wide-ranging across a variety of media, no specifically 'Nordic' concept of the Anthropocene emerges, beyond the usual Eurocentric ideas of the state of the planet.

The rest of Europe

Results from the Irish Republic indicate solid media interest in the Anthropocene. The highest-circulation newspaper, *The Irish Independent*, sets the tone with 'Scientists Hope to Reverse Damage to Planet' (2014), reporting a lecture at the Royal Irish Academy by the environmental historian, John McNeill, who portrays the Anthropocene as the way scientists deal with problems caused by humanity, including climate change. He is quoted: 'within 50 years, they could be geoengineering'. 2015 brings another eminent scholar into the debate, 'Ahead of His Time: James Lovelock on Radical Hope', a very long and positive discussion of his new book *Rough Ride to the Future* (the book is not as pessimistic as the title). 'Is Paris Treaty All That Stands Between Us and Mass Extinction?' (2015) documents vanishing frogs, bats, rhinoceroses, coral, etc. 'in our Anthropocene age, named for the impact of humans on the planet'; and 'World Facing Biggest Mass Extinction Since Dinosaurs – With Two-Thirds of Animals Wiped Out in 50 Years' (2016) is informative but, as usual, the extinction of humanity is not broached. The other major newspaper in Ireland, *The Irish Times* was the most prolific source of Anthropocene-related items, around 30. 'The Modern Darwin' (2014) by the environmental writer Paddy Woodworth, profiles E.O. Wilson. In the course of this long and original article, Wilson's criticism of the ecologist Erle Ellis is highlighted. Woodworth quotes the now infamous statement made by Ellis: 'Nature is gone … You are living on a used planet. If that bothers you, get over it. We now live in the Anthropocene, a geological epoch in which Earth's atmosphere, lithosphere, and biosphere are shaped primarily by human forces'.[25] Woodworth suggests that Ellis is not alone in holding this view, but gives Wilson the last words: 'Ultimately, the "Anthropocene ideology" will lead to artificial ecosystems, designed for our pleasure – you know, like a good swimming pool. That's essentially what's on their mind: we can do almost anything, and we don't need old-fashioned nature or the pain and expense of trying to retain and restore it'. It would have been helpful if this stimulating article had acknowledged more clearly that Wilson and Ellis represent only two of many interpretations of the Anthropocene.

The *Irish Times* tells us that more celebrities came to Dublin in 2016 to discuss these issues. 'Naomi Klein Argues Climate Change is a Battle Between Capitalism and the Planet', reports two lectures in Dublin, by Klein and the Marxist sociologist John Bellamy Foster. Foster's subject was 'the Anthropocene and the crisis of civilization: climate change and capitalism'. Seasonal fun in 2017, with the paper's Christmas quiz lightening the mood. Question 32 is 'What scientific name has been coined for the new, human-dominated era of the planet?' This, at least, tells us that the *Irish Times* accepts its educational mission. The Cork-based *Irish Examiner* asks in 2015: 'Anthropocene: Are Humans Destroying Planet

Earth?, beginning with general warnings about the risks, but concluding, neutrally, on start dates. A surprise awaits us from *RTE* (Irish state TV) with 'The Future of Retail: 15 Destinations Coming to a Shop Near You' (2017) predicting 'the recent trend of online retailers opening physical stores means the in-store experience looks set to progress and evolve for the digital age. As we enter the new geological epoch of the Anthropocene the age of consumerism still looks set to continue to dominate'. The link takes us to 'The Anthropocene Epoch: Scientists Declare Dawn of Human-Influenced Age' (2016), a thoughtful and thorough survey by Damian Carrington, environmental editor of *The Guardian*, but with little indication that an existential crisis may be looming.

In the Netherlands, the national broadcaster *VPRO* reports the Anthropocene from time to time, most notably in 2017 with 'The Earth Can Best Exist Without People, But Not the Other Way Around'. The influential left-leaning magazine *Vrij Nederland* presented: 'Admit it: Once Again, When You Hear Strong Climate Resolutions, the Eyelids Become Heavy'(nd); this implies 'eco-fatigue', a conclusion not borne out by the sparse Dutch coverage of the Anthropocene, though climate change is mentioned frequently in the Dutch (and global) media. The centrist *De Volkskrant* melodramatically reports: 'Climate change is not Apocalyptic in Nature' (2015), a positive endorsement of eco-modernism, quoting Ted Nordhaus of the Breakthrough Institute, who claims that the environmental movement is 'elitist and far too negative'.

Anthropocene media coverage in Luxembourg consists mainly of six items in the national weekly *Le Jeudi* (now apparently defunct). In 2017, it printed a despairing article 'God, Consultants, and Functionaries ...All Relegated to Oblivion', explaining that there is no place for religion in the age of the Anthropocene. The multi-lingual green weekly *Woxx* and the tabloid *Le Quotidien* each provide one uninformative cultural reference to the Anthropocene.

Conclusion

The Anthropocene, as reported in the media of western Europe, is a very mixed bag, with many differing opinions displayed. Media coverage of the Anthropocene is overwhelmingly neutralizing and/or reassuring, all the way from simple announcements that some geologists felt the need to argue that the Earth System had left the benign conditions of the Holocene and entered the uncertain Anthropocene (maybe caused by human profligacy, maybe not) to an acknowledgment that the new epoch did present considerable risks, though humans were likely to be able to cope with these risks. Our interpretation of these results from our media searches is best characterized as 'Business as usual', despite some apparently radical changes (best exemplified by the gradual exit from coal in Europe and elsewhere) and, in general, the optimistic belief that renewable sources of energy, as well as technological innovation, will allow us to keep living and consuming more or less as we do now. This conclusion rests on reading between the lines and finding that the old adage 'it is easier to imagine the end of the world than the end of capitalism' often subtly undermines the unwelcome hints that there really is an existential problem for humanity. This is apparent from more pessimistic to more optimistic transitions (or vice versa*)* between the top and the bottom of many Anthropocene stories. The analysis in Chapter 14 of this book suggests that this is quite common in media coverage in all the regions, though it seems particularly marked in western Europe. No one can rule out the possibility that adaptation and technological innovation will allow us to deal with the consequences of the Anthropocene. But it requires a considerable leap of faith to imagine that the very institutions that caused the problems will also provide the key to their solution.

Notes

1 Mario Brincat, Alex Damianos, Medine Duvarci, Louisa di Felice, Can Mert Kökerer, Magnus Örn Sigurðsson, and Anya Verkamp also contributed to this chapter. We gratefully acknowledge research support on the German language material by Vanessa Bittner, Stefan Galley, Viviane Nolte, Patrik Rist, and Patrick Zähringer from the University of Konstanz.

2 As suggested above, under current circumstances, genuine recycling, renewables, and conservation, while all welcome, will have only marginal impacts on global warming and existing degradation of ecosystems in the short to medium term. See Jacobson (2017, 2020) for a more optimistic 'Wind–Water–Solar' scenario, and the previous critique by Trainer (2013).

3 See Hart and Leiserowitz (2009) on the impact of this film.

4 *Libération* reported discussions on ecocide in the National Assembly in 2019, for example, 'France must recognise ecocide in the penal code', but with no mention of the Anthropocene. See also Higgins et al. (2013) and Crook and Short (2014) which does relate it to the Anthropocene.

5 The reference is to a 'nonsense' poem by Edward Lear. The idea that most people have heard of the Anthropocene is a delusion that still afflicts some scientific and cultural elites.

6 As the poster makes clear, the meeting addresses a wide variety of issues. The blank panel to the left of Figure 9.2 is due to copyright issues regarding three pictures (a cityscape, a stone carving, and an idyllic landscape).

7 Re-reading this in 2020, when much of the world is self-isolating, we might ponder the irony.

8 This is one of the very few references to climate change 'sceptics' in the West European results.

9 Lynas chronicles his shift in *The New Statesman* in an article in February 2004. Castree (2015a) and Foster (2016) provide shrewd analyses of how the Anthropocene challenges leftist politics.

10 For an interesting study of media–climate science relations in Germany, see Ivanova et al. (2013).

11 These exhibitions are analyzed in more detail in Chapter 13 of this book.

12 For the cover of *Der Spiegel* No. 47 (1981) see https://www.spiegel.de/fotostrecke/die-besten-spiegel-cover-von-1977-bis-1986-fotostrecke-144417-5.html.

13 German-language media make up roughly two-thirds of the total western European market, of which about one-quarter is in French. However, most of our results come from the Francophone media.

14 On degrowth, see D'Alisa et al. (2014), Burton and Somerville (2019), Sol (2019) and Figure 14.2.

15 As for many other excellent Anthropocene images, we were unable to obtain permission to reproduce it in the book.

16 Decisively rejected by the electorate, the Swiss environmental movement considered the referendum as a victory.
17 See Frisch (1979: 48). An English translation by Geoffrey Skelton, *Man in the Holocene*, was published in *The New Yorker* (May 19 1980).
18 A video of Latouche in action at the Sorbonne in 2018, with Jean-Baptiste Fressoz, 'Survivre dans l'Anthropocène', is available at https://www.youtube.com/watch?v=IYCPP8J5dMM
19 In April 2020 (in the midst of the Coronavirus pandemic), the renowned radical film maker, Michael Moore, released a documentary 'Planet of the Humans', outing some prominent environmentalists, and claiming that more-or-less all renewables were, in fact, based on fossil fuels. This attracted much media (social and otherwise) attention. For one of many point-by-point assessments of the evidence, revealing many serious errors in the film, see https://portside.org/2020-04-26/skepticism-healthy-planet-humans-toxic-critical-review
20 This magazine has closed and been reincarnated on several occasions.
21 See also Mau (2017) in Danish, but it is well worth the effort to obtain a translation. It contains a graph on the use of the term 'Anthropocene' in Danish.
22 AURA (Aarhus University Research on the Anthropocene), was a six-year transdisciplinary and transnational project; see https://anthropocene.au.dk/
23 However, see Lyytimäki and Tapio (2009) on climate change reporting.
24 This is the Gibson (Katherine Gibson) of the celebrated feminist authorial duo of economic geographers Gibson-Graham (see Gibson-Graham 2011). The media article focuses on her step-by-step 'seeds of a good Anthropocene' strategy.
25 See Ellis 'Stop Trying to Save the Planet' in the US magazine *Wired* (2009) at https://www.wired.com/2009/05/ftf-ellis-1/ See also Ellis (2018: 105–7) on the 'Pristine Myth'.

References

Bonneuil, C. & J.B. Fressoz (2017) *Shock of the Anthropocene*. London: Verso.

Burton, M. & P. Somerville (2019) Degrowth: A defence. *New Left Review* 115: 95–104.

Castree, N. (2015a) Unfree radicals: Geoscientists, the Anthropocene, and left politics. *Antipode* 49(S1): 52–74.

Crook, M. & D. Short (2014) Marx, Lemkin and the genocide–ecocide nexus. *The International Journal of Human Rights* 18(3): 298–319.

D'Alisa, G. et al. eds. (2014) *Degrowth: A Vocabulary for a New Era*. London: Routledge.

Editors (2017) Anthropocène. *Les Annales* 72(2): 263–265.

Ellis, E. (2018) *The Anthropocene: A Very Short Introduction*. Oxford: Oxford University Press.

Foster, J.B. (2016) Marxism in the Anthropocene: Dialectical rifts on the left. *International Critical Thought* 6(3): 393–421.

Frisch, M. (1979) *Der Mensch Erscheint im Holozän – Eine Erzählung*. Frankfurt: Suhrkamp.

Gibson-Graham, J.K. (2011) A feminist project of belonging for the Anthropocene. *Gender, Place and Culture* 18: 1–21.

Hamilton, C. et al. eds. (2015) *The Anthropocene and Global Environmental Crisis: Rethinking Modernity in a New Epoch*. London: Routledge.

Hart, P.S. & A.A. Leiserowitz (2009) Finding the teachable moment: Analysis of information-seeking behavior on global warming related websites during the release of 'The Day After Tomorrow'. *Environmental Communication* 3(3): 355–366.

Higgins, P.D. et al. (2013) Protecting the planet: A proposal for a law of ecocide. *Crime, Law and Social Change* 59: 251–266.

Ivanova, A. et al. (2013) Is there a medialization of climate science? Results from a survey of German climate scientists. *Science Communication* 35(5): 626–653.

Jacobson, M.J. (2020) *100 Percent Clean, Renewable Energy and Storage for Everything*. Cambridge: Cambridge University Press.

Jacobson, M.Z. et al. (2017) 100 percent clean and renewable wind, water, and sunlight (WWS) all-sector energy roadmaps for 139 countries of the world. *Joule* 1: 108–121.

Lewis, S.L. & M.A. Maslin (2018) *The Human Planet: How We Created the Anthropocene*. London:

Lyytimäki, J. & P. Tapio (2009) Climate change as reported in the press of Finland: From screaming headlines to penetrating background noise. *International Journal of Environmental Studies* 66(6): 723–735.

Mau, S. (2017) The stratigraphic signatures of capital. *Eftertryk (Reproduction)*. www.eftertrykket.dk/2017/05/01/kapitalens-stratigrafiske-signaturer

Pálsson, G. et al. (2013) Reconceptualizing the 'Anthropos' in the Anthropocene: Integrating the social sciences and humanities in global environmental change research. *Environmental Science & Policy* 28: 3–13.

Sol, J. (2019) Economics in the Anthropocene: Species extinction or steady state economics. *Ecological Economics* 165. https://doi.org/10.1016/j.ecolecon.2019.106392

Trainer, T. (2013) 100 percent renewable supply? Comments on the reply by Jacobson and Delucchi to the critique by Trainer. *Energy Policy* 57: 634–640.

10 The Anthropocene in Middle Eastern media

Invisible oil?

Baran Alp Uncu and Ramzi Darouiche[1]

As a region, the Middle East is one of the major contributors to the human activities responsible for the advance of the Anthropocene. Containing almost half of the crude oil reserves on the planet, the region remains the largest producer of global energy, adversely impacting the Earth system, as well as probably being 'the world's most unequal region'.[2] It is also a region that faces severe risks posed by the Anthropocene, especially as a result of climate change.

Compared with the other regions investigated in this book, the media in the Middle East display one of the lowest rates of Anthropocene coverage. We searched 230 national and local media and pan-Arabic news sources from 14 Middle Eastern countries, of which around 150 had no coverage of the Anthropocene at all. In the remaining 80 sources, the total number of articles that refer to the Anthropocene is relatively low, around 200. Except for *Gulf News*, based in Dubai (22 stories), no Middle East source referenced the Anthropocene more than 20 times over the search period of almost two decades. These findings show that most media in the Middle East have been paying very little attention to the Anthropocene. Furthermore, even if some readers in the region come across the Anthropocene concept in news stories, the information they will gather from Middle Eastern media would often be limited, leaving little room for readers to develop a comprehensive understanding.

In general, Middle East media frame the Anthropocene as a scientific issue without highlighting its social, economic, and political aspects. Presented as a new geological epoch (which is yet to be recognized officially), only around one-third of our results highlight problems and risks attributed to the Anthropocene. Furthermore, an overwhelming majority of these 'alarming' accounts claim optimistically that detrimental impacts of the Anthropocene can be mitigated through scientific advances and corrective measures within existing political and economic systems. Information on the Anthropocene relies heavily on sources outside the Middle East, particularly from Europe, North America, and Australia. With few exceptions, the Anthropocene is not localized with respect to its impacts on the region or to contributions made by actors in the Middle East. More than half of the news articles printed about the Anthropocene are borrowed from western media outlets. Among these, *The Guardian*, *The New York Times*, *The Independent*, Reuters, *Deutsche Welle*, Associated Press, and Agence France Presse (AFP) stand out as the sources that are mostly used by the Middle East media. Pan-Arabic media, such as *Al Jazeera*, *Lebanon Daily Star*, and *Gulf News* rely almost exclusively on these western sources.

An overview of media in the Middle East

It is important to situate the media in the Middle East and North Africa (MENA) in the context of transformations in the political landscape, following the so-called 'Arab Spring', especially in the case of Tunisia. The evolution of MENA media follows a somewhat similar pattern to that in the global south. We can identify three phases of Arabic media, namely the colonial, post-colonial, and 1990s phases (Lahlali 2011). Following the post-colonial and nationalist eras, total state control over media emerged at the latest during the 1970s and 1980s, when direct and indirect censorship of radio, extensively used for the Nasserist agenda, passed over into government control. At the same time, Saudi Arabia, experiencing significant oil wealth and seeking to extend its hegemonic role over the region's media, undertook great efforts to bring broadcasting in the region under its control (Zayani 2012). Newly emerging media technologies in the second, post-colonial phase resulted in a spike of news outlets with great penetration of audiences in the MENA region and beyond. Following the CNN monopoly over Gulf War broadcasting, the availability of direct broadcasting channels allowed outlets in the MENA region to target audiences directly. However, despite increased opportunities for audiences in the region to access information and to articulate discontent, the alleged trend towards liberalization did not go hand in hand with democratization of the media environment. By the mid-1990s, the Qatar-funded *Al Jazeera* filled this gap and shifted regional politics onto a transnational level. However, alternatives to *Al Jazeera* emerged, such as the LBC-Al Hayat partnership, Abu Dhabi TV, Dubai's and later Saudi Arabia's *Al Arabiya*, as well as Hezbollah's *Al Manar* (see Lynch 2006). Whereas radio and then television became the main agents of national integration in the post-independence era across the MENA countries, the Internet has become increasingly important as a source of news as Soffer (2015) points out in the case of Israel (see also Kraidy 2011, Mellor et al. 2011, Gunter and Dickinson, 2013, Morgan 2019).

Media coverage of the Anthropocene

There is a lively but not extensive body of research on media and environmental issues in the Middle East (see Reinisch 2010, Uzelgun and Castro 2014, Freeman 2016, Schapiro 2017), none of which explores the Anthropocene.[3] When the Middle East media do engage with the Anthropocene, almost all the stories concentrate on other regions for describing and exemplifying detrimental impacts. In other words, the local impacts of the Anthropocene rarely find their way to the Middle East media. For example, in 2016, several media cover the Living Planet Report of the World Wide Fund for Nature (WWF). In *Gulf News*, we find 'Wildlife Population Falls 60% Since 1970', in the popular Turkish dailies *Hurriyet* and *Yeni Şafak*,

'Two-Thirds of the Living Species Population Could Become Extinct by 2020', and *Sözcü*, 'WWF Says Wake Up!' The original WWF report refers to breaches of planetary boundaries as the main cause for the escalated levels of biodiversity loss, and the conditions of different regions of the world are evaluated. According to the WWF report, all of Turkey, Syria, and Yemen, in addition to some parts of Saudi Arabia, the Gulf countries, and Iraq are identified as places where soil is significantly degraded. However, none of the MENA media coverage of the WWF report talks about these findings. The articles cited, borrowing from Western sources, present the general results. Few articles in MENA media explain the links between ecological conditions in the region and the Anthropocene.

However, we do find media reporting about local aspects of the Anthropocene in cultural items that introduce exhibitions and artwork, the subject matter of which is related to the Anthropocene. For instance, *Gulf Daily News* has 'Fragile Ecology Amid Changing UAE [United Arab Emirates] Landscapes' (2016), an interview with the British photographer Richard Allenby-Prat who documents the decay of the 'surprisingly rich' but also 'fragile' ecology of the UAE by drawing attention to the negative impacts of high consumption levels. Similarly, the Iranian English-language daily *The Financial Tribune*, with 'Contemporary Art Annual at Niavaran' reports on an exhibition at the Persbook Art Centre in Tehran, in which exploitation of fossil fuel resources in the Middle East is mentioned. Though art reviews give readers ideas about some of the local impacts of the Anthropocene, to the extent that they are expressed through artworks, they lack information regarding the future trajectory of the Anthropocene, what to do about it, or the possibility of controlling it. Two articles in the Dubai-based *The National* present contrasting attitudes to ecological distress in the region. The first, 'Our Mental Health is Linked To Our Environment' (2016) by Zayed University Professor Justin Thomas, draws attention to the possible impacts of the Anthropocene-led extinctions on the animal species endemic to the region as follows: 'When the hammour [Gulf fish] is no more, how will you feel? When the Arabian leopard goes extinct, how will you feel? When the only place you can see an Arabian oryx is in a zoo, how will you feel?' More problematic is 'Little Wonder That, According to Geologists, We Have Entered the Anthropocene Age: an Age Defined By Man's Massive Impact on the Planet' (2011), by Edward Glaeser, professor of economics at Harvard. He argues: 'To many this is cause for alarm. Because when it comes to the environment, everybody knows "urban" is a dirty word – right?' Cities, he declares, are the "greatest achievement" of mankind, making us not only richer, smarter, healthier and happier, but also greener'. It may come as a surprise to learn that Glaeser cites Dubai to clinch his argument. His depiction of Dubai hides massive ecological and socio-economic problems, notably in relation to water, energy use during its construction, maintenance, and operation as a city and, of course, inequalities. Indeed, it could be argued that Dubai and other Gulf cities stand as exemplars for many of the ills associated with the Anthropocene, ignored by *The National* and most other global media outlets.[4]

Another story that localizes the Anthropocene appears in the Israeli daily, *Haaretz* 'The Watershed: A Journey Along the Polluted Ashalim Stream [in the

Negev Desert]' (2016). Given the importance of water issues in the Middle East, it is surprising that, for example, we found no discussion of environmental issues across Israel, Jordan, and Palestine around the Jordan River, where important cross-border civil society work is being carried out by Eco Peace Middle East (see Figure 10.1).[5]

Nevertheless, there is much of interest in Middle East media coverage of the Anthropocene. The earliest media item from the Middle East that explicitly mentions the Anthropocene is in the Saudi-based Arabic-language daily *Al Riyadh*: 'The Environment is Heading To the Brink of Disaster' (2008), which introduces the Anthropocene concept by mainly focusing on climate change. It explains the basics of the Anthropocene, the start date (industrialization and the Great Acceleration) and potential 'catastrophic' outcomes, for both the planet and the region. Unsurprisingly, for a pro-government publication, this original (if confusing) op-ed piece by government spokesman Dr Mohammed Hussein Al Askar questions the general tendency to hold energy producers responsible for the rise in greenhouse gases (GHG), claiming that their contribution is 'not scientifically proven'. He calls for recognition of the responsibilities of 'customers', who are described as 'the owners of their choice and activity', accusing the powers controlling world politics of ignoring the warnings of scientists and failing to implement the solutions that are available: 'The shouts and warnings of scholars fall on the deaf ears of leaders and decision makers, because each country pushes the other

Figure 10.1 Palestinian, Jordanian, and Israeli teenagers campaign together to save the Jordan © Eco Peace Middle East.

to suspend the bell, as if humanity is unconsciously subjected to self-destruction'. The highly influential pan-Arab media takes up the story in the Qatar-based *Al Jazeera* with 'Climate Change Leads to a New Geological Era' (2008), arguing: 'the Earth has entered a new geological era' due to human-caused changes in climate, with the land, and oceans carrying potential risks. This article is translated verbatim from the British daily *The Independent*, based on a research project by a team of western geologists led by Jan Zalasiewicz. In the following years, the Middle East media's interest in the issue remained low.

It could be argued that Israeli media were the first to cover Anthropocene-related issues in the region, without using the term. Two articles report a conference, 'Global Warming. Is Our Future in Peril?' at Tel Aviv University in December 2006, namely 'High Fever' in *The Jerusalem Post* and 'Artificial Cooling to the World' in *NRG*. Both articles link climate change to human activities and refer to the work of Paul Crutzen, who participated in the conference. In 2007, we find in *Ynet*, the online outlet of *Yedioth Ahronoth*, the largest daily in Israel, a story reviewing research on methane emissions in relation to climate change, naming Paul Crutzen, Will Steffen and Jan Zalasiewicz. The total number of items on the Anthropocene increases sharply in 2016 and, in spite of a subsequent slight decrease, it still remained higher in 2017 than in the pre-2016 period. More than half of the total number of articles on the Anthropocene in Middle East media are published in 2016–2017. In other words, the total number of items on the Anthropocene in these two years (2016 and 2017) amounts to more than the total number of articles published between 2006 and 2015. This is due to the almost nine-fold increase in the number of 'neutral' media items covering the Anthropocene. Around three-quarters of the articles employing 'neutral' frames alone or in combination with 'optimistic' frames were published in 2016–2017.[6] Rising interest in the Anthropocene appears to parallel intensification of the international scientific debates revolving around the Anthropocene.

The Anthropocene as a scientific phenomenon

The most common approach in the Middle East media is to present the Anthropocene as a scientific phenomenon, without referring to its negative/positive consequences or emphasizing solutions to tackle the risks it poses. Articles based on this 'neutral' framing of the Anthropocene constitute almost one-half of the total number of Anthropocene-related coverage. When stories that carry both 'neutral' and 'optimistic' elements are included, this ratio exceeds one-half. The pan-Arabic media, as well as the media in Egypt, Turkey, Lebanon, Saudi Arabia, and the UAE offer neutral frames in their coverage of the Anthropocene.

Within the neutral frame category, less than one-third of the media stories specify a start date for the Anthropocene. Of those that provide a start date, more than three-quarters point to the mid-20th century as the beginning of the Anthropocene. Whereas most of these accounts name only the time period, without elaborating the reasons, a small number of news articles discuss processes and dynamics of the proposed time period, that gave rise to the Anthropocene, notably the 'Great Acceleration'. For example, the Iranian English- language daily *The Financial Tribune*, 'New Geological Age' (2016) indicates that, even though the 'formal' start date of the Anthropocene is yet to be determined, some scientists believe that it is the 1950s when the 'Great Acceleration' was identified, with an immense increase in consumption and population. Coverage of this issue results in ambivalence with respect to which specific social and economic systems and their embedded activities have most influence on the risks associated with the Anthropocene.

All neutral, as well as optimistic and pessimistic frames, offer a definition of the Anthropocene, except for a small amount of art-related media coverage. These definitions range from short descriptions, condensed in a phrase, to comprehensive accounts of what the concept refers to. Whether brief or detailed, all the neutral accounts provide similar definitions of the Anthropocene, through their emphasis on the transformative impacts of human activities on the planet and nature. Some of these definitions make reference to the geological age during which humans became the dominant influence on our planet's environment. For example in 'Anthropocene Epoch, Meaning That Human Activity is Now the Dominant Influence on Climate and the Environment' (*Gulf News*, 2016), 'How Man has Tipped Earth into a New Epoch … Age of Man' (also in *Tehran Times*), 'Climate Change Endangering 432 N. American Bird Species … New Age of Man' (*Akhbaar*, 2017), 'Welcome to the Anthropocene! … the Present Geological Age, Whose Distinguishing Feature is the Way We've Altered the Planet (*Lebanon Daily Star*, 2016), and 'On the Aesthetics of Extinction … a Proposed Era That Begins When Human Activities Begin to Have a Major Global Impact on the Geology of Earth and its Ecosystems' (*Al Jazeera*, 2016).

The neutral frame category can be sub-divided into two forms of presentation: first, by exploring the Anthropocene as a whole, and second, by exploring it in an issue-specific manner. These neutralizing accounts commonly describe the Anthropocene as a geological period that is shaped and dominated by humans and the workings of which will be determined by the scientific debates. Accordingly, they usually consist of lengthy accounts, providing basic scientific information on the causes, indicators, and (to a lesser extent) outcomes, including scientific debates revolving around the Anthropocene. Given that most of the news coverage is being borrowed, directly or indirectly, from western news sources, almost all of the scientific studies, reports, and debates introduced by Middle East media come from North American, European, or Australian scientists. Renowned Anthropocene scholars such as Paul Crutzen, Will Steffen, and Jan Zalasiewicz are cited in many of the neutral accounts. For example, the article 'How Man Has Tipped Earth into a New Epoch' (2016), cited above, reports on the efforts of the Anthropocene Working Group (AWG) for the official recognition of the Anthropocene by providing a comprehensive account of how the concept was coined and what it involves. In a report from AP, the English-language daily *The Jordan Times* 'With Their Mark on Earth, Humans May Name Era, Too' (2015), underlines that the Anthropocene is not limited to climate change. This informative article summarizes many indicators, including ozone loss, disruption of nitrogen and phosphorus cycles that is causing dead zones, changes in water, acidification of the ocean, endocrine disruptors, and deforestation. The reference is to Will Steffen, described as 'one of the leaders of the Anthropocene movement'. The second category of neutral coverage concentrates on specific ecological issues, notably extinction/biodiversity loss, waste, plastic pollution, or climate change, linked separately to the larger framework of the Anthropocene. Most of the media accounts that combine neutral and optimistic frames belong to this group. Whereas the Anthropocene is described as a scientific issue, questions about what can be done to address specific threats are also raised, such as, in *Gulf News* 'The Hot Topic of Climate Change' (2015), *The National* 'End of the World As We Know It is Just Ahead (2016), *Iran Daily* 'Human Impact Has Pushed Earth into the Anthropocene' (2016), *Independent News Agency* 'Have We Entered The "Anthropocene" Or The Human Geological Era' (2013), *Hürriyet* 'Two-Thirds of Living Species Might Be Extinct By 2020' (2016), and *Al Araby* 'Pollution Kills Nine Million People Annually and Threatens Human Societies' (2017).

Optimistic narratives for a 'good' Anthropocene

Most of our results can be classified into the neutralizing or the optimistic frame, while acknowledging risks, or a mixture of the two. Until 2014, optimistic Anthropocene stories printed each year by the Middle East media totalled between one and four. This figure doubles after 2014, coinciding with some initiatives at the international level that heightened hopes and expectations in relation to climate change and sustainability such as the Paris Agreement and the UN Sustainable Development Goals (SDGs), respectively.[7] Optimistic accounts of the Anthropocene are found most frequently in the national and pan-Arab media based in Qatar, Saudi Arabia, and the UAE, as well as in Israel and Jordan. The pan-Arab media has a greater tendency to use optimistic frames for covering the Anthropocene, in comparison with local and national media in the region.[8]

As is the case elsewhere, optimistic accounts appearing in the Middle East media essentially describe the Anthropocene as a new geological age in which human activities dominate the planet's ecological system and engender serious ecological problems. In contrast to the neutral frames, the argument in these accounts is that the Anthropocene and its various outcomes can be controlled and managed by human ingenuity. To this end, the solutions suggested tend to involve mobilization of science and technology, as well as political and behavioural changes. Nevertheless, these solutions are usually vague, without requiring radical changes in the existing political, economic, and social systems. These accounts report the Anthropocene as an open-ended process that can be managed by humans dealing with the risks. For example, we find a quasi-alarmist optimistic framing in the English-language *Daily News Egypt* (reproduced from *Deutsche Welle*) under the title 'Are We Really Heading for an Anthropocene Apocalypse? (2017), which investigates the most appropriate approach to address the impacts of the Anthropocene. Building on a Proceedings of a National Academy of Sciences (USA) (PNAS) report on mass extinction, the Anthropocene is described as human impact on the planet producing adverse effects in a wide range of spheres. The article takes an optimistic turn, referring to biologist/journalist Christian Schwägerl, author of a popular science book on the Anthropocene. *Daily News Egypt* (and Schwägerl) argue that emphasizing catastrophe by using 'apocalyptic language' leads people into inaction, culminating in the minimization of any possibilities of tackling challenges posed by the Anthropocene. Schwägerl argues that nature and economy should be reconciled in order to prevent the 'misuse of the Earth's resources', thereby enabling a 'more harmonious relationship with our planet' but fails to show how this could be achieved under existing political and economic systems.

In optimistic narratives, emphasis is commonly put on humans' capabilities and capacities to control the Anthropocene. An AFP-sourced article in 2011 in two English-language dailies (*Tehran Times* and *Lebanon Daily Star*) under the same title, 'Extreme Makeover: are Humans Reshaping Earth?' clearly illustrates this approach. The article provides a detailed scientific account of the Anthropocene and its hazards. Zalasiewicz, Crutzen, and Steffen are cited, as is Erle Ellis, professor of geography and ecology at the University of Maryland, who argues that the Anthropocene can evolve into something 'good, even better' as long as humans change their ways of behaving and thinking, by 'taking over ownership of the planet'.[9] The article concludes by tasking scientists with determining what to do in order to tackle any undesirable outcomes of the Anthropocene: 'At the same time, however, the concept forces us to ponder whether humanity's outsized impact on the planet could lead to undesired, possibly uncontrollable, outcomes, and what, if anything, humanity should do about it. That leaves scientists, who may be more comfortable classifying rocks than rocking the boat, in a tricky position'. In a similar vein, *Iraq Independent News Agency*, with the familiar title 'Have We Entered the "Anthropocene" or The Human Geological Era' (2013), argues that humans have transformed 'nature' irreversibly. It is no longer nature as we know it since it is inseparably intertwined with 'technical influences' of humans, but humans can eradicate Anthropocene-led problems such as increasing temperatures, droughts, and waste. In two similar articles in 2013, *Al Jazeera* 'What Does the Future Hold for the Global Economy?' and in *Al Arabiya* 'Writing the Future for the Global Economy', Jeffrey Sachs tells readers that technology can minimize ecological as well as socio-economic problems caused by the Anthropocene. 'Writing the Future' (from Project Syndicate) disseminates a feeling of hope by claiming that humanity does possess the necessary know-how and techniques for producing beneficial results such as making food production more environment friendly, improving public health ('both for rich and poor'), and emitting lower levels of greenhouse gases. Sachs concludes with the message that humanity should set bold goals, built on 'human values' and 'creativity', in order to make the planet a more sustainable place: 'In our generation, sustainable development will be the test, encouraging us to use our creativity and human values to establish our path to the sustainable well-being of our densely populated and endangered planet'. Nowhere is the question of how to reconcile economic growth with the SDGs seriously addressed.

Al Jazeera, with 'Problem Solving in the Anthropocene Era' (2016), offers a somewhat more critical perspective, arguing that previously successful 'solutions', such as the Green Revolution, are unlikely to address anticipated food insecurity. Instead, 'a smarter approach to strategic decision making' and 'a broader understanding of innovation' are required, in order to cope with challenges posed by the Anthropocene. However, the solutions remain vague. Rather, what is suggested is that the tools of the already existing system, such as GDP, are redefined by incorporating parameters of economic, natural, and social capital in combination with the 'dignity of the ecosystems and social structures that underpin this output'.

Another theme that recurs in optimistic narratives is the emphasis on the necessity of increasing the ecological awareness of humanity as a whole, implying recognition of ownership of the planet and its responsibilities, as well as the development of appropriate norms and values. *Yemen Today* (2010), for example, portrays Yemen as a 'victim' of climate change by highlighting local environmental problems, such as floods in Hadhramaut, arguing that Yemen is not responsible for the Anthropocene, but it is exposed to all present and future hazards. In *Khaleej Times*, we find 'Vatican-Appointed Panel Warns of Climate Change" (2011), a scientific report on the causes and effects of mountain glacier retreat occurring as a result of climate change. It is unusual for an article to cite a religious authority with regards to Anthropocene-related issues.[10] The article highlights the need for all countries to implement 'fair' and 'effective' policies to tackle climate change. By exemplifying measures taken by the Vatican under the leadership of 'green' Pope Benedict XVI, such as installing photovoltaic roof solar panels and solar cooling units at the Vatican facilities, as well as participating in reforestation, it is indicated that, in order to achieve a sustainable world, humanity should converge on a shared understanding, 'living in the same home', a key Anthropocene theme. According to these accounts, it is imperative to curb destructive and negative behaviours, like 'greed', and to develop shared values and norms that will guide human action to a more sustainable model. Jeffrey Sachs, director of the UN Sustainable Development Solutions Network is a prominent participant in these debates. His 2012 article 'Heat is On To Solve the Global Climate Crisis' (Project Syndicate) appears in *The National*. After evaluating climate change as a 'human-induced' problem, taking place in the context of the Anthropocene, Sachs states that 'shortsighted greed' as well as 'political timidity' and 'anti-scientific corporate propaganda', which put the planet in the present risky situation, should all be replaced with scientific solutions in order to ensure human survival. These are fine words, but, like so much Anthropocene commentary, they fall short of concrete action for radical change. Like Sachs and mainstream opinion, the Saudi *Arab News* cherishes the concept of sustainable development. This is expressed in another op-ed by the director of the Potsdam Institute for Climate Impact Research, the eminent scientist Johan Rockström (plus two more Project Syndicate articles by the same author in *Lebanon Daily Star* and *Jordan Times*). Rockström introduces an important (and emotional) element into the debate with 'Leaving Our Children Nothing' (2015), emphasizing technology as a prerequisite for overcoming problems created in the Anthropocene. However, he argues that technology-based solutions are insufficient unless they are harnessed with appropriate values, resting on the responsibility of humans to eradicate 'massive human and economic costs'. Anticipating the impacts of Greta Thunberg and others, Rockström reminds adults that what we do will have serious consequences for the ability of future generations to survive on planet Earth, but again, he offers no specific radical proposals for change.

Sustainable development in one form or another is frequently proposed as a remedy for the risks associated with the Anthropocene by the 'optimistic' media in the Middle East, as it is in the rest of the world. Examples of this are found in *Khaleej Times*, 'When Unchecked, Growth is Bad For Our Health (2017), *The National* 'End of the World As We Know it is Just Ahead' (2016), *Jordan Times* 'The Climate Pope' (2015) and *Jordan Times* 'The Health Costs of Environmental Change' (2015). A reformulated version of 'sustainable development' is also proposed by *Al Jazeera* in 'Redefining Sustainable Development' (2013), a critique of the conventional form of sustainable development, originally defined as 'development that is able to meet the needs of the present without compromising the ability of future generations to meet their needs'. The author, the Australian David Griggs, stresses that this conventional understanding of the concept is neither effective nor adequate in the Anthropocene. However, his alternative is the usual rhetoric of the need for new values, which falls short of exposing the unsustainable dimensions of our current consumption and ways of living. Only around one-quarter of the optimistic articles take the Anthropocene as a multi-dimensional issue by referring to all its aspects. The rest of the optimistic news coverage, as shown above, focuses on specific ecological issues.

As in all other regions, climate change appears as one of the most discussed Anthropocene-related topics, with around one-half of the optimistic Anthropocene items in the Middle East media focusing on this issue. Media interest in climate change has increased since the 2015 Paris Agreement, seen by some as the most concrete step taken to date at the international level towards tackling climate change, in spite of its shortcomings and defects. Accordingly, of the 28 optimistic articles, a majority that made a mention of climate change were published after 2014. For example, in 2016, *Egypt Independent* publishes '2015 a "Tipping Point" for Climate Change', defining the Paris Agreement as a 'rare moment of unity' among 195 states in the fight against climate change. Though uncertainty is also expressed, regarding whether the agreement will achieve its goals, the article hopefully says that the world has finally decided to take action, instead of endlessly debating climate change. In 'Gaia Vince's Voice of Hope Amid Climate Change' (2016), an interview published in *Gulf News*, the popular science writer claims that humanity already has the necessary technology to stop the devastating impact of climate change that is implied as the major problem of the Anthropocene. Vince explains her optimism: 'I have had people saying it is unnecessarily positive... If you are human, if you want to see a future, you have to see a future that is hopeful. You have to work towards that. And I am optimistic. I think we are very resilient, we are very adaptable. We are very ingenious as a species and, hopefully, we will get through'. Likewise, renewable energy is often presented as the main solution for tackling climate change in much media coverage. In an op-ed in *Gulf Daily News* 'The Hot Topic of Climate Change' (2015), originally published in *The New York Times*, Curt Stager, an American professor of natural sciences, argues, 'A switch from finite fossil energy to cleaner, renewable energy sources is inevitable. We are only deciding how and when to do it'.

A few articles take philosophical or sociological paths, such as connecting extinction and biodiversity loss, conservation, and human life. *Gulf Times*, in 'How To Mend the Conservation Divide' (2014), reprints a debate from *The New York Times* between two American ecologists, Emma Marris, author of the influential book *Rambunctious Garden*, and Greg Aplet of the Wilderness Society. This is an excellent example of media engagement with how environmental humanities contribute to Anthropocene studies (see Chapter 12 of this book). Readers are told that 'fantastic diversity of life' can be secured through preserving sites where wildlife still exists and through scientific innovation in already-altered places. Conservation remains a divisive site of struggle. One of the rare examples, suggesting that humans are also exposed to the risk of extinction, appears in *Gulf News* with 'Humans "Will Be Extinct in 100 Years" Scientist Blames Population Explosion' (2010). Borrowing from the UK tabloid, *The Daily Mail*, readers are told that humans will face the risk of extinction due to 'population explosion' and 'unbridled consumption'. A similar argument (also borrowed from a western source) comes in *The Tehran Times* in 'Mass Extinction Threat: Earth on Verge of Huge Reset Button?' (2010), promoting conservationist policies and measures, and *Gulf News* (borrowing again from *The New York Times*) reprints 'Why the Tide is Shifting Towards Marine Conservation' (2017), profiling the research of the North American environmentalist Emma Marris. We also find reporting of engagement by social scientists in Anthropocene studies. For example, *Khaleej Times* 'Towards a Global Treaty on Plastic Waste', and *Jordan Times* 'Let's Get Rid of Plastic Before It Chokes Cities' print the same op-ed on plastic pollution and the Anthropocene in 2017, sourced from Project Syndicate. The authors of this article, Nils Simon and Lili Fuhr (Berlin-based researchers at Adelphi Research and the Heinrich Böll Foundation, respectively) identify plastic pollution as the defining problem of the Anthropocene, arguing in great detail that it can be addressed through available tools and mechanisms, with simultaneous use of technology, political measures, and behavioural change.

Pessimistic radical change narratives

As is the case in all other regions, these stories are rarely reported in Middle East media coverage of the Anthropocene. In general, articles that utilize the pessimistic frame address the Anthropocene either directly, in terms of a general process, as in *The National* 'The Decade' (2010), and *Lebanon Daily Star* 'Rio Summit: Scientists Warn of Emergency on Global Scale' (2012), or, indirectly, by focusing on one major ecological problem, usually anthropogenic climate change. Pessimistic media representations commonly draw attention to the detrimental impacts of the Anthropocene, or a specific Anthropocene-related ecological crisis, on specific species as well as on the ecological balance in general. In that regard, imminent risks to life on the planet are raised, even though human life is not explicitly specified. *Mako*, a daily newspaper in Israel, poses the question, 'Have We Asked For Forgiveness From Future Generations?' (2016).

Among all the Middle East media sources, *Haaretz* from Israel and *Gulf News* from UAE stand out as the only two that print pessimistic stories more than once. *Haaertz*, which is the oldest quality daily newspaper in Israel (with liberal and left-wing tendencies), prints three pessimistic Anthropocene articles in total, whereas *Gulf News* prints two. Although most pessimistic items are translations of stories published in the Western media, a very limited number of original items expressing pessimistic views also appear. For example, an op-ed piece in *Haaretz*, written by the Israeli historian/academic Ofri Ilany, 'Let the World Burn: Welcome to the Trumpocene' (2017), argues that detrimental impacts of the Anthropocene will not go away, but will only be delayed at best, even if tools and mechanisms that are expected to repair the damage, such as the Paris Agreement, are put into effect. Ilany cites an article 'We are Entering the Trumpocene' by Graham Readfearn (*The Guardian*, 2016), where the author claims that we live in a time period when politicians like Trump 'open the gates of hell', while humans in general refrain from accepting the fact that they are destroying the planet by their own hands. Declaring that nature is already dead, Ilany indicates that humans who are born into the Anthropocene are trapped into 'Holocene thinking', a term borrowed from the Australian ethicist Clive Hamilton, carrying the false expectations of returning the planet to what it was during the Holocene period. The author clearly enunciates the bleak end towards which the planet is inevitably heading but offers no radical alternative. Another original op-ed piece that takes a pessimistic view of the Anthropocene is from *Cumhuriyet,* the oldest daily newspaper in Turkey, and known to have a centre-left political stance. Written by columnist Hakan Kara, it draws attention to the Anthropocene's various detrimental impacts on living species. The author makes a grim future projection by claiming that all that will be left behind from our times will be human-produced hazardous substances. 'The content of that thin soil layer, which will be left behind from our civilization, makes me think: "Radioactive materials, plastic, concrete, fertilizers, and ash ... I wonder what the future geologists will think about our civilization. As one of my beloved friends says: "You leave either a trace [*iz* in Turkish] in life or smut [*is*]"'.

Yet, pessimistic accounts are not uniform with respect to how they approach the problem of the Anthropocene. An alternative pessimistic narrative is built on a more substantial criticism of the existing economic and social system. Whereas this type of narrative, underlining the necessity for radical changes in the social and economic systems in order to stop the disastrous impacts of the Anthropocene, is rarely to be found in Middle East media, it is not entirely absent. For example, *Kassioun* (2013), the daily newspaper of the communist People's Will Party in Syria, argues that new production methods are required in order to stop 'accumulation of rubbish', which is singled out as the main indicator of the Anthropocene. In a similar vein, an op-ed piece from *Haaretz*, 'Turning On the Air Conditioner in the Burning House' (2015), defines climate change as the major ecological crisis of the Anthropocene era, in which everything is 'artificial'. Not only the political leaders, but also regular citizens are accused of living without changing their consumption habits and lifestyles, as well as giving their consent to the dominant economic system, capitalism. Therefore, the article attributes direct responsibility for participation in the capitalist system as the major cause of climate change: 'in the present age, everything natural can be changed, except for the concentration of capital by a tiny group of people. Fight against it? It is not possible. "That's the way it is in nature", says the obedient citizen and shrugs...It's politics'.

Middle East media sources mostly refrain from identifying actors who are responsible for increasing Anthropocene risks. Rather, humanity in general is responsible for the Anthropocene, regardless of class polarization, gender inequalities, or imbalances within decision-making processes. In other words, human activities and their impacts are mostly considered as uniformly and equally distributed across time and space among humans. However, two 'pessimistic' articles printed in *Gulf News* are exceptions. The first (re-printed from *The Guardian*), reproduced with the title 'Climate Deniers Have Won' (2014), associates the Anthropocene with 'man-made global warming' and 'man-made mass extinction of species' and argues that climate change deniers have been successful in their efforts to resist any significant changes that could stop anthropogenic climate change. Citing corporate lobbying and the negligence of politicians seeking financial support, the article argues that climate denialism has taken the form of a 'cult', whose followers conveniently believe the opposite of scientific findings, in order to justify expanding consumption and resource exploitation habits: business as usual. The author, British columnist Nick Cohen, argues that it is an impossible task to convince societies to pursue another way of living. He concludes: 'But the task feels as hopeless as arguing against growing old. Whatever you do or say, it is going to happen. How can you persuade countries to accept huge reductions in their living standards to limit (not stop) the rise in temperatures? How can you persuade the human race to put the future ahead of the present?'

Gulf News 'Extinctions Fueled by Global Warming' (2014) is a review of Elizabeth Kolbert's influential book, *The Sixth Extinction: An Unnatural History*, reproduced verbatim, again from *The Guardian*. After providing a brief account of human-induced hazards and their devastating impacts on species, the reviewer explains Kolbert's claim that extinction is the outcome of our choices, even if unintentionally, as humans. In response, the reviewer refuses the implication that attributes responsibility to all of humanity and instead points to 'individuals who are spending hundreds of millions of dollars to keep people in the dark about climate change and who are blocking moves to a green economy'. As indicated, it is difficult to find any pessimistic narratives that put forward concrete proposals for radical change that could deal with the potential risks of the Anthropocene.

Conclusion

In 2018, *Al Araby* published an op-ed 'What We Will think?', expressing surprise at the Arab world's neglect of the concept of the Anthropocene, mentioning that Wikipedia does not have an Arabic entry for 'Anthropocene', despite it being one of the most-spoken languages in the world and despite 39 other languages having an entry for it. In 2020 it appears a word has emerged, *al-anthrapoussin* (الأنثروبوسين). It remains to be seen if the media in the Middle East are now paying more attention to it.

Perhaps the most striking finding from our results is not what is written, but what is not written, hence the sub-title of this chapter: 'Invisible oil?' In about 200 Anthropocene media stories, there were only two passing mentions of oil, both in the same item, sourced from *The Guardian*, one mention of petroleum (in the English-language *Arab News*), referencing a government minister in Saudi Arabia, in, as discussed above, an article by a Saudi government spokesman, blaming consumers not producers. In a report by the Lebanese-American international journalist Dana Moukhallati on the World Energy Congress in Abu Dhabi, we find ' "Demon Oil" on the defensive over climate change' (2019): 'At the dawn of an era scientists have dubbed the Anthropocene, driven by human impact on the planet, the energy industry's four-yearly gathering was forced onto the defensive on climate change. With the burning of "demon" fossil fuels blamed for playing havoc in the "age of man", many agreed that, after decades of energy wars, future conflicts would be driven by competition for clean water as the glaciers recede and rivers dry up'.[11]

Reporting on the Anthropocene does not fit easily into media agendas in the region, with most countries mainly focusing on questions of economic development (especially the Arab OPEC members), and, of course, there are more immediate political and military crises, as in Syria, Yemen, and Iraq for example. As topics of media interest, the Anthropocene and related ecological issues, such as climate change, face structural barriers.

Notes

1 Elya Milner, Sarah Morris, and Elizabeth Schmitz also contributed to this chapter.
2 See *Measuring Inequality in the Middle East, 1990-2016: The World's Most Unequal Region?*", https://wid.world/news-article/new-paper-inequality-middle-east/ and https://www.theguardian.com/environment/2019/oct/09/revealed-20-firms-third-carbon-emissions)
3 On environmental policy in the Arab world, see Djoundourian (2011), and Ouis (2002) for a wide-ranging case study of the UAE.
4 See https://dubaization.com/post/176821563973/sultans-of-green-arab-gulf-cities-and-the-new for a summary of the complexities of this issue.
5 The 'Environment and Climate in the Middle East' media digest at <mideastenvironment.appa.01.york.ca> provides regular reporting, as does WAFA (Palestinian News Agency), but not on the Anthropocene.
6 In what follows, 'neutral' and 'optimistic' are used in the senses outlined in Chapter 1 of this book, and 'optimistic' suggests the 'good' Anthropocene narrative.

7 In 2015, The Sustainable Development Goals (SDGs), consisting of 169 targets in 17 fields, were set by the United Nations General Assembly as part of the 2030 Agenda.
8 On pan-Arab media, see Yaghi (2017) and Sarnelli and Lomazzi (2019).
9 As indicated in previous chapters, Ellis has become an influential and controversial figure in Anthropocene studies.
10 See also 'The Climate Pope' in *Jordan Times* (2015) and *Lebanon Daily Star* (2015), and Chapter 12 of this book.
11 See https://phys.org/news/2019-09-demon-oil-defensive-climate.html

References

Djoundourian, S. (2011) Environmental movement in the Arab world. *Environment, Development and Sustainability* 13: 743–758.

Freeman, B. (2016) Protecting the gulf: Climate change coverage in GCC print media. *Cogent Arts & Humanities* 3. http://dx.doi.org/10.1080/23311983.2016.1212690

Gunter, B. & R. Dickinson eds. (2013) *News Media in the Arab World: A Study of 10 Arab and Muslim Countries*. London: Bloomsbury.

Kraidy, M.M. (2011) The emergent supranational Arab media policy sphere. In Mansel, M. & M. Raboy eds. *The Handbook of Global Media and Communication Policy*. Oxford: Blackwell, chap. 18.

Lahlali, E.M. (2011) *Contemporary Arab Broadcast Media*. Edinburgh: Edinburgh University Press.

Lynch, M. (2006) *Voices of the New Arab Public: Iraq, Al Jazeera, and Middle East Politics Today*. New York: Columbia University Press.

Mellor, N. et al. (2011) *Arab Media: Globalization and Emerging Media Industries*. Cambridge: Polity Press.

Morgan, A. (2019) What to expect from the post-pan-Arab media. *Conflict Research Programme Blog*. https://blogs.lse.ac.uk/crp/2019/01/23/what-to-expect-from-the-post-pan-arab-media/

Ouis, P. (2002) Greening the emirates: The modern construction of nature in the United Arab Emirates. *Cultural Geographies* 9: 334–347.

Reinisch, L. (2010) Environmental journalism in UAE. *Arab Media & Society* 11: 1–19.

Sarnelli, V. & V. Lomazzi (2019) The end of pan-Arab media? National, transnational media and identity in Morocco, Tunisia and Jordan after 2011. *International Communication Gazette* 81(5): 418–444.

Schapiro, M. (2017) The unique burden of covering climate change in the middle east. *Pacific Standard* (14 June).

Soffer, O. (2015) *Mass Communication in Israel: Nationalism, Globalization, and Segmentation*. New York: Berghahn.

Uzelgun, M.A. & P. Castro (2014) The voice of science on climate change in the mainstream Turkish press. *Environmental Communication* 8(3): 326–344.

Yaghi, M. (2017) Media and sectarianism in the middle east: Saudi hegemony over pan-Arab media. *International Journal of Media & Cultural Politics* 13(1 & 2): 39–56.

Zayani, M. (2012) Transnational media, regional politics and state security: Saudi Arabia between tradition and modernity. *British Journal of Middle Eastern Studies* 39(3): 307–327.

11 Oceania

Big islands, small islands, and the Anthropocene

Leslie Sklair[1]

Coverage of the Anthropocene in Oceania is, unsurprisingly, dominated by the media in Australia (about 100 items from 91 sources, 42 of which had no results). New Zealand produced around 40 results, achieving moderate coverage and picking up on many of the Anthropocene themes found in larger media environments, whereas searches of media in Indonesia produced fewer results. The coverage of Anthropocene-related creative arts events is notable. Several small Pacific Island territories gave us a few results, signifying that the term is recognised but not afforded much attention in comparison to (in some cases) substantial coverage of local indicators of anthropogenic climate change in places already at peril. The three media sources we were able to search in Papua New Guinea (PNG) yielded nothing at all on the Anthropocene. The narratives in half of the items found in Oceania were classified, all falling into neutral or optimistic 'good' Anthropocene narrative frames, with the latter being slightly more frequent than the former. Practical proposals for radical change to cope with risks (where recognised) are almost totally absent. Adaptation and mitigation strategies (though not often using these terms) were common.

Australia

Media ownership in Australia is divided between two dominant groups, namely Rupert Murdoch's NewsCorp (N/C) and what was, until 2018, the Fairfax Group, now 9 Media (9/M).[2] Each group has one national newspaper that circulates widely throughout this vast territory, namely *The Australian* (N/C) and the *Australian Financial Review* (9/M; self-described as a business tabloid, hereafter *AFR*). NewsCorp is recognised as a supporter of climate change scepticism, whereas 9 Media is recognised as somewhat more mainstream. Both have many titles in and around the main cities, encouraging a thriving media environment. Their reporting of the Anthropocene does not always resort to type.

Two well-known Australian Anthropocene notables, the Earth System scientist Will Steffen and the ethicist Clive Hamilton, both feature in the media globally. Will Steffen was actually born in the United States but has worked in Australia for much of his professional life as executive director of the Australian National University Climate Change Institute and a member of the Australian Climate

Commission until its dissolution in 2013. From 1998 to 2004, he was the executive director of the International Geosphere–Biosphere Programme. Along with Johan Rockström, at the Stockholm Resilience Centre, he popularized the ideas of planetary boundaries and safe operating spaces for humanity as ways to interpret and mitigate the risks of the Anthropocene. His publications on these topics are widely cited by scientists and others writing on the Anthropocene. His name appears about 20 times in the results from Oceania[3] (and around 100 times in all regions). After Paul Crutzen and Jan Zalasiewicz, Steffen is the most widely cited scientist in our results from global media. His name occurs in most of the major print media in Australia and New Zealand, the first instance being in The *Courier-Mail* (Brisbane, N/C) in 2008, 'Species Wipe-Out Human Induced', where Steffen is quoted to the effect that the planet is already in a 'human-induced mass extinction event', which is defining 'a new geological age known as the Anthropocene'. Steffen argues that human impact had been pronounced in Australia, due to the highly variable climate, unique wildlife, and poor soils. He concludes with a warning that human history is littered with examples of civilisations that have collapsed because of their inability to adjust to environmental change. 'The best course of action is to mitigate climate change and reduce greenhouse gas emissions as soon as possible'.

Four articles in *The Melbourne Age* (9/M) confirm Steffen's place as the pre-eminent Anthropocene scientist from Australia. The first, 'Winds of Change Brewing' (2012), discusses the role of Andy Revkin (creator of the dot.Earth blog in *The New York Times*) at the 'Planet Under Pressure' conference in London (3,000 specialists, more than two million tweets) which 'reflected a shift away from "solving the climate crisis" ... Instead, it focused on strategies to manage climate risk by enhancing the capacity of populations to withstand extreme conditions, while not giving up on efforts to stem emissions'. The article concludes with a quote from Steffen: 'There is quite strong evidence that we're moving into a new state of the Earth system called the Anthropocene ... But it is a dynamic state, and we don't know where it is going to end up'. In 2015, borrowing from *The Guardian* (the origin of many articles in all regions), *The Age* (9/M) cites research 'led by Prof. Will Steffen of the Australian National University, on safe operating levels', highlighting Steffen's prominence in the international community of Anthropocene scientists. In 2016, *The Age* printed an article based on research by Steffen and others, originally published in the prestigious journal *Science*, arguing: 'Humans have produced enough concrete to pave every square metre of the Earth's surface with a kilogram of the material, and carbon dioxide emissions are rising at 100 times quicker than the fastest pace recorded in the past 800,000 years'. This was followed up by another piece along the same lines in 2017, reproducing a long, detailed, and sobering article Steffen wrote with Owen Gaffney (who is identified as an Anthropocene analyst and communicator and co-founder of the Future Earth Media Lab). Under the title 'Introducing the Terrifying Mathematics of the Anthropocene'[4] where human impacts (concrete, plastics, technofossils) are documented: 'it is the scale that humans have altered Earth's life support system that is the most concerning ... So, can the Anthropocene equation be solved?' (see

Gaffney and Steffen 2019). The current rate of change must return to around zero as soon as possible. It cannot continue indefinitely. Either humanity puts on the brakes or it would seem unlikely a global civilisation will continue to function on a destabilized planet. The choice is ours'. These articles offer examples of how extreme risks can be made clear within reassuring, optimistic narratives, offering no specific solutions, implying that humanity can 'put on the brakes'. Steffen is cited with similar messages in *The Australian* (N/C) in 2015 'Humanity Heading Into the Danger Zone'.

In 2016, in a long and very informative piece titled 'Towards Clear Blue Skies' by the journalist Graham Lloyd, Clive Hamilton is also referenced. This article concludes, enigmatically: '[ecological] footprint research suggests the struggle is not against capitalism per se but to lift populations above lower-middle-income-levels'. A letter to *The Newcastle Herald* (9/M) in 2016 implores us: 'Listen to Science' and reject climate change deniers, quoting the ABC television programme *Catalyst* in which Will Steffen went so far as to warn that 'we have to leave nearly 90 per cent of existing coal reserves in the ground, and leave half of the gas reserves in the ground, and about 35 per cent of oil reserves in the ground. So, no Galilee Basin, no new gas, no new oil. And that's to have a 50/50 chance (of saving the Earth)'.[5] These results suggest that the views of Steffen and others who promote the optimistic 'good' Anthropocene narrative sit more or less comfortably in a variety of media with contrasting ideological positions.

Clive Hamilton, often portrayed as an alarmist on climate change and the effects of the Anthropocene, is frequently cited globally in the Anthropo-scene literature, and sometimes in the more expansive Earth System literature that goes beyond the science. His name appears around 50 times in all regions. Although there are few references to him in the media results from Oceania, his views tend to be reported at length, usually in reviews of his several books aimed at the general public. *The Canberra Times* (9/M) has 'Scary Steps' (2013), a review of his book *Earth Masters* and, in 2017, a long excerpt from his most apocalyptic book *Defiant Earth* in 'The Great Climate Silence: We are On the Edge of the Abyss But We Ignore it'. This book was also covered in *The Hobart Mercury* (N/C) in 2017: 'The Awful Truth Of Our Mastery Over Nature', which ends: 'Hamilton posits the possibility of a second civilization emerging from the planetary ashes of the old one' and concludes, 'these new humans would look at those ashes and declare "Never again".' So says Hamilton the rationalist. I'm not so sure'. Hamilton is clearly pessimistic but offers no concrete radical proposals. He is not entirely alone in this – such views are generally ignored or otherwise neutralized in the media, Oceania being no exception.[6]

Popular science, sceptics and the Anthropocene

Australia stands out as a place where popular science books on climate change and the Anthropocene are reported in the media. For example, a review from 2005 in *The Advertiser* (Adelaide): 'This, the Anthropocene period, [Tim] Flannery says, is likely to be Earth's shortest era unless governments, corporations and individuals rein in their profligate use of fossil fuels and bring down the levels of carbon in the atmosphere. The way we are heading is a downward spiral of climate change – extinction of species, hurricanes and cyclones, floods, plagues, and rising seas'. Flannery is a distinguished scientist, with many publications in peer-reviewed journals and is a leading figure in the efforts to raise consciousness about the risks posed by climate change.[7] In 2011, *The Australian* (N/C) prints a review by Flannery of *Deep Future,* a book by Curt Stager (musician and academic specialist in lake ecology and palaeoecology, who has aired sceptical views on climate change). Flannery concludes his review: 'If all "climate sceptics" were as truly sceptical as Stager, the media debate between the climate scientists and their opponents would be far more interesting. As it stands, it's a sterile argument between scientists and opinionists that serves only to mislead the public'. The editor adds: 'Tim Flannery is Panasonic Professor of environmental sustainability at Macquarie University in Sydney. His most recent book is *Here on Earth: An Argument for Hope*'. Another example of this popular science genre is an opinion/essay published in the *AFR* in 2011, where Paul Gilding, author of *The Great Disruption*, argues: 'The global ecosystem has not only changed, it has now started a planetary-wide shift into what an increasing number of scientists are calling the Anthropocene ... . This is not the end of the world, nor the end of nature, but it is a scale of change so profound that we need to face up to what it means. Human behaviour now determines the fate of the global ecosystem'. These articles are representative of a much larger number that would appear to take the situation implied by anthropogenic ecosystem damage seriously, but not too seriously, usually planting the idea that humanity can cope with what lies ahead.

The Australian (N/C) also prints more sceptical views. In 2008, 'Scientists Argue for Redraw of Earth's Time Line' sensationalizes preparations for the 33rd International Geological Congress meeting in Oslo: 'A heated scientific row is brewing as British geoscientists lead a push to establish a new chapter in the history of Earth – one based on human activity. Led by geologist Jan Zalasiewicz, of the University of Leicester, the rabble-rousers argue that changes wrought since the Industrial Revolution are so profound they are now visible in the physical and living fabric of the planet'. This is the only time Zalasiewicz has been dubbed a 'rabble-rouser' or criticized at all in our results. In 2015, *The Australian* reinforced this theme of scientific irresponsibility in 'Take Panic Out Of Climate Change', a report on Andy Revkin's reassuring 'good' Anthropocene message (based on innovation and transformation) to a Global Integrity Summit in Australia. This attack on climate science is taken to new levels in four articles in *The Herald Sun* newspaper in Melbourne (N/C), all by Andrew Bolt, a Sky News 'shock jock' (in the language of tabloid journalism; see Bacon 2013: 4.3 and 4.6). In 2013, an article with the inflammatory title ' "Climate Refugee" Found – From an Island Growing, Not Sinking' tells the story of Ioane Teitiota from Kiribati, a tiny atoll (coral) island in Micronesia, who applied for asylum in New Zealand as a climate refugee due to rising sea levels that threatened to drown his family. *The Herald Sun* reported his deportation.[8] Three further pieces from the same source followed. In 'Warmist Journalists Drown in Untruths About Sinking Islands' (2015),

Bold invokes research in the journal *Anthropocene* to suggest that hundreds of coral reef islands in the Pacific and Indian Oceans may have remained stable or increased in size. In 2017, we have 'No One Leaving This "Sinking" Island [Marshall Islands]. So, What Gives'? However, subsequent research (Esteban et al. 2019), which involves one of the researchers cited by Bolt (Paul Kench) –far from showing that sea level rises are not occurring in small islands, focuses on the struggles of those who live on these endangered islands to adapt to the perils of rising sea levels, sometimes successfully, sometimes not.[9]

The Hobart Mercury (also a NewsCorp title) proclaimed in 2015: 'Don't Raise the Wilderness White Flag', a critique of eco-modernism. 'The ideals of conservation are under attack by what are termed the green modernists, a group of influential, mainly American thinkers, who take a pragmatic stand to argue that the human species has become so dominant in its control of Mother Nature that we might as well regard the Earth as a giant garden. We live in what's now termed the Anthropocene Epoch – the age of man'. Also in 2015, *The Australian* reinforced this message in 'Pragmatic Environmentalists Pin Hopes On a Technological Fix', presumably more to do with the local politics of conservation than acceptance of the facts on climate change. Waving the green flag of taking the Anthropocene seriously, three items in *The Canberra Times* are notable. The first reprints an article by Jim Robbins from *The New York Times* in 2014, 'Building an Ark For the Anthropocene: We are Barrelling into the Anthropocene, the Sixth Mass Extinction in the History of the Planet'. Taking off from a study published in the journal *Science* (which is often cited in our results), Robbins reports that 'governments, scientists, and non-profit organisations try to build a modern version of Noah's Ark … a patchwork quilt of approaches, including assisted migration, seed banks, and new preserves and travel corridors, based on where species are likely to migrate as seas rise or food sources die out'. Many working scientists are cited in this long and wide-ranging piece to the effect that the urgency of the situation is compelling. The paper prints another long, accurate article by the American journalist Faye Flam in 2015, focusing on Zalasiewicz's much-reported research on plastics, explaining that the idea that people could permanently change the planet was recently considered nonsense, but not now in the Anthropocene era. The third item raises an important but neglected issue: 'Why CSIRO [Australia's national science agency] Climate Cuts Could Threaten Security?' (2016). The global significance is clear: 'Understanding how climate change is unfolding in different countries and regions in terms of rate of change, exact locations, and specific impacts lies at the heart of security planning in this new epoch of the Anthropocene'.[10]

Some popular science magazines in Australia also cover the Anthropocene, but not in great numbers of items or in much detail. *Australasian Science* in 2016, for example, with 'Welcome to the Anthropocene', concludes that 'the concept of the Anthropocene is immediately beyond the expertise of natural scientists alone'. *Australian Geographic* discusses extinction in the Anthropocene in 2013 and, again in 2016, with greater urgency, providing a link via Google Scholar, to 101 related papers (both magazines publish seven articles on the Anthropocene between 2000 and the end of 2017). *Cosmos* (eight items) covers a wider range of Anthropocene-related issues, such as carbon storage and plastics.

New Zealand

In 2018, the New Zealand government blocked a plan to merge ownership of the *New Zealand Herald* (and many local papers and radio stations) and Fairfax Media, which owns the country's most popular news website, *Stuff*,[11] as well as three daily newspapers in the largest cities. Most of the main newspapers in New Zealand cover the Anthropocene but none to any great extent. *The New Zealand Herald*, with 11 items, leads the pack in print media. The first mention appears in 2010, 'End Of an Era', dealing with the debate around the Anthropocene in the Geological Time Scale, followed in 2015 with: 'Experts Haggle Over Start Date For Earth's Era of Human Domination', and a report from an article in *Science*, 'Earth Pushing Planetary Boundaries ...It becomes more and more like living on a spaceship than living on a planet'. Also, in 2015, *National Weekend Radio* announced 'The Sunday Feature: Resilience in the Age Of the Anthropocene: A Panel Discussion on Climate Change', one of many metonymic references, substituting the part for the whole. *The Dominion Post* (four items) features an op-ed by Professor Tim Naish: 'Dawning of Age of Anthropocene' in 2016, another 'Welcome To the Age of the Anthropocene', ending on a note of guarded optimism. *The Press* (six items) starts its coverage in 2016: 'It's Life on Earth ... But Not As the World's Known It' – a satirical discussion between "the glacier" and "*Homo sapiens*"'. Several New Zealand magazines were also searched. *The North & South* has no result, *The Listener* has one item: 'The Dawning Of the Age of Anthropocene', another op-ed arguing that the concept will be useful 'for societies to plan for, to mitigate against, possible changes' (2011).

National Business Review reviews a new book by the local journalist Rod Oram (*Three Cities: Seeking Hope in the Anthropocene*) in 2016, on Beijing, London, and Chicago. In New Zealand, the only searchable popular science magazine we found was *New Zealand Geographic*, a beautifully illustrated publication with four well-researched articles, each mentioning the Anthropocene in passing.

The other islands of Oceania

Indonesia is said to consist of at least 13,000 islands, of which fewer than half are inhabited. The first mentions of the Anthropocene were found on the website of the Indonesian arm of *D/W* (Germany's leading organization for international media development): 'The Holocene Age Ends, The Anthropocene Age Begins' (2012). Anthropocene translates as *zaman pengaruh manusia* (the impact of man era). A more philosophical piece in 2013 concludes: 'Today, there is no longer a realm of nature that is opposed to human culture and that developed itself as it once was. Now the Earth and its environment are massively influenced by humans'. These were followed by an article in 2015 from *National Geographic Indonesia* (an offshoot of the US-based organization), summarizing the theory that the Anthropocene began in 1610 with the Columbian Exchange; another in the popular science magazine *Geologinesia* (2016), 'When Did the Anthropocene Begin?', a thorough discussion, citing the main publications; and *Viva News*, also in 2016: 'The Description of Anthropocene'. Finally, in 2017, in the cultural/literary website *Cantrik*, we find, 'Discourse about Anthropocene' by an Indonesian philosophy student, Unies Ananda Raja, critiquing the 'Assumption of Natural Stability', and concluding: 'From a geological point of view, the Earth is never in a "stable" or "harmonious" condition'. The popular Indonesian women's magazine *Femina* offers an explanation of Steffen and Gaffney's (2017) Anthropocene Equation in a long and well-informed article in 2017: 'Latest Research! Humans Destroy the Environment 170 Times Faster Than Nature' quoting Steffen: 'The results show us, as many as 9 billion people in the world, estimated that this number will occur in 2050, can still live, while reducing greenhouse gas emissions at the same time … That way, we still have time to prevent climate change that endangers the environment'.

The *Jakarta Globe* has one item on the Anthropocene, in 2016, 'Wildlife Populations Plunge Almost 60 Percent Since 1970'drawing this conclusion: 'yet another sign that people have become the driving force for change on Earth, ushering in the epoch of the Anthropocene'. Borrowing from Reuters, the piece references Professor Ken Norris, Director of Science at the Zoological Society of London ("London Zoo"), and Deon Nel, World Wide Fund for Nature (WWF) global conservation director. Nel says: 'I don't speak at all about doom and gloom – we do see a lot of positive signs ... Last month, the giant panda was taken off an endangered list after a recovery in China'. A reassuring message to take the edge off 'doom and gloom'. While it is not mass media, even in our inflated use of the term, *Jurnal Balairung,* written by the aforementioned Unies and others from the University Gadjah Mada, Yogyakarta (Indonesia), merits attention both for its evocative cover art (see Figure 11.1) and for its contents.

Space limitations preclude more than a listing of topics, in this unusual collection (see Unies ed. 2018), referencing local and foreign debates from social sciences and environmental humanities, as can be seen from the following: 'The Human in Nature's Disequilibrium: Critique of Saras Dewi's Ecophenomenology'; 'Complexity and Risk Perception in the Anthropocene'; 'The Regime of Deniers: The Impact of Climate Change Deniers on the United States Environmental Policy'; 'Power and Exclusion: REDD+ as "Climate Leviathan" and the Land Use Conversion in Indonesia'; 'The Changing Consumption of Water from Piyaos [free drinking water] to Pepsi in India'; and 'Interactive Interview with Jan Zalasiewicz: "The problem is not too little evidence but too much"'. *Jurnal Balairung* is mostly in Indonesian but the references are largely in English. It will be interesting to see if the media in Indonesia follow this lead on what many consider is the most important global issue of our times.

Coverage of the Anthropocene in searchable media in Micronesia and Polynesia is sparse (though coverage of climate change is much more frequent). A long and, at times, poignant article emanating from the Pacific Islands News Association (*Pina*) in 2014: 'Island Issues, Global Problems' grabs our attention in its opening

Figure 11.1 Anthropocene special issue, *Balairung* (Indonesia), (2018) © *Jurnal Balairung.*

paragraph. 'The problems affecting Small Island Developing States (SIDS) are as far removed from the idyllic tropical holiday as the picture-perfect postcard that is sent home by the tourist. But how does one capture, in a photo for a postcard, the rising seawaters that are affecting our aquifers, the increasing intensity of extreme weather events, such as cyclones, or the land-based pollution destroying the livelihoods of people in small fishing villages?'[12] Gyan Chandra Acharya (the United Nations Under-Secretary-General and High Representative for the Least Developed Countries, Landlocked Developing Countries and Small Island Developing States) warns: 'if this SIDS reality is not addressed now, it could very quickly become a reality for countries across the globe. Speaking to *The Samoa Observer*, Acharya said the islands are on the frontline for the many issues affecting the world… The reason these issues must be discussed globally is because of what he called Anthropocene'. *The Solomon Star* also had one item in 2014 (suggesting that if one source publishes on a neglected topic regionally, others might follow) 'On Mother Earth Day, UN Urges Protecting Planet From Heavy Hand of Humankind'. UN General Assembly President John Ashe is quoted: 'today's lifestyles have ushered in what scientists are now calling the Anthropocene Era… encouraging delegations to consider the relevant information on the UN Harmony with Nature website, as they consider ways to further build a knowledge network on sustainable development, the efforts of which will serve to produce a balanced paradigm for the planet and for people'.

The online news site *Matangi Tonga* offers three items on the Anthropocene, the first in 2011, reporting on research from the Arctic, predicting sea level rises of five feet by the end of the century. This references the Vatican publication *The Fate of Mountain Glaciers in the Anthropocene*, a report that circulated around the world. In 2017, *Matangi Tonga* commented on another much-discussed publication, from *The Lancet*: 'The Health Costs of Environmental Change ... Scientists Believe We Have Entered a New Geologic Epoch – the Anthropocene'. And we find 'Toward a Global Treaty on Plastic Waste' (2017), picking up on the theme that future geologists will find plastic fossils ... 'Plastic pollution is a defining problem of the Anthropocene'. Media in French Polynesia produce one item, in *La Depeche de Tahiti*, 'Take a Bad Turn, Be in the Dictionary' (2015) on the decision of the publishers Larousse and Robert, the two main guardians of the French language, to include 'l'anthropocène' (competing for attention with 'décroissants', 'les climatosceptiques' and 'décarboner'). No doubt, alarming for some.

Creative arts, scholars and the Anthropocene in Oceania

Media coverage of how the creative arts have engaged with the Anthropocene in Oceania is multi-dimensional. For example, the *AFR*, in 'Age of Nostalgia' (2015), argues that theories of the sublime and nature from the 18th century feed into art about the Anthropocene; *Sydney Morning Herald* 'Apocalypse Soon' (2011) describes an exhibition, 'Anthropocene', by the 'Awfully Wonderful' Hayden Fowler 'where the artist lives, caveman-like, inside a hollow dome and eats only tins of beans in a fake garden, doesn't offer a lot of hope, although it does look amazing'. Perhaps most original is a review in the *Australian* in 2014 of a biography of the iconic W.G. Sebald: 'Labyrinthine Look At an Enigmatic Emigrant'. It concludes: 'Sebald may be the first writer of the Anthropocene: this terrifying era in which we are beginning to understand humans as agents of planetary change'.[13] In 2016, *AFR* began to report on Burtynsky's multi-media 'Anthropocene Project' (Chapter 13 of this book). *The Illawarra Mercury* in 2017 features the artist Arthur Apanski (originally from Belarus) creatively expressing his views on Anthropocene extinction: 'I see this as my personal duty before I'm gone from this planet, to tell the truth the way I see it'. And, more bizarrely, *The Newcastle Herald* (2017) announced the launch of a new wine 'Anthropocene Epoch Mencia' ($29 a bottle). In New Zealand, the weekly *National Business Review* (2017) reports on a photographic exhibition (Chapter 13 of this book). A final example is from the *Jakarta Post*, 'Rookie Artists Question Indonesia's Maritime Identity' (2016), including a discussion of an exhibit labelled 'Anthropocene Guidelines Chapter III: Heading towards a decade of coral bleaching'. This is the work of Gadis Fitriana (a student at Jakarta Institute of Art at the time).

Debates among Australian scholars on questions of the Anthropocene are extremely lively. For example, the historian Alison Bashford (2013) engages creatively with the issues of deep time and Big History in the Australian imaginary, writing geography as history and history as geography. There is little hint of such ideas in media coverage of the Anthropocene. Bashford also draws attention to the work of Libby Robin, a leading proponent of Environmental Humanities in Australia and internationally (Robin 2017). None of this appears in our results, perhaps it is behind the paywalls.[14] Media coverage in New Zealand, while more limited in a much smaller population, is similar in content to Australia (not surprising since Fairfax Media has a large share of both markets). An interesting similarity is the presence of some Anthropocene items in the business press of both countries; not many, but notable compared with other media environments in which business publications tend to ignore the topic. Results from Indonesia were problematic as Anthropocene searches gave several items that were related to anthropogenic changes and risks but did not actually mention the Anthropocene. This was not uncommon in general, but, as there were few genuine Anthropocene items in the Indonesian media, it is difficult to come to robust conclusions on coverage. Added to this complication, the German *D/W* media distribution platform appears to be the main source of articles.

In their article on 'Pacific Islands in the Anthropocene', Vitousek and Chadwick (2013: 1) argue: 'islands provide a useful model for understanding how coupled human and natural systems experience the Anthropocene, and perhaps for how they can manage its impacts'. Though their main focus is on Hawaii, the idea that the thousands of inhabited islands of Polynesia and Micronesia could teach us something in the present about how to cope with the Anthropocene is intriguing. The *Pina* article referred to above, which references the SIDS organisation, is an eloquent testimony to this sentiment. *Matangi Tonga* also addresses the problems caused by rising sea levels for these small islands, as well as more general issues of health in the Anthropocene and the scourge of plastic waste, especially toxic for beaches all over the world. Also significant in Anthropocene reporting in French Polynesia is the information in *La Depeche de Tahiti* that the French have officially accepted the term Anthropocene – linguistically, but with no recognition of its potentially alarming consequences. Again, with the exception of the occasional article on rising sea levels and other ecological risks, the story of the Anthropocene in most of Oceania is one of reassurance.[15]

Conclusion

As noted earlier, two media groups dominate news and opinion in Australia – Fairfax and NewsCorp. Although the latter is notable for the platform it gives to contrarian voices, apart from a few cases, there is not a great deal of difference between the contents of the coverage from the two groups. Recent research shows that Australian print media defined climate change in economic rationalist terms, influenced by the dominant political actors (see Das 2019: 139). Nevertheless, media coverage of the Anthropocene in Australia and New Zealand is wide ranging in terms of themes, notably start dates, potential, and actual impacts. Optimistic 'good' Anthropocene narratives rub shoulders with extinction, climate, and other ecosystems, denial, security, Anthropocene ‘celebrities’ from Australia (Steffen and Hamilton), and Australian exceptionalism. We have quoted Steffen’s view that human impacts on the planet have been pronounced in Australia, due to the highly variable climate, unique wildlife, and poor soils. However, despite this and similar views, there is no genuine attempt to provincialize the Anthropocene in the media in Australia, apart from some localising references, such as the drying up of the Murray river (Chapter 13 of this book). The one Anthropocene reference to this occurs in the context of a dance performance. Surprisingly, there is only one (passing) reference to drought in all of our results.[16]

Notes

1 Cameron Muir and Jahnnabi Das also contributed to this chapter.
2 Media ownership is complicated, so these labels simplify. Bacon (2013) provides a critical and still useful overview.
3 Unrestricted access to all media in Australia (without reference to Anthropocene) would probably show dozens if not hundreds of results for both Steffen and Hamilton.
4 The digital article is embellished with many useful links, providing a basic primer for the Anthropocene (originally published in *The Conversation.*)
5 Fossil fuels left in the ground are increasingly referred to as ‘stranded assets’. For a brief analysis of the global fossil fuel lobby see: https://www.theguardian.com/environment/2019/oct/10/vested-interests-public-against-climate-science-fossil-fuel-lobby, and Brulle (2020) and Wetts (2019, 2020).
6 Chapter 12 of this book discusses Hamilton’s global impact.
7 See *AFR* in 2016: ‘Tim Flannery speculates that whole extinct species could be resurrected’.
8 Whereas the United Nations High Commissioner for Refugees (UNHCR) officially addressed the problem at the end of 2018, the concept of ‘climate refugees’ is not yet recognised in international law. The UN refers to such unfortunate people as ‘persons displaced in the context of disasters and climate change’. See https://www.unhcr.org/uk/climate-change-and-disasters.html. Scholarly interest in climate refugees is rapidly expanding (for example, Bettini 2013, Hingley 2017, and White 2019). In 2020 the *New York Times magazine* began publishing a series of articles on this topic (see Lustgarten (2020).
9 There are over 200 results for ‘warmist’ on *The Australian* website. For documentary evidence of how islanders adapt to rising sea levels, see: https://www.youtube.com/watch?v=gwaS9hlNv5M. Wodak (2018) shows how Bob Dylan highlights rising sea levels in ‘The times they are a-changing’. See also Lipset (2017) for an original approach.
10 This article cites ‘Climate Change Adaptation and Resilience Directive’, the comprehensive climate policy from the US Department of Defense (see also Hynes 2015 on ‘Pentagon pollution’).
11 The website *Stuff.co.nz* gives about 20 results for ‘Anthropocene’ via Google Search. *Stuff* appears to be a subsidiary of Australia’s Fairfax Media Limited, operating nine daily newspapers throughout New Zealand, as well as one weekly paper, *The Sunday Star-Times*.
12 See note 9 above.
13 See Malkmus (2019) on Sebald and the Anthropocene.
14 See Debrett (2011) on climate change reporting, and Bardsley and Wiseman (2016) on learning from remote indigenous communities, in terms of the Australian Anthropocene. For analysis of the environmental movement in Australia, see Tranter and Lester (2017) and, in New Zealand, O’Brien 2015.

15 It is interesting to compare our results on reporting of the Anthropocene in the Pacific Islands media with Jackson (2010) and Shea et al. (2020), on climate change reporting in and on the region.

16 In 2012 the ABC released a film which tells the remarkable story of Lake Eyre, drought, and water. At the time of writing (2020), the main issue is wildfires; see 'From Fire Evacuation Rooms, "Diary of a Wombat" Author Pens Her Message to Australia' by Jackie French (*Sydney Morning Herald*, 9 January 2020). French writes: 'Unless we keep up the rage [at the lack of political leadership], the passion and compassion, our children and our children's children will die in more climatic disasters, from winds to cyclones, floods, tornadoes, bushfires, and storm surges: the "new normal" of the Anthropocene'. The Coronavirus pandemic has created another 'new normal'.

References

Bacon, W. (2013) *Sceptical Climate Part 2: Climate Science in Australian Newspapers*. Sydney: Australian Centre for Independent Journalism, University of Technology (30 October).

Bardsley, D. & N. Wiseman (2016) Socio-ecological lessons for the Anthropocene: Learning from the remote indigenous communities of central Australia. *Anthropocene* 14: 58–70.

Bashford, A. (2013) The Anthropocene is modern history: Reflections on climate and Australian deep time. *Australian Historical Studies* 44(3): 341–349.

Bettini, G. (2013) Climate Barbarians at the gate? A critique of apocalyptic narratives on 'climate refugees'. *Geoforum* 45: 63–72.

Brulle, R.J. (2020) Denialism: Organized opposition to climate change action in the United States. In Konisky, D. ed. *Handbook of Environmental Policy*. Northampton, MA: Edward Elgar, chap. 24.

Das, J. (2019) *Reporting Climate Change in the Global North and South: Journalism in Australia and Bangladesh*. London: Routledge.

Debrett, M. (2011) Reporting on climate change: An Australian perspective. *International Journal of Science in Society* 2: 149–160.

Esteban, M.L. et al. (2019) Adaptation to sea level rise on low coral Islands: Lessons from recent events. *Ocean and Coastal Management* 168: 35–40.

Gaffney, O. & W. Steffen (2017) The Anthropocene Equation. *Anthropocene Review*. https://doi.org/10.1177/2053019616688022

Hingley, R. (2017) 'Climate Refugees': An oceanic perspective. *Asia and the Pacific Policy Studies* 4(1): 158–165.

Hynes, P. (2015) Pentagon pollution. *Climate & Capitalism*. https://fortheleft.blogspot.com/2015/02/pentagon-is-worlds-bigest-polluter.html

Jackson, C. (2010) *Staying Afloat in Paradise: Reporting Climate Change in the Pacific*. Reuters Institute Fellowship Paper. Oxford: Reuters Institute.

Lipset, D. (2017) Masculinity and the culture of rising sea-levels in a mangrove lagoon in Papua New Guinea. *Maritime Studies* 16(2): 1–21.

Lustgarten, A. (2020) The Great Climate Migration. *New York Times magazine* (23 July).

Malkmus, B. (2019) The Anthroposcene of literature: Diffuse dwelling in Graham Swift and W.G. Sebald. In Wilke, S. & J. Johnstone eds. *Readings in the Anthropocene*. New York: Bloomsbury, chap. 12.

O'Brien, T. (2015) Social control and trust in the New Zealand environmental movement. *Journal of Sociology* 51(4): 785–798.

Robin, L. (2017) Environmental humanities and climate change: Understanding humans geologically and other life forms ethically. *WIREs Climate Change*. doi:10.1002/wcc.499

Shea, M.M. et al. (2020) Representations of Pacific Islands and climate change in US, UK, and Australian newspaper reporting. *Climatic Change*. https://doi.org/10.1007/s10584-020-02674-w

Tranter, B. & L. Lester (2017) Climate patriots? Concern over climate change and other environmental issues in Australia. *Public Understanding of Science* 26(6): 738–752.

Unies Ananda Raja, M. ed. (2018) Antropocen. *Jurnal Balairung* 1(1).

Vitousek, P. & O. Chadwick (2013) Pacific islands in the Anthropocene. *Elementa Science of the Anthropocene* 1. http://doi.org/10.12952/journal.elementa.000011

White, G. (2019) 'Climate Refugees' – A useful concept? *Global Environmental Politics* 19(4): 133–138.

Wodak, J. (2018) Shifting baselines: Conveying climate change in popular music. *Environmental Communication* 12(1): 58–70.

Part III

From the Anthropocene to the Anthropo-scene

12 Media coverage of the Anthropocene in the social sciences and environmental humanities[1]

Viviane Riegel

Introduction

This chapter focuses on how media all around the world have reported the ways in which social science and environmental humanities scholars discuss the Anthropocene. While none of these scholars appear to use the term 'Anthropo-scene' they are usually not doing research on Earth System science as such. Obviously, many scholarly articles connect what it means to be human with the Anthropocene, but it may be surprising to some to discover that mass media of several different types (quality nationals, big-city newspapers, tabloids, small-town papers, specialist community, general magazines) also engage with these issues, presenting and discussing some of the theories, research, and thoughts we find in academic books and journals. Social scientists and environmental humanities scholars are represented in the media from most of the regions in our study. As will be seen from the excerpts reproduced below, our results display a range of opinions. In terms of exegesis, we find that some media reporting of the contributions of social scientists and environmental humanities scholars is reasonably accurate and some inaccurate, but all is understandably incomplete. Media coverage in these spheres focuses on relatively few scholars, and here we present the material found in media reporting the Anthropocene related to Bruno Latour, Timothy Morton, Maristella Svampa, William Cronon, Jason Moore, Donna Haraway, Sverker Sörlin, Alf Hornborg, Kathleen Morrison, Dipesh Chakrabarty, Amitav Ghosh, Joan Martínez-Alier, Kyle Whyte, Eduardo Viveiros de Castro and Débora Danowski, Leonardo Boff, Clive Hamilton, and Robert Macfarlane.

Anthropo-scene scholars in the media

Bruno Latour, probably the most famous Anthropocene-related social scientist (certainly in the Francophone world), argues that the Anthropocene resonates deeply with social theory because it builds upon several of the same fault lines as those upon which social science had established its fragile tenets for many decades.[2] First, the idea of the Anthropocene places human agency at the centre of attention. Second, the Anthropocene is often interpreted as defining human agency by drawing on a range of entities, some related to the natural sciences while others relate to what ethnographers have learned to register throughout their field work, like patterns of land use, migrations of plants, animals and people, city life, trajectory of epidemics, demography, inequalities, classes, and state policies. With a rather neutral reading of his views, the French magazine *L'Express*, in 'Those French People Who Renew Thought' (2015), labels Latour as one of the most famous contemporary French intellectuals, widely translated abroad, highlighting his spectacular exhibition, 'Anthropocene Monument', which was taking place at that time. The historian of science Elena Zheltova even discusses the connection between Russian literature and Latour's Actor Network theory in *Gazeta* (2015).[3] An article published in *Le Monde*, also in 2015, 'Civilization Challenged By the Anthropocene', quotes Latour: 'humans should make science compatible with other devices, political and cultural, particularly, to deal with global warming … The Anthropocene would then cease to be the somewhat desperate discovery that humans have become a geological force, to become the index of a completely different composition: that of a possible civilization'. In a similar tone, the Norwegian *Aftenposten* 'More Than One Way To the Truth' (2013), translates Latour's acceptance speech for the Holberg Prize, which mentions his ideas on the Anthropocene. Introducing a local reference, Latour argues for cosmopolitan science, 'all science, from agronomy to atmospheric physics, from microbiology to volcanology, from computer models to good, old-fashioned systematics and natural history'. The local reference is that the meteorologist Vilhelm Bjerknes was born in Bergen. Latour's views are also represented positively, in *El Pais* (Spain), with 'It was Not Written That Ecology was a Party: Bruno Latour Has a Bitter and Provocative Perspective on Society And the Environment' (2013). This article outlines his thoughts on the current green political movement, suggesting that the concept of the Anthropocene is very useful for overcoming philosophical blocks to green politics, by making it impossible to separate ecology from human activity. It also briefly mentions the concept of degrowth as a solution in the Anthropocene, as a narrative for possible radical change. Latour's pre-eminence in Anthropocene studies (at least in France) is illustrated on the cover of a recent book (see Figure 12.1).

Timothy Morton, born in the UK, an English (literature) professor in Texas, has attained a level of celebrity, due to the fact that celebrities have been attracted to his writings. He is influential in both social science and environmental humanity discussions of the Anthropocene. Morton argues that the Anthropocene puts an end to the idea that nature is a stable, nonhuman background to (human) history. He affirms that what the Anthropocene actually names is mass extinction, an inconvenient truth for scholars convinced that any hint of talk about reality smacks of reactionary fantasy. A long profile in *The Guardian* (23 August 2017), 'A Reckoning For Our Species: the Philosopher Prophet of the Anthropocene' explores his ideas and their appeal, for example his view that anarchist thought would enhance Marxism.[4] In the Danish paper, *Kristeligt Dagblad* (2017), he is quoted: 'we must stop romanticizing nature …instead, we should dissolve the concept of "nature". The perception of nature as something different, and different from man and culture is the very basis of the plundering of the planet that has

Figure 12.1 *L'Atlas de l'Anthropocene'* (2019) © Sciences Po Presse.

caused the climate catastrophe'. His most-discussed concept is the 'hyper-object', as reported in the *Gulf News* from Dubai in a thorough summary of extinction in the Anthropocene: 'The extinction crisis is so vast and complex that it almost repels thought. It is what the cultural critic Timothy Morton calls a hyper-object' (2014).

The Argentine sociologist Maristella Svampa joins the debate over what has been labelled the Capitalocene, adding a political economy dimension to what are often purely descriptive or philosophical accounts of the Anthropocene. The article 'Key Questions To Understand the World to Come' in the Argentine paper *La Nación* (2016) discusses her ideas about alternatives to development, post extractivism,[5] and the Anthropocene. She is a proponent of the Capitalocene, arguing that this is the greatest debate of the 21st century, both for the nation and for humanity as a whole. The article takes a neutral position on Svampa's radical perspective, representing it fairly but, unsurprisingly, unable to devote the space to explain it fully. This is a problem with most of the material analyzed in this chapter.

William Cronon is a leading environmental historian from the University of Wisconsin–Madison in the United States. In 2013, the environmental journalist Keith Kloor wrote a two-part series on the Anthropocene for the US magazine *Discover*. 'Is the Anthropocene Doomed?' set the scene with 'It's not often that an aging social movement gets a chance to redefine and reinvigorate itself. Environmentalism has that opportunity now, with the Anthropocene, which *National Geographic* has dubbed, *The Age of Man*' neatly summarizing the state of the Anthropocene debate. In the second part, 'What Should the Anthropocene Look Like?' Kloor highlights the work of Cronon, in particular, his 'heretical claim' that the time has come to rethink wilderness (see Cronon 1996). Cronon's complex argument about wilderness and nature itself for the Anthropocene is summarised as follows: 'Wilderness hides its unnaturalness behind a mask that is all-the-more beguiling because it seems so natural. As we gaze into the mirror it holds up for us, we too easily imagine that what we behold is Nature when in fact we see the reflection of our own unexamined longings and desires. For this reason, we mistake ourselves when we suppose that wilderness can be the solution to our culture's problematic relationships with the nonhuman world, for wilderness is itself no small part of the problem'. Kloor concludes: 'We are supposed to be in the business of changing the world ... The question is: Are we? ... That's the million-dollar question the Anthropocene suggests to me'. Cronon, it must be said, rarely mentions the Anthropocene in his work, though it is clear that this is what he is writing about.[6]

Jason Moore is a Marxist sociologist at Binghamton University in the United States. His widely cited edited book *Anthropocene or Capitalocene?* focuses on the capitalist world-system and the Earth System, with several contributors proposing their own alternative '-scenes'. The US-based magazine *Slate*, in 'Why Climate Change Discussions Need Apocalyptic Thinking' (2017), highlights the spread of the term Anthropocene outside university gates into popular magazines, like *Slate* itself. It mentions Moore's ideas on the Capitalocene in the context of a provocative discussion of why 'hope is dangerous', a clear example of climate change as a metonym for the Anthropocene. The Danish daily *Dagbladet Information* mentions Moore's work in the article 'The Anthropocene Ideology' (2016), comparing the term Anthropocene unfavourably to Capitalocene, identifying capitalism as the main reason humanity is facing such an environmental challenge in the new era. This often-contentious discussion continues in the academic literature but is rarely explored seriously in the media.

Donna Haraway (emerita of the University of California, Santa Cruz, PhD in biology) is a major figure in the worlds of feminist theory and science and technology studies. Her many publications, notably ongoing research on 'Staying

With the Trouble' are widely regarded as foundational contributions to post-humanist interpretations of the Anthropocene, which she famously rejects in favour of the Chthulucene, defined as processes of composting, a path towards something that might possibly have a chance of living on, as we collect up the waste of the Anthropocene. It is not surprising that the regular mass media mostly fail to engage with her work in much detail. She is referenced in several publications, including in *The Conversation* (2015) as a proponent of the Capitalocene (not strictly accurate),[7] in the Nicaraguan *La Voz del Sandinismo* (2017), in an extensive review of a book on eco-feminism in the Danish *Information* (2018), and in the Danish *Dagbladet Politiken*, a rather neutral description of her ideas as a way to 'think about what it's like to live in a damaged world'. *Liberation* (2015) locates her in eminent company: 'One can only follow Latour, Haraway or Viveiros Castro, for whom the worst will be avoided only if the "earthlings" (those who know that they belong to the Earth) prevail by their struggles and by the right over the "Modern" (who believe that the Earth belongs to them)'. Finally, and perhaps more significantly, a *New Scientist* report on an Anthropocene-themed arts festival in Norway (2017) cites Fabrizio Terranova's film 'Donna Haraway: Story-Telling For Earthly Survival', with a picture showing her despair at fellow intellectuals, 'mesmerized by the smartness of the latest analysis of capital'.

A debate in 2017 between two Swedish scholars, Sverker Sörlin and Alf Hornborg, appears in the Scandinavian media. Sörlin, an environmental historian at the prestigious Swedish Royal Institute of Technology, published an article in *Aftonbladet*: 'Now a new story is being written about the planet's future', highlighting his idea of 'the Anthropocene of hope', with a focus on democracy as the most fundamental environmental issue. He argues that most of the regulations governing our societies are based on principles from early capitalism, which are completely in conflict with sustainability. Alf Hornborg, a Swedish anthropologist and human ecologist, responds in *Dagens Nyheter*. Hornborg affirms that, although we must try to feel hope, the Anthropocene is not just a condition for experiencing that our lives have meaning. It must also be the moment for an in-depth analysis of the structures that drive hyper-consumption and the consequent concern about the climate change caused by consumption. Hornborg argues for more responsible efforts to build an interdisciplinary approach to the Anthropocene. For this, he emphasizes the need for different perspectives from the South and East, those regions that are often invisible in discussions of the Anthropocene.[8]

The contribution of scholars from India

As indicated in Chapter 7 of this book, three of the most important contributions on the Anthropocene in social sciences and environmental humanities come from scholars who have strong connections with India. These works have resonated all over the world, even in the media. The first and most influential (at least in terms of scholarly citations) is the essay by the Indian historian Dipesh Chakrabarty, whose article 'The Climate of History: Four Theses' (2009) could be considered to be the foundational text for the study of the Anthropocene in the social sciences and environmental humanities.[9] In 2016, two articles citing Chakrabarty in the context of the Anthropocene appear in two leading Indian magazines, *Outlook* and *India Today*. In 'Quo Vadis Earth?' Pranay Sharma, the editor of *Outlook,* describes how the term Anthropocene came into use. The reader will certainly come away with much accurate information about the geological aspects of the Anthropocene, but, with two brief mentions of Chakrabarty in a long and informative article, Sharma hardly engages with his Anthropocene work. However, *Outlook* has regularly published other articles by and on Chakrabarty (not mentioning the Anthropocene but significant in understanding current debates), notably 'In the Waiting-Room of History' (2004), by the novelist and academic Amit Chaudhuri. This is a penetrating and sympathetic critique of Chakrabarty's important book, *Provincializing Europe*, which acknowledges the influence of European thought on both Chakrabarty and Chaudhuri (as Indian intellectuals). The title of the article refers to the 'not yet' interpretation of modernity in Asia (particularly India). Drawing an interesting quasi-parallel between Chakrabarty's 'provincializing' and Edward Said's 'Orientalism' (with no reference to ecological issues), the pre-Anthropocene Chaudhuri introduces an original complexity to what was to follow. *India Today*, 12 years later, publishes 'Civilisation and Madness' by the sociologist Amita Baviskar, illuminating the connections between the historian Chakrabarty and the novelist and eco-critic Amitav Ghosh. Starting with his 'Climate of History', Baviskar explores the impact of Chakrabarty's unexpected intervention into environmental debates on the social sciences and literary studies (especially, by 2016, debates around the Anthropocene).[10] Baviskar argues: 'If social science writing on the environment is a muffled minority confined to a ghetto, the situation of literary fiction is no better'. The latter is one of the three major themes of Ghosh's book *The Great Derangement*, 'a powerful explanation for this collective amnesia'. Baviskar highlights Ghosh's controversial argument that colonialism held back industrialization in India and China, thus reducing fossil fuel-induced ecological damage and, by implication, letting capitalism off the hook. Engaging critically with the willingness of Ghosh to applaud the interventions of Pope Francis and other religious leaders in the politics of climate change, she concludes: 'If this deeply thoughtful book offers that as our only hope, we are doomed indeed'. As a coda to Baviskar's challenging contribution, *India Today* adds an interview with Ghosh. He reflects on 'the beating heart of bourgeois seriousness – because the bourgeois is nothing if not serious. And you realise that the seriousness itself is a kind of derangement'.

The Great Derangement is reviewed at length in *Outlook* as well as in *Daily Asia Age* (2017), *The Telegraph* in India (2017), *Nepali Times* (2017), and *The London Review of Books* (LRB) (2017). In 'Horror Story Of a Climate Calamity' (2017), Kunda Dixit, the publisher of *Nepali Times* argues: 'the key message of his book, about the nature of empire in an age of globalization that is driven by populism and climate denial, is eerily prophetic'. He is less enthusiastic about Gosh's views on why novelists don't write about climate change. Dixit sees the book as already outdated and calls for a new edition. The online Bangladeshi

Daily Asia Age (2017) prints 'Stranger than Fiction' by Siddhartha Deb (who also reviews for *The New York Times*). Deb accepts that 'in his marvellous recent book, *The Great Derangement*', Ghosh is correct to point out the shortcomings of modern fiction in a time of 'climate crisis', but the Anthropocene hardly merits a mention. The critique of colonialism is discussed in a 4,000+ word article in *The London Review of Books* in 'If On a Winter's Night a Cyclone' (2017) by Thomas Jones, editor of the LRB blog. Jones, perhaps inspired by Ghosh's insightful stories of ecology and colonialism in the book and in his novels, adds a few of his own. He quotes Ghosh: 'Who could possibly make a convincing case … for the poor to make sacrifices so that the rich can continue to enjoy the fruits of their wealth?' Jones notes that Ghosh ponders the similarities between New York and Mumbai, the two port cities vulnerable to extreme weather conditions, though the arguments over the extent to which the rich can avoid the worst and who, exactly, is to blame, drags on. Jones concludes that, whether novelists write about climate change or not, will make little difference in the long run. The article in *Asian Age* (2016) 'Like a Train Hurtling Towards a Cliff', by the journalist Kusumita Das, best transmits the sense of urgency in *The Great Derangement*, a sense that readers might find neutralized by the last section of the book (on literature) and some of the other commentary. Das usefully picks out the various responses that Ghosh cites as to why the 'climate crisis' is so often not taken seriously. She quotes him: 'Often the impacts of climate change and man-made impacts are very hard to distinguish from each other. For instance, in Chennai, when we had this terrible deluge, the effects of rainfall were greatly worsened by mismanagement…Politics is more thrilling to talk about. As regards climate change, with many, it's just the feeling that it's too big to do anything about, let's focus on these more local things, where I can blame someone. Who do you blame when it comes to nature?' This review also highlights the importance of individual actions for Ghosh, and consumer choices in general, citing Michael Pollan (intriguingly identified as Professor of Practice of Non-Fiction at Harvard University): 'the single most important thing you can do for the environment is to cook your own food' (we might add that cooking is good, growing and harvesting then cooking is better). Also mentioned here is the usually ignored Jevons effect. Ghosh explains: 'efficiency in energy consumption leads to greater consumption. If you have cheaper light bulbs, you will only buy more light bulbs. Will it solve the problem? It is just a different way of keeping the same system, making it go even faster'.

The writings on the Anthropocene of both Chakrabarty and Ghosh have received extensive, though mixed, treatment in the scholarly literature. For example, Emmett and Lekan (2016) reproduce the text of an international conference on Chakrabarty at the Rachel Carson Centre in Munich; and Boscov-Ellen (2018), from a Marxist perspective, argues that we must move beyond Chakrabarty's apparently contradictory idea that European concepts are 'both indispensable and inadequate' but that that we can learn from Chakrabarty's mistakes. Ghosh has a *Journal of Asian Studies* Round Table (Thomas et al. 2016).

This summary of media coverage of the works of Chakrabarty and Ghosh reveals four things about how the media deal with the complexities of the Anthropocene in the work of Anthropo-scene scholars. First, as usual, climate change is mostly used as a metonym for the Anthropocene.[11] Second, the arguments around colonialism, climate change, and the Anthropocene are rarely clearly reported in ways that would allow robust conclusions about the world today. Third, the Anthropocene is most often mentioned in passing, rarely discussed as a key topic.[12] And finally, there is almost no direct discussion of the issues around provincializing and localizing of the Anthropocene. The only exception to this occurs in *Outlook* (2016) which publishes an interview with the distinguished anthropologist and environmental historian of South India, Kathleen Morrison (University of Pennsylvania). In 'The Anthropocene Concept is Oddly Eurocentric',[13] Morrison explains that by 'oddly Eurocentric' she means that the insistence on points in history as times of major transition are actually '*times of change in western Europe*, not global transitions' (italics added). Models of anthropogenic land cover change (ALCC) tend to be based on European agriculture, which is not always applicable elsewhere, particularly not in Asia, as Morrison and her co-researchers are currently documenting (see Morrison 2015).

Colonialism, 1610, and the Anthropocene

Although largely ignored by the media, Chakrabarty's central idea of 'provincializing' is a hotly contested concept – constructive criticism of this has come from Pernau (2016).[14] Though the media rarely challenge Eurocentrism, we find many indirect critiques of colonialism, especially settler colonization, in coverage of the Anthropocene. This most often occurs in the context of Latin America, with the thesis that the origins of the Anthropocene can be found in what Lewis and Maslin conceptualize as the Columbian Exchange, which attempts to achieve the crossover between Earth System science and socio-environmental history. Media coverage of the Anthropocene usually reports this in terms of 1610 (an alternative start date for the Anthropocene).[15] *Público* with 'Anthropocene: What If We Have Forever Changed the Earth's Geological History?' (2015) explains the importance of colonization of the Americas: 'by 1610, almost all of Eurasia's arable land had already been transformed by agriculture, although this had not yet happened in most of America'. Simon Lewis (not in the Anthropocene Working Group, AWG) argues against the AWG-preferred 1945 start date. It is precisely because everyone is using the Anthropocene concept in an informal way that Maslin says 'clear definitions are needed to understand each other'; *Diário de Notícias* (2015) '1610, the Year That Man Changed the Planet', concludes: 'It is in this context of intense discussion that the group led by Simon Lewis of University College London (UCL) today published in the journal *Nature* its proposal for the beginning of the Anthropocene in 1610. The date was found after the researchers reviewed the last 50,000 years of human occupation on the Earth, to identify environmental episodes with a different anthropogenic mark' (reprinted in *Grande Reportage* in the following year). All these sources are from Portugal, which is not so surprising when we recall that most of the ships bound for the 'new world' left from the Iberian Peninsula. However, the story of 1610 and the article in *Nature* is told by media in all regions, for example in the Russian paper *Lenta* 'For the Earth, the Discovery of America was Considered More Important Than Nuclear Explosions' (2015) – referencing the later theory that the Anthropocene starts with nuclear testing. The article repeats the story that the Anthropocene started in 1610, following the effects of migration of Europeans to America, based on research at UCL that was published in *Nature*. From Bangladesh, *NTV* asks, 'Is the Anthropocene World Different From the Stable Holocene Epoch of the last 11,700 Years That Allowed Human Civilization to Develop?' (2016), adding one more layer to the debate. Jan Zalasiewicz is reported as telling the BBC, 'It adds positively to the overall debate on the Anthropocene, and to the growing number of suggestions about where it should start. ... The 1610 suggestion clearly reflects a historically important event, though it would need more evidence, I think, whether the criteria they suggest would work better than the multiple signals now known to be associated with the mid-20th century Great Acceleration [the basis for the proposed 1945 start date]'; Chile's *La Tercera* offers 'British Research Fixes the Beginning of the Anthropocene in 1610, the Year in Which Man Begins To Change the Earth. Study Reopens Debate on Date of Beginning of New Geologic Era' (2015), revisiting Paul Crutzen's famous spontaneous declaration of the 'Anthropocene' at a conference in Mexico; and finally *Surging News* in China with 'New Earth Geology, "Anthropocene" will likely be accepted next year? (2015), reports that some scientists are saying that it began in 1610. This article draws attention to the geological criteria involved in deciding these issues, noting that the AWG hopes to resolve the question soon. In the United States, *The Minneapolis Post* does not mention 1610 but gives us one of the most detailed and accurate articles we found on the topic in 'From Climate Change to "Technofossils", We're Revising Earth's Geologic History' (2016) by the environmental journalist Ron Meador, who explains: 'Proposals for marking the start of the Anthropocene have included (i) an "early Anthropocene", associated with the advent of agriculture, animal domestication, extensive deforestation, and gradual increases in atmospheric carbon dioxide (CO_2) and methane (CH_4) levels thousands of years ago; (ii) the Columbian Exchange of Old World and New World species associated with colonization of the Americas; (iii) the beginning of the Industrial Revolution at ~1800 CE; and (iv) the mid-20th century "Great Acceleration" of population growth, industrialization, and mineral and energy use'. In 'The Most Intriguing Environmental Stories of 2015', *Scientific American* provides more specific detail: 'A terrible earthquake, massive drought and nuclear power captured the imagination this year... a new proposal pegs the start of the Anthropocene to the little ice age and the Columbian Exchange', and staff writer David Biello, in 'Mass Deaths in Americas Start New CO_2 Epoch' (2015), reports succinctly that placing the Anthropocene at this time [1610] highlights Lewis and Maslin's idea that colonialism, global trade, and the desire for wealth and profits began driving Earth towards a new state. Lewis is quoted: 'We are a geological force of nature, but that power is unlike any other force of nature in that it is reflexive, and can be used, withdrawn, or modified'.[16] It is estimated that, in 1492, between 54 and 61 million people inhabited the Americas; by 1650, the population had declined to around six million (Lewis and Maslin 2018, chap. 5). Reflecting on this catastrophic loss, *BBC News* 'America Colonization Cooled Earth's Climate' (2018) argues 'the Great Dying of the Indigenous Peoples of the Americas led to the abandonment of enough cleared land that the resulting terrestrial carbon uptake had a detectable impact on both atmospheric CO_2 and global surface air temperatures', highlighting the research of Maslin and his colleagues'. Lewis and Maslin's book, *The Human Planet* (2018), is one of the few on the topic to be priced for a mass market.

As noted above, with very few exceptions, mainstream media fail to discuss colonialism directly. However, an increasing number of scholars have started to make connections between indigenous knowledge and the Anthropocene. Some of this has been reported in specialist media. For example, *The National Observer* (Canada) in 'Urgency in Climate Change Advocacy is Backfiring, says Citizen Potawatomi Nation Scientist' (2019) reports the path-breaking research and activism of Kyle Whyte,[17] who argues that the Anthropocene is the deliberate enactment of colonial processes that refuse to acknowledge specific relations between humans, the land, and our other kin. The damming of rivers, clear-cutting of forests, and importation of plants and animals remade the worlds of America into the vision of a displaced Europe, fundamentally altering the climate and ecosystems. Furthermore, the forced displacement that many tribal communities suffer involves adaptation to entirely new environments, to new climates, new ecosystems, new plants and animals. These processes of environmental transformation and forced displacement can be understood as climate change, or more broadly, a preview of what it is like to live under the conditions of the Anthropocene. Speaking for Indigenous Peoples, Whyte is quoted: 'The best ways climate scientists can relate to us and work with us, is to rebuild and strengthen our own knowledge systems'. Whyte is also quoted in the digital indigenous news site *Indian Country Today* (nd), in 'This is not the sixth extinction, it's the first extermination event: What we are witnessing is not a passive geological event but extermination by capitalism, says Justin McBrien', a passionate, uncompromising critique of the destructive role of capitalism in anthropogenic ecosystem degeneration. McBrien argues: 'Focusing on a dystopic future allows the privileged to ignore the dystopic horror that already exists today for a great many people on this planet. As philosopher and environmental activist Kyle Powys Whyte writes, many indigenous peoples have long lived in a dystopian "Anthropocene" –it is here, now, yesterday. They have also long fought an existential war against it' (see also Whyte 2018).

The logical conclusion of this line of reasoning is that, to adequately address climate change and other environmental catastrophes, we also need to seriously think through and enact processes of decolonization. This involves self-governance for Indigenous Peoples, the return of stolen lands, and reparations for the descendants of captured Africans, but it also fundamentally questions the bounds and the legitimacy of the nation-state structure itself. In 2017 the Spanish periodical *Ecologia Política* discusses 'Centenary Ritual [Nego Fugido] in Bahia'

[Brazil], and the struggle for freedom with origins in the slave system in the cane plantations, enhanced by the advance of extractivism in the globalized economy. This article cites the work of two important theorists, Joan Martínez-Alier et al. (2016) and Achile Mbembe (2014), who expose the depoliticizing tendencies of the Anthropocene, especially their generalizing and homogenizing effects (as this and other chapters in this book illustrate, these criticisms are beginning to resonate with most Anthropo-scene scholars, and some Earth System scientists). However, as Mbembe (2014) argues, the article questions whether the violence of colonialism and slavery should be considered as part of the Anthropocene. Research on indigenous knowledge uncovers many ways to consider nature and contributes to recasting global environmental studies in the Anthropocene. Indigenous conceptions of nature vary greatly, as each group has its particular way to conceive nature and understand the relations established with it. However, one can say that there is something in common among all of them, in that the 'natural world' is primarily a wide network of inter-relations between and among agents (human or nonhumans). As noted above (and in Chapter 5 of this book), some media did cover the book by the Brazilians Deborah Danowski and Eduardo Viveiros de Castro (2016), which analyzes the traditions of indigenous groups in Latin America and relates this to current debates around the Anthropocene. For example, *El Pais Brazil* 'From the Anthropocene To the Earth Age' (2014) explores their ideas about the planet and Brazil, and the degradation of life caused by climate change, presenting the possibility that indigenous communities may have valuable lessons about surviving the end of the world.

Religion and ethics in the Anthropocene

We also find in the media in some countries, especially in Latin America, references to the work of theologians who see connections between religion and the Anthropocene. For example, the major Mexican newspaper *El Universal* in 'Boff: the Church is Barely Engaging with Environmental Causes' (2010) profiles Leonardo Boff, a Brazilian priest and one of the founders of Liberation Theology, developed during the 1970s in Latin America. While not referencing the Anthropocene directly, his critique of capitalism and consumerism and his call for an end to the obsession with growth and recognition of the need for radical change (and the role of science) indicates his grasp of the central issues of anthropogenic ecological degradation. In 2014, the Bolivian newspaper *El Diario* prints 'An Emerging Paradigm', a rather abstract article reflecting on humanity. Boff writes: '[humanity can] work beyond the logic of its own interest and in favour of the interests of the weakest beings, as it can also attack nature and destroy species (new era of the anthropocene)'. Boff also appears in *Jornal do Brasil* 'The Future Of the Earth Does Not Fall From the Sky', a warning about the possibility of a Sixth Extinction, and in the Costa Rica edition of *El Pais* (both in 2018) with 'The Geological Age of the Anthropocene Versus the Ecozoic'.[18] Boff's critique of Anthropocentrism also attracts coverage in Latin American media, but not in the context of the Anthropocene. The majority of religion-related Anthropocene stories refer to *Laudato Si*, the encyclical 'on care for our common home' of Pope Francis in 2015, widely seen as a call to action on climate change and the Anthropocene. This is reported in *The Manila Times, The Conversation, MDZ* online (Argentina), *The Vancouver Courier, O Clarim* (Macao), *The Irish Times, Lebanon Daily Star, New Dawn Liberia, Malay Mail* (all in 2015), and in *Excelsior* (Mexico) and *Borneo Bulletin* (both in 2017). Of particular interest (though it does not mention the Anthropocene) is the long, polemical article in *The Manila Times* 'Is the Climate Change "Crisis" a Fraud?' (2015) by the former right-wing Philippine politician and journalist Francisco Tatad. Writing from Washington DC, Tatad acknowledges that many Catholics have been moved by the encyclical: 'But many devout Christians, Catholics faithful to the Church Magisterium, feel Laudato Si's paean to "Mother Earth" as the source of all life is poetry, not theology or science, and a possible slipping toward paganism. As a Catholic layman, I quote this objection boldly without endorsing it; I simply want a clarification'. Associated with the Schiller Institute, an obscure think tank (proponent of various conspiracy theories and anthropogenic climate change denial, with outposts in Washington DC, Germany, and elsewhere), Tatad recommends a report calling the Paris conference 'satanic' and a 'swindle'. This is probably the most uncompromising example of climate change denial we found in all our searches, highlighting the dilemmas that face Catholics (and other religious denominations) when their leaders draw attention to the issues of potential anthropogenic ecosystem collapse.[19]

The Australian ethicist, Clive Hamilton, is one of the most prominent Anthropocene voices in the media, warning of the existential dangers to human survival it poses. He has engaged energetically in several key Anthropocene debates, summed up in his articles in *Nature* and elsewhere, and in his books, notably *Defiant Earth* (2015), where he launches a ferocious attack on 'good' Anthropocene narratives. His own summary of this book in *Scientific American* 'The New Environmentalism Will Lead Us to Disaster. So-Called Ecopragmatists Say We Can Have a "Good Anthropocene"' (2014) raises ethical and moral issues forced on us by the idea of the Anthropocene. His work can be found in media all over the world, for example, in *The Conversation* (2013) 'Climate Change Signals the End of the Social Sciences', challenging the social sciences. In *Conservation Africa* 'The Anthropocene Belongs to Earth System Science' he engages with Earth System scientists on the significance of the Anthropocene. *MIT Technology Review* (2017) hosts a debate between Hamilton and a prominent proponent of geoengineering – Hamilton finds the idea 'chilling'. More mainstream media have also found him newsworthy; *The Belfast Telegraph* 'Scientists to Launch Global Hunt For "Line in the Rock" Making New Man-Made Epoch' (2016) prints his declaration (also in *Nature*): 'Some scientists even write: "Welcome to the Anthropocene". At first, I thought they were being ironic, but now I see they are not. And that's scary. The idea of the Anthropocene is not welcoming. It should frighten us. And scientists should present it as such'. Several sources cite *Defiant Earth* in 2017, including Karachi's *News International* (2017), *The Telegraph* (UK), and *The Guardian* (UK), which prints an extract from the book. Hamilton also features in Andy Revkin's contributions in *The New York Times,* in lively and friendly arguments about the risks of the Anthropocene.[20]

Last, but by no means least, in 2016, *The Guardian* publishes a remarkable, long article 'Generation Anthropocene: How Humans Have Altered the Planet For Ever' (2016) by Robert Macfarlane, the British author of many critically acclaimed books on nature. This newspaper article has attracted its own lengthy, literary Roundtable discussion (Sykes et al. 2017 reprints the article and commentary).

Conclusion

Media coverage of the contributions of social sciences and environmental humanities suggests that the media share many of the same concerns about the prospects of the Anthropocene (and the Anthropo-scene) and the challenges it presents to theory and practice in different parts of the world. Unsurprisingly, mass media offer limited representation of how these scholars deal with the Anthropocene, especially its risks, compared with wider academic media (peer-reviewed journals, books, etc.), although the mass media do rely on scholarly publications for many of their stories. It is difficult to avoid the conclusion that, in the case of the Anthropocene, social scientists and environmental humanities scholars are mainly talking (often arguing) among themselves with little attention paid to their arguments outside their own narrow specialisms, despite regular appeals for inter-disciplinarity.

The perspectives of media in regions from the Global South, especially South Asia and Latin America, tend to be critical about the role of western and northern countries in terms of the contemporary ecological issues faced by humanity, but also point to different possibilities, with divergent perspectives of how life can be regarded on the planet. Whereas media reports on issues around postcolonialism and decolonialization in the context of the Anthropocene are sparse everywhere, we do find some critique of capitalism as a major factor in the deterioration of the environment. This is one of the main results found, that gives some visibility to academic research on the Anthropo-scene by social scientists and environmental humanities scholars attempting to unveil and think through the risks of the ecological challenges we face.

Notes

1 Leslie Sklair also contributed to this chapter.
2 See https://blogs.commons.georgetown.edu/engl-595-spring2016/files/2016/01/LatourAgencyattheTimeoftheAnthropocene.pdf
3 *War and Peace* is referenced in his famous essay on 'agency' (Latour 2014).
4 https://conversations.e-flux.com/t/timothy-morton-the-philosopher-prophet-of-the-anthropocene/6708
5 Hamouchene (2019: 5) explains: 'Extractivism refers to activities that overexploit natural resources, destined particularly for export to world markets. As such, it is not limited to minerals and oil: it extends to productive activities which overexploit land, water, and biodiversity, such as agribusiness, intensive forestry, industrial fish farming, and mass tourism'. See also Kartha et al. (2018).

6 It is notable that, in both articles, Kloor rejects 'horror stories' of the Anthropocene.
7 Haraway (2016: 52-3) levels seven objections to the Anthropocene concept, but concludes: 'in so far as the Capitalocene is told in the idiom of fundamentalist Marxism, with all its trappings of Modernity, Progress, and History, that term is subject to the same or fiercer criticism'.
8 Hornborg (2019) argues that, to deal with ecological crisis, we need to move away from capitalism, especially its conception of money.
9 The charge of sexual harassment made against Chakrabarty at the University of Chicago appears not to have tarnished his scholarly reputation.
10 In addition to the ecological themes in most of his novels, Ghosh published a book on 'petrofiction' in 2005. For details, see the triple review of *The Great Derangement* in *Boundary 2* at https://www.boundary2.org/?s=ghosh.
11 In his review of Ghosh, Jones suggests that climate change is used as a synonym for the Anthropocene.
12 For example, we discovered an article on Ghosh in the *Hindustan Times* only because of a comment by Ghosh that a short story by Kanishk Tharoor ('Elephant at Sea') reminded him of the Anthropocene.
13 There are several earlier versions of this important essay; for its latest incarnation, see Morrison (2018).
14 See Ghosh and Chakrabarty (2002) for a deeply informative pre-Anthropocene discussion, ideas that might suggest what Baldwin (2020) labels the 'racial Anthropocene'.
15 See Lewis and Maslin (2018), especially chap. 5. The year 1610 references the Orbis Spike (ibid.: 318–21), geological and social marker of decisive human impact on the Earth system.
16 The only outright rejection of the Columbian Exchange thesis we found was an article by the Russian scientist Dmitriy Alekseev, in *Moskovskiy Komsomoletz* (2015); see Chapter 8 of this book.
17 Whyte has also been quoted in *Newsweek*, 'Threatened by Climate Change, Native American Tribes to Honour Paris Accord' (2017), and in *The Guardian* 'Fierce Storms and Rising Seas are Causing Existential Angst For Communities in Alaska, Arizona, and Other States On the Front Line Of the Climate Crisis' (2019). Though neither of these articles directly mentions the Anthropocene, the implication is clear enough.
18 On the ecozoic and an interesting alternative timeline for the Anthropocene, see https://ecozoictimes.com/what-is-the-ecozoic/what-does-ecozoic-mean/
19 Limitations of space preclude a more thorough analysis of some interesting coverage of the Anthropocene in the religious press (Christian, Islamic, and Jewish). See also Turner (2017) and Deane-Drummond et al. (2018). Barnett (2015) provides more historical perspective.
20 For Australian media coverage of Hamilton, see Chapter 11 of this book.

References

Barnett, L. (2015) The theology of climate change: Sin as agency in the enlightenment's Anthropocene. *Environmental History* 20: 217–237.

Boscov-Ellen, D. (2018) Whose universalism? Dipesh Chakrabarty and the Anthropocene. *Capitalism Nature Socialism*. https://doi.org/10.1080/10455752.2018.1514060

Chakrabarty, D. (2009) The climate of history: Four theses. *Critical Inquiry* 35: 197–222.

Cronon, W. (1996) The trouble with wilderness: A response. *Environmental History* 1(1): 47–55.

Danowski, D. & E.V. de Castro (2016) *Is There Any World to Come?* Cambridge: Polity Press.

Deane-Drummond, C. et al. eds. (2018) *Religion in the Anthropocene*. Cambridge: Lutterworth Press.

Emmett, R. & T. Lekan eds. (2016) Whose Anthropocene? Revisiting Dipesh Chakrabarty's 'Four Theses'. *Transformations in Environment and Society* 2. (RCC Perspectives, whole issue).

Ghosh, A. & D. Chakrabarty (2002) A correspondence on provincializing Europe. *Radical History Review* 83: 146–172.

Hamilton, C. (2017) *Defiant Earth: The Fate of Humans in the Anthropocene*. Cambridge: Polity Press.

Hamouchene, H. (2019) *Extractivism and Resistance in North Africa*. Amsterdam: Transnational Institute.

Haraway, D.J. (2016) Staying with the trouble: Anthropocene, Capitalocene, chthulucene. In Moore, J. ed. *Anthropocene or Capitalocene? Nature, History, and the Crisis of Capitalism*. Oakland, CA: PM Press, chap. 2.

Hornborg, A. (2019) *Nature, Society, and Justice in the Anthropocene: Unraveling the Money-Energy-Technology Complex*. Cambridge: Cambridge University Press.

Kartha, S. et al. (2018) Whose carbon is burnable? Equity considerations in the allocation of a 'right to extract'. *Climatic Change* 150: 117–129.

Latour, B. (2014) Agency at the time of the Anthropocene. *New Literary History* 45(1): 1–18.

Lewis, S.L. & M.A. Maslin (2018) *The Human Planet: How We Created the Anthropocene*. London: Pelican.

Martínez-Alier, J. et al. (2016) Is there a global environmental justice movement? *The Journal of Peasant Studies* 43(3): 731–755.

Mbembe, A. (2014) *Crítica da razão negra*. Lisboa: Antígona.

Morrison, K.D. (2015) Archaeologies of flow: Water and the landscapes of Southern India past, present, and future. *Journal of Field Archaeology* 40(5): 560–580.

Morrison, K.D. (2018) Provincializing the Anthropocene: Eurocentrism in the earth system. In Cederlöf, G. & M. Rangarajan eds. *At Nature's Edge: The Global Present and Long-Term History*. New Delhi: Oxford University Press, chap. 1.

Pernau, M. (2016) Provincializing concepts: The language of transnational history. *Comparative Studies of South Asia, Africa and the Middle East* 36(3): 483–499.

Sykes, R. et al. (2017) Contemporary studies network: Responding to Robert Macfarlane's 'Generation Anthropocene'. *Open Library of Humanities* 3(1). http://doi.org/10.16995/olh.153

Thomas, J.A. et al. (2016) JAS round table on Amitav Ghosh, the great derangement: Climate change and the unthinkable. *Journal of Asian Studies* 75(4): 929–955.

Turner, B.S. (2017) Ritual, belief and habituation: Religion and religions from the axial age to the Anthropocene. *European Journal of Social Theory* 20(1): 132–145.

Whyte, K. (2019) Too late for indigenous climate justice: Ecological and relational tipping points. *WIREs Clim Change* e603. https://doi.org/10.1002/wcc.603

Whyte, K.P. (2018) Indigenous science (Fiction) for the Anthropocene: Ancestral dystopias and fantasies of climate change crises. *Environment and Planning E: Nature and Space* 1(1–2): 1–18.

13 Media coverage of Anthropocene-related creative arts

Leslie Sklair[1]

Introduction

One of the most surprising findings of the Anthropocene Media Project is the volume and scope of reporting of Anthropocene-related arts events. The largest numbers of these items appear in North American and Western European media, ranging, as a proportion of total results, from around 20% in North America, 15% in Western Europe, and somewhat less elsewhere. As will be seen from the excerpts reproduced below, some of these articles express very definite opinions about the Anthropocene, though most appear neutral as to its consequences. This is almost certainly due to the fact that some art exhibitions (especially in the United States) attracted multiple short announcements in local media. Another issue is one of interpretation, here more so than in the general results, as reporting the arts is probably as likely to reproduce the interpretations of the journalist as the intentions of the artist. This was forcefully brought home to me when I participated in an event in Brighton, supported by Arts Council England, in 2019, in which a group of local artists invited the public to 'talk about the Anthropocene' and visit a multi-media exhibition in which the Anthropocene was interpreted aesthetically in many different ways. When artists in any medium take the Anthropocene as their subject matter, they must confront at least two enormous problems, first, what Timothy Morton calls the 'hyper-object' that it is (Morton 2013),[2] and second, what we can do about it.

References to the Anthropocene in media coverage of the arts display a variety of attitudes. It is usually explained in stock phrases like 'Age of Humans' and 'man's impact on the planet' but there are also more overtly dystopian sentiments implied, as will be seen. A good guide to the historical and performative complexities of trying to express the Anthropocene artistically is provided by Robin and Muir (2015) who introduced the challenging idea of 'Slamming the Anthropocene', illustrated graphically with images of environmental hazards and reviews of events. I begin with the most prominent examples of media coverage of Anthropocene-related arts events and then review a selection of other items, organized by genre.

The Anthropocene in museums

Two major multi-media exhibitions in Germany, ‘The Anthropocene Project’ at the Berlin museum Haus der Kulturen der Welt (HKW), and ‘Welcome to the Anthropocene: The Earth in Our Hands’ in the Deutsches Museum (DM) in Munich, are outstanding examples of how the Anthropocene Working Group (AWG) embraced the arts and how the arts embraced the Anthropocene. It is already documented that HKW worked closely with the AWG, for example, by providing a meeting place for its deliberations. An early item on HKW appears in *Edición Impresa* (Argentina) in 2013 ‘Nature Claims For Justice’, by the German journalist/poet Timo Berger. He interviews Paulo Tavares, urbanist and architect, born in Brazil, and teaching at Goldsmiths College in London.[3] Berger tells us that Tavares was one of the experts invited to HKW in Berlin for an international meeting to debate the future of our planet. *Business Report*, a South African online news site, has an article in 2014 by the Washington-based environmental journalist Seth Borenstein, ‘Anthropocene, the age of humans: we’re changing the Earth. There is no question about that, I’ve seen it from space’ (a version of this also appears in *China Post* in 2016). Borenstein reports that the American Association for the Advancement of Science is hosting an art exhibition, ‘Fossils of the Anthropocene’, and that the AWG ‘ramps up its efforts to change the era’s name with a meeting at a Berlin museum’ (HKW). In 2015, *The Conversation* prints ‘The Anthropocene Has Been Shaped by the Media And Our Digital Lives’ by the communications theorist Jussi Parikka. He writes: ‘the Anthropocene has given new inspiration to contemporary artists’, citing the exhibition ‘Rare Earth’ at the Thyssen-Bornemisza Art Contemporary in Vienna, and in Berlin, ‘HKW recently finished a massive multi-year project on the Anthropocene that curated exhibitions, talks, publications, and a public visibility for a concept that has unified the humanities, earth and climate science’. The *Egyptian Gazette* (no date available) provides links to Anthropocene Curriculum | HKW in Berlin, and to the ongoing multimedia ‘Anthropocene Project’ by Edward Burtynsky and his colleagues (see below). The German daily *Bild Zeitung* covers the HKW exhibition several times, such as a preview in 2012 which reports on the upcoming two-year project by artists and scientists ‘to reflect on human’s forming of nature’, and again in ‘Experts Want To Introduce New Age of the Earth (*Anthropozan*)’ in 2016. In Singapore, *The Straits Times* (2014) picks up the story from Reuters in Berlin with ‘Scientists Prepare For Change Of Epoch to Reflect Humankind’s Deep Impact On the Planet’. This article tells us that the AWG, being in no hurry to make important decisions, will report its conclusions in August 2016 to the geology congress in Cape Town. ‘Berlin’s HKW, Where the Talks Took Place, is Holding an Exhibition Of Related Artworks, Videos, and Seminars’. We are fortunate in having several commentaries on the exhibition by some involved scholars (Swanson et al. 2015:152–54, discusses this and other academic events). An extensive digital archive exists, searching for the Berlin ‘Anthropocene Project HKW’ on the Internet in 2020 bringing up more than 400,000 results. As noted above, the Munich exhibition announced itself with a more direct title, illustrated in the cover of its book (see Figure 13.1).

Figure 13.1 Special exhibition on the Anthropocene at the Deutsches Museum Munich (2014) © Deutsches Museum.

El Clarín, in 'Technology is Our Second Skin' (2014), profiles Helmuth Trischler, a historian at the Deutsches Museum (DM) and director of the prestigious Rachel Carson Centre for Environment and Society in Munich, who discusses androids, and the origins of the Anthropocene in the agricultural revolution thousands of years ago. *Frankfurter Allgemeine Zeitung* (2014) 'We Have Arrived At a New Geological Age … Traces of Man are Everywhere' and *The Sydney Morning Herald* 'Lightning Rod for Science' (2015) both announce the Anthropocene exhibition in the DM. Also in 2015, *The Newcastle Herald* (Australia) reports on DM (dubbed the world's largest museum of science and technology), and comments: 'Having been around for so long, there's a certain tension between its historic exhibitions and those depicting the current state of science. On one hand, vast halls are populated by maritime and airborne craft of the past; on the other, temporary exhibitions such as "Welcome to the Anthropocene"'. *Abend Zeitung* (2013) reports that, for thousands of years, people have been shaping the environment through agriculture and that the special exhibition is dedicated to 'Anthropocene, the Age of Man'. The article tells us that the exhibition gives a sense of how much we humans have changed and adapted our planet in recent millennia and how powerful the influence of technical innovations has become. Visitors are invited to create visions of the future on a field of 1000 paper flowers' (this article is reprinted in 2015). *Süddeutsche Zeitung* profiles the exhibition 'Welcome to the Anthropocene' and explains what kind of technology is shown. 'We live in man's vulnerable age. The Great Exhibition Anthropocene in Munich shows how mankind changes the globe far down to Earth's crust' (2015). Also in 2015, the US-based magazine *Slate* (which covers the Anthropocene regularly) publishes 'Bringing Postnatural History into View', a scholarly article on the domestication of species of plants and animals that mentions DM's 'Welcome to the Anthropocene' in passing. More informatively, the *New Zealand Geographic* (2016) celebrates a long-lost recording of the call of the now-extinct huia bird, 'part of Munich's Anthropocene exhibition helps us appreciate loss in the Anthropocene. But man also changes himself'. In *The Guardian*, Robert Macfarlane cites this poignant story of the huia in his comprehensive article 'The Anthropocene Generation' (2016). He writes: 'The DM in Munich is currently hosting "An Anthropocene Wunderkammer", which it calls "the first major exhibition in the world" to take the Anthropocene as its theme'. Macfarlane argues that 'Anthropocene art is, unsurprisingly, obsessed with loss and disappearance'. *Welt Am Sonntag* is one of the few media sources to mention Berlin and Munich together and put them into a broader context. In a thoughtful article in 2014 'Forget New Year's Eve! A New Era Has Begun', Andreas Rosenfelder, head of the Arts section of the paper, ranges far and wide to give meaning to the Anthropocene, 'the hottest scientific debate these days', referencing the major exhibitions at HKW and DM. Scholarly opinion on these two exhibitions is divided. The conclusion reached by two historians of technology, reviewing the DM exhibition for the leading journal in their field, is highly critical: 'the Anthropocene idea has prompted many scholars and activists to point out the radical environmental injustice of the epoch and to critique the capitalism that has led to it. We see neither in this exhibit. Instead, it offers a relatively benign vision of a changing planet. The change is not pictured as threatening, in spite of being rapid. The exhibit says Welcome to the Anthropocene, not Goodbye to the World You Knew' (Jørgensen and Jørgensen 2016). However, other scholars (for example, insiders Keogh and Möllers 2014 and Robin et al. 2014) present much more positive evaluations of the Berlin and Munich events and, in general, the challenges that museums face to represent the Anthropocene aesthetically and symbolically.

Of the many other museums that hosted Anthropocene exhibitions, two others also receive significant coverage in the media. Calatrava's 'Museum of Tomorrow', whose opening exhibition had strong Anthropocene themes, was labelled in 2016 by the Mexican paper *El Reforma* as 'The New Iconic Building of Rio de Janeiro', also reported in *Extra El Clarin*, *Revista*, *Epoca*, *Le Soir*, and *The Hong Kong Economic Times* (see Chapter 5 of this book). The Smithsonian in Washington made the news when it officially recognized the Anthropocene. The museum official John Kress is quoted by *The Scranton Times-Tribune* (from an AP release reprinted in several other papers in the United States): 'the Smithsonian is embracing the term because … for us it kind of combines the scientific and the cultural in one word', music to the ears of the AWG. The Smithsonian also appears in similar press stories in media as far afield as *Los Andes*, *Borneo Bulletin*, and *The China Post.*

Anthropocene Project: Burtynsky, Baichwal, de Pencier

Perhaps the longest-running arts-related Anthropocene project is the mission of the Canadian photographer Edward Burtynsky to photograph human impacts on the planet, notably images of land, mining, energy, water, and industrialization. *The Houston Chronicle* (2016) reports that 'FotoFest 16' will tackle the world's toughest environmental issues with its 16th Biennale, featuring 34 leading photographers and mixed-media artists from nine countries. Burtynsky is highlighted but we do not learn much about his work, and it is squeezed into a busy programme. *The Guardian* (2016), with 'Edward Burtynsky: The Photographer Finding Art In Rivers Of Toxic Waste', tells us a great deal more, for example, that his 'subversive activism … is now more important than the artistry' and that a new film [the third in a series] called 'THE ANTHROPOCENE: The Human Epoch', which Burtynsky made with the Canadian filmmakers Jennifer Baichwal and Nicholas de Pencier, will investigate a current scientific theory, arguing that we are now living in the Anthropocene epoch, because 'mankind's activities are shaping – and destroying – the Earth's ecosystems'. *Maclean's* magazine in Canada in 'The Unjudging Eye of Ed Burtynsky's Camera' (2016), by Brian Bethune, declares 'One of Canada's finest photographers considers his craft'. This is a glowing assessment of Burtynsky's work, illuminated by spectacular photographs. Burtynsky is quoted: 'How long can we be the top predator acting without concern for all the rest?' Notably, for the first time, Burtynsky is photographing animals, illustrated by an iconic pyre of elephant tusks. *The Guardian* (2017), 'In This New Human Age, the Anthropocene' quotes Burtynsky again: 'We've reached peak everything' and follows up in 2018 with 'The Human Signature: Edward Burtynsky's Anthropocene – In Pictures', more spectacular images online.[4] There was more coverage of the 'Anthropocene' film after its release in Canada in 2018, for example, in *The Ottowa Star* (which had previously given space to climate change deniers). Articles documenting the evolution of the project also appear in *The Toronto Star* and *The Ottawa Citizen.* In *The Citizen,* Burtynsky explains that Jennifer [Baichwal] said, 'Why don't we take this word, Anthropocene, and build a film and book and everything around that? Burtynsky was sceptical ...Every time I spoke to audiences I would say, has anybody heard this word? and maybe four hands would go up. Now it's on the side of buses in Toronto'. In *The Toronto Star*, we find: 'Sprawling Anthropocene Project Shows Humanity's Enormous Impact On the Planet'. The art critic Murray Whyte says: 'Burtynsky's pictures have always held a terrible beauty. His compositions veer close to the abstract in their capturing of horrendous damage ... They're gorgeous first, horrendous later, and that's surely the point'. De Pencier is quoted: 'Someone called us the three horsemen of the apocalypse ... I really hope that's not the case. But we can't claim neutrality anymore. We used to say this is not a polemic, and you can draw your own conclusions'. Whyte (who is not entirely uncritical of some aspects of the exhibition) continues: 'For Burtynsky, who has worked along thematic lines tied to specific industries—railway lines carved into the landscape; mines; farming; urban sprawl– *Anthropocene* presents as an unintended catch-all term for his entire career'. This is a thought-provoking conclusion that might also fit many environmental humanities scholars. *The Daily Mail* (UK) in 2019, with 'Tyre Towers, Mesmerizing Mines and Deforestation', enthuses on Burtynsky's amazing aerial photos, showing how humans reshaped the planet'.[5] The 'Anthropocene' exhibition in the National Gallery of Art in Ottawa and the Art Gallery of Ontario in 2018 gave visitors an opportunity to express, in a word on a digital board, how the exhibition made them feel: 23% felt worried, 17% angry, 9% unconcerned or suspicious. It is difficult to keep track of the many voyages of this multimedia project. 'The Anthropocene' film premiered at the prestigious Toronto International Film Festival and has been screened at many theatres and galleries across Canada, the US, France, Italy, the UK, and elsewhere.[6] The project (film, photographs, and installations) has travelled around North America, Flowers Gallery in London, and to Bologna in 2019. A complementary coffee table book was published by Steidl (enhanced digitally), with commentary from two AWG scientists (Jan Zalasiewicz and Colin Waters) and new work from Margaret Atwood, and another serves as a catalogue for the exhibition.[7]

La Repubblica: Anthropocene and the arts in the Italian media

A distinctive feature of coverage in the Italian media (over 100 items found) was the prominence of Anthropocene-related arts events in the centre-left daily *La Repubblica* (27 of 46 items found). We find items covering many art forms. 'One Hundred Films From All Over the World' (2005), reports on the Ecovision film festival in Rome. The Anthropocene reference is for Nobel prize-winner

Paul Crutzen, famous for his studies on the ozone layer, presenting his new book. The title of the event is 'Welcome to the Anthropocene'. In 2013, the first in a series of Anthropocene-related musical events is announced, 'Leu Music in Casalborgone' reporting, rather obscurely, 'hypothetical finding of a black box dispersed in the Anthropocene that brings to light extinct sounds, images of lost places, voices of daily life and excerpts of conversations'. Several articles announce events involving the multi-media musician Marco Paolini (one of whose concerts, 'The Anthropocene' at the Roma Europa festival in 2017, is available on YouTube), and at a music festival in Barcelona, 'Dance and Sounds of the Earth: At the Sonar You Dance the Era of Anthropocene' (2016). The reporter wryly observes: 'Welcome to the era of Anthropocene. The sentence was not written on any door or access gate here at the Sonar in Barcelona, yet'; an 'Anthropocene' recital by the eminent Russian pianist, Arkadij Volodos, at the Carlo Felice theatre in Genova is announced (2016); 'Giraldi Between Man and Nature' (2017) is a sympathetic review of the exhibition 'cosmetics of the Anthropocene' featuring Piero Giraldi, one of Italy's leading politically engaged artists, his 'nature rugs' attracting great critical interest. Anthropocene-related museum events also feature in the coverage in *La Repubblica,* notably a wide-ranging review of a major exhibition 'Maxxi in Rome, Art And Science Of the Cosmos' in 2017. The catalogue explains: 'From gravitational wave detectors to an installation on the world after the Anthropocene: on display are instruments and works by scientists and artists. Closer than you think'. A subsequent event in 2018: 'Gravity. Imagine the universe after Einstein' is previewed, 'the Cosmic Concert' by Tomàs Saraceno features an enormous image of a spider that moves on the canvas to reflect the texture of space.[8] The curator observes that art and science are closer realities than you think. Saraceno, for example, was the first artist with access to the CERN laboratories, and his work in the Maxxi museum on *Aeroke* (reflecting spheres suspended in the air, surrounding the Cassini antenna) represents how the artist imagines the world after the Anthropocene. The final selection from the Anthropocene arts archive in *La Repubblica* reports the last stage of an exhibition at the science museum in Trento. 'The Ecosystem And Its life cycles' (2016) is dedicated to the Anthropocene, apparently labelled 'Extinctions. Stories of Catastrophes and Other Opportunities'. The following section is a small selection from a large archive, organized by genre.

Genres

Art exhibitions

The third 'End of the World Biennale' in the city of Ushuaia, in southern Argentina, was reported in *El Universal* (Mexico) and in the Argentine papers *Ambito Financiero*, and *La Prensa* in 2011. Works by 100 artists from all five continents were exhibited, inspired by the idea of the Anthropocene. Despite the ambiguous title (Ushuaia markets itself to tourists as the city at the end of the world) the consensus is that the exhibition is a great opportunity to demonstrate that another world is possible if compromise and common sense prevail. The article concludes that the Anthropocene concept is not just gaining attention in the scientific spheres, but also within artistic circles. The growing number of art exhibitions related to the topic might be showing the beginning of a social process, where arts help to pose new questions about our common future'.[9] *Hangzhou Daily* 'Taipei Biennale, There is a Big World in the Small Play' (2014), reports that the theme is 'drama and acceleration', and the proposition is 'Anthropocene'. The works of 52 artists from around the world are on display here. The Luxembourg tabloid *Le Quotidien*, 'Casino: Activities Continue During Building Works' (2016), tells us that 27 young artists from Strasbourg exhibit their works on the theme of the anthropocène, or rather on its effects as a principal actor in climate change. *Daily Star* (Bangladesh, 2016) 'Fragments of the Anthropocene: Bengal Art Lounge' describes the works of two artists from Dhaka, from their exhibition about the loss which the onset of the Anthropocene has caused. In 2012, *The Boston Globe* prints several items on the artist known as Swoon, in front of her installation 'Anthropocene Extinction' at the prestigious Boston Institute of Contemporary Art. The artist is pictured speaking at the museum's Teen Night with some Boston high school students. In Argentina *Página 12* 'Nature: Refuge and Resource for Humans' (2017) briefly summarises an art exhibition including 'the origin of an idea, the Anthropocene' along with utopias-dystopias, sustainability, ecological engagement, activism, ecosystems, new worlds, and new natures. The item concludes: 'More evidence on the growing number of art exhibitions and events placing the Anthropocene as a central axis of curatorial projects'. Finally, in Scotland, *The Greenock Telegraph*, reports 'Grove Park Hosting Green Residency This Weekend' (2016), where eight Scotland-based artists will spend a weekend looking at the connections between the arts and a more environmentally sustainable society. The event is the third annual artist residency organised by Creative Carbon Scotland (CCS). CCS director Ben Twist is quoted: 'The Anthropocene marks a significant shift in how we understand humanity's role and responsibility in relation to issues such as climate change ... Exploring this topic during the residency offers an important opportunity for Scotland's artists to consider how their practices could contribute to a wider cultural movement towards a low-carbon future'. Also quoted is Lex Braak, director of the Van Eyck Institute in Maastricht: 'I am sure my understandings of the Anthropocene will change and deepen from the engagement. Residencies are wonderful settings in which to create shared understanding across disciplines and practice settings'.

In addition to the images reproduced in the preceding chapters, many other images express rather different aspects of the Anthropocene. Robin Wood represents in dramatic (perhaps horrific) terms human impacts on other sentient animals (Figure 13.2) and Barcelona-based street artist Pejac symbolizes the sheer waste of the Anthropocene, as our home planet literally 'goes down the drain' (Figure 13.3).

Figure 13.2 Destroying nature is destroying life (2018) © Robin Wood/ Illusion CGI Studio.

Photography

In 2017, *The Washington Post* reported 'Climate Change Exhibit to Open Near Trump's House', profiling the work of photographer Justin Brice Guariglia, who brings the catastrophic results of climate change to Trump's own doorstep, in Florida. The exhibition 'Earth Works: Mapping the Anthropocene' was covered all over the world, for example in *Egypt Today* and Taiwan's *United Daily Times*, featuring images of melting glaciers that Guariglia collected while on a NASA mission to Greenland. *The New York Times* profiles his work in 'A Man on an Eco-Mission in Mixed Media' (2017). He tells the *Art Newspaper*: 'One of the unique things an artist can do is to help shape realities and, at the same time, have a political and social message. Artists today really have to use art to carry an important message forward'. This is powerfully illustrated with reference to 'Capitalist Ruins' in Figure 13.4. What is distinctive about Guariglia's art is its blend of photographic vision and the use of new technology to create an original aesthetic.[10]

The Canadian magazine *Maclean's*, with 'Postcards From the Edge Of the Earth' (2016) reviews a photography exhibition, chronicling the environmental movement from the 1960s to the present. The curator says that the title suggests falling off a cliff, perhaps one that mankind blasted itself. 'Maybe human beings are not the strongest species on Earth. It's something to say, beware'. *La Crosse Tribune* (Wisconsin), 'UW-L Photographer's Winona Exhibit Comments On a Precarious Environmental Future' (2016), reviews 'Anthropocene', pictures

Figure 13.3 Stain, Spain, Santander (2011) © Pejac @pejac art.

of mass-produced disposable goods such as candy and cheap trinkets. The artist explains: 'the aesthetic conveys the turmoil of an inevitable ending that we are aware of on some level, and that breeds a very complex anxiety'. *National Business Review* of *New Zealand* reports how a collection of photographs of the industrial past is transformed by Chris Corson-Scott into an exhibition, 'Dreaming in the Anthropocene' (2017).[11]

Figure 13.4 Capitalist Ruins, acrylic, one of four high-density polystyrene panels, epoxy, hand-carved text (2020) © Justin Brice Guariglia.

Music

In *The Seattle Times* 'Say Goodbye To 2016 With a Song' surveys the year in music with reference to Nick Cave's song 'Anthrocene', which 'felt appropriate because it wraps up the loss felt this year and the severe impact of humans on the environment'. The writer explains that the word 'Anthrocene' was first used, instead of the now more common Anthropocene, in 1992 by science writer Andrew Revkin in his book *Global Warming* to describe a 'geological era significantly impacted by humans'. In England, *The Oxford Mail*, 'Death of Hi-Fi Create a "Squall" of Sound' (2012), profiles a local group and their new album, 'Anthropocene'. Lead vocalist Andy says: 'There's a beauty in darkness in some ways and I get a lot of solace from it. I was probably the only person to read Cormac McCarthy's '*The Road*' and find it uplifting'. The new album 'Miss Anthropocene', by the enigmatic Canadian pop star Grimes, is reviewed widely in the media in 2020.[12] In the Danish *Politiken* 'Classical Music In Antropocæn time' (2016) connects the "Voice of the Whale" from 1971, a key work in environmentally conscious composition music' with the Anthropocene. Finally, several media in the US report that the Norton museum in Florida had hosted musical events inspired by the photography of Justin Brice Guariglia (see above).

Opera

DV (Iceland, 2016) reports the contemporary opera 'UR' by Icelandic composer Anna Þorvaldsdóttir, based on the idea that Man considers himself above the environment, affecting the Earth in an unprecedented way. She is quoted: 'Because of this, theories have emerged that have us living in the Anthropocene [*mannöld*], an epoch of human impact in Earth history. This can maybe be seen most clearly in places where people have lived for thousands of years in a very close relation to the harsh natural forces, like on Greenland, where the concept work for the opera was partially carried out. The glaciers are melting, the oceans are acidifying, and the livelihood of animals and human societies is crumbling'. In 2017, *La Repubblica* revisits the musician Marco Paolini, reporting on the programme for the new opera season: opening with Wagner 'we close with an extraordinary appointment [*appuntamento*]', Paolini's '#Anthropocene'. The Scottish Opera company production 'Anthropocene' featuring an unfrozen ice maiden was reviewed by several Scottish newspapers in 2018.

Dance

Media reporting of Anthropocene-related dance is generally uninformative, usually listing title of performances but not their content. The Bangladesh *Daily Sun* (2016) reports on the Jahangirnagar University dance troupe, and its eleven-day-long festival to celebrate three decades of its journey. The Anthropocene connection is obscure. *Salzburger Nachrichten* (2016) advertises 'Anthropozän', an urban dance performance, based on the influence of man on the Earth, while *Süddeutsche Zeitung* (2017) 'In the Crosshairs Of the Gods' reviews a ballet that deals with the end of humans on Earth and their self-destruction as a metaphor for the Anthropocene. In Australia, *The Canberra Times*, 'Art Gets in On the act of climate debate' (2009) reports on a climate change and sustainability forum and the performance of Mirramu Dance Company's 'River', implicating the Anthropocene in the drying up of the Murray River. An optimistic activist is quoted: 'It is fixable ... Clear vision is more important than deep anger'.

Sculpture

Grapevine, the Icelandic English-language weekly and online news site, profiles the sculptor Anna Rún Tryggvadóttir in 'Everything is in Motion, Nothing is Solid' (2017), a long and especially instructive article on artistic practice in the Anthropocene by the Berlin-based writer and performer, Steindór Grétar Jónsson. Tryggvadótti is also based in Berlin, her exhibition 'Garður' (Garden) is at the Reykjavík Art Museum. The main components are Icelandic rocks the artist sculpts and treats. Suspended above these rocks, containers of coloured liquid are rigged to drip every hour, watering them and changing their appearance. Slowly, they transform both in colour and in texture, as chemical compounds on the rocks, like salt, react to the stimuli. Her central philosophy is rooted in the state of absolute authority over a situation, while being completely powerless to control it at the same time. The situation of the artist, she says, involves 'how we position ourselves toward nature and how society constantly demands that we encroach upon it ... It's a twisted spot we're in. Scientists are calling this era the anthropocene—this period of time from when mankind started to significantly impact the world. We tend to define everything from the perspective of humans and how we understand and perceive things. I try to mirror our function down to a material or even a molecular level, through the use of matter we consider as being inanimate'. Her techniques inevitably lead to the artwork morphing from day one up until the final day of the exhibition. Reflecting on her continued connection with Iceland, she explains, 'Even though Icelanders are raised in a very European society, we face wild and crazy weather and unstable elements, it's so different from living in a big city, a manufactured area for human existence, built to service our needs. We're at a point in human history where we have to balance this with the idea of conserving the resources needed to create this kind of reality. In Reykjavík, you can look around and see nature, the mountains, the ocean, almost 360 degrees. In Berlin, I don't have any of this. Being here has stirred something in my subconscious'. This article is quoted at such length because it is almost unique in showing how the work is manifestly influenced by and centred in the Anthropocene. *The State Journal-Register* (Springfield, Illinois, 2014) is one of many media to report on the art of the British sculptor, Jason de Caires Taylor: 'The only way to see this incredible museum in Mexico is by scuba diving'. The installation is of 500 underwater sculptures off the coast of Mexico's Riviera Maya, via the Museo Subacuático de Arte. One exhibit is called 'Anthropocene' a sculpture of a Volkswagen car made to house lobsters.[13]

Theatre

Le Soir (France, 2017) gives few details in 'The Yes Men: Spreading Some Joy At la Fabrique de Théâtre à Frameries, Sponsoring a Studio Dedicated To the Anthropocene'. *Süddeutsche Zeitung* (2017), under the rather obscure title 'This is the Molecular Genetics', profiles the German dramatist Konstantin Kuspert, who is known for his dystopian (and sometimes utopian) plays and controversial topics. Kuspert talks about the Anthropocene, saying: 'it is outrageous how it is exploited by the media but there seems to be no change in people's mentality'. *Canberra Times* (2016) reviews playwright Hannie Rayson's 'Extinction', explaining 'The sixth great dying – the so-called Anthropocene extinction – is the one we're living through now. It covers the loss of biodiversity since humans emerged as a super-predator, and, although scientific consensus on key questions remains elusive, there's broad agreement our ecological impact on the planet is global and severe'. But nothing, the reviewer concludes, 'offers relief from the playwright's thin take on a problem thick with dramatic potential'. This dramatic potential appears more realized in *The Calgary Herald* review of 'Extremophiles' (2018), Georgina Beaty's forward-looking apocalyptic and absurd solo show, inspired (if that is the word) by resource extraction in Alberta, set in a desertified and abandoned Canadian North in the year 2020. She is quoted: 'I am fascinated by the Anthropocene Age, the time in which we are living, where humans have permanently altered the earth's natural systems. Climate change is an unavoidable undertone in my work'. Once again, we see climate change as a metonym for the Anthropocene, understandable in the context of Alberta's tar sands and wildfires, localizing the Anthropocene. Also from Canada, *The Edmonton Journal* 'Swallow Returns To Edmonton To Ask Hard Questions About Our Future' (2016), gives us a lengthy report on the Frente (as in 'out in front') Theatre Collective. Playwright Leslea Kroll's play, originally conceived as a comedy for a melting globe, chronicles the demise of a fictional island entirely swallowed up by the rising ocean. Kroll's sense of urgency is intensified on a day when the city was full of Fort McMurray evacuees and wildfire smoke. The Portuguese magazine *Sabado* in the curiously entitled 'Your Majesties, Welcome To the Anthropocene' (2015), reviews a theatrical performance, 'Talk to the Demon', in which Wim Vandekeybus works with a team of six performers and a child. Vandekeybus explains: 'We humans are desperately busy with giving sense to everything that happens between two mysteries – birth and death ... We all create demons inside ourselves to fight out our inner struggles. Good can't exist without bad'. The reviewer says 'Talk to the Demon is a piece without music. In that sense it's very scary, very naked, but at the same time also pure and touching'.

Poetry

In the South African media platform, *Netwerke 24*, we find 'T. T. Cloete's Latest Collection is a Fascinating Study of the Relationship Between God, Human and Nature' (2014) by Andries Visagie (Professor in Afrikaans at the University of Pretoria). This is a poetry review, interesting because it is an early reference to the Anthropocene in Afrikaans literature. The work is an attempt to reconcile science and religion, using nature as a lens through which to experience humanity's effect on the world, 'seeking meaningful ways to live their faith in the world'. The section on the Anthropocene outlines an oblivious image of modern humans, who have to sacrifice quality of life because of its rapid growth and its insatiable urge for consumption'. *Smoky Mountain News* (Waynesville, North Carolina) publishes 'Book of Poetry Has a Disturbing Beauty' (2017), a thoughtful review of 'Anthropocene Blues' by local author John Lane, 'these poems are told from the perspective of a traveling geologist as he sings the blues of our epoch… about how we humans are taking ourselves down along with everything around us. This is not an apocalyptic book about the collapse of human existence on the planet, but it is a wakeup call, which has been the role of the poet for many millennia, as the harbinger of good and bad news'. Local landscapes feature strongly in these poetic reflections, echoing the 'localizing of the Anthropocene'.

The year 2016 was a bumper year for poetry in the Anthropocene, beginning in *The Telegraph* (London) with a review of 'Poetry and the Anthropocene: Ecology, Biology, and Technology in Contemporary British and Irish poetry' by Sam Solnick (an academic at University of Liverpool): 'This book asks what it means to write poetry in and about the Anthropocene, and argues that the Anthropocene requires poetry and the humanities to find new ways of thinking about unfamiliar spatial and temporal scales, about how we approach the metaphors and discourses of the sciences, and about the role of those processes and materials that confound humans' attempts to control or even conceptualise them'. In 'Poems Help Comprehend the Incomprehensible' the popular Danish newspaper *Politiken* argues that 'Poetry in the Antropocæne [sic] era deals with both dystopian and utopian dreams, with future disasters and inhumane scale conditions – in short, everything that science can describe, but does not make sense'. *Süddeutsche Zeitung* brings the anthology 'Poetry in the Anthropocene' edited by Anja Bayer to the attention of its readers. The *Vancouver Observer* opens its pages to the local academic and poet Stephen Collis, in 'Poetry Against the Pipelines'. Collis eloquently explains the relevance of poetry to real life, 'I've written about the desire for change in a world that is already rapidly changing – in ways that alarm us, or at least should alarm us'. He discusses the curious fact that his poetry was used as 'evidence' in an Air India terrorism trial, and his book of poetry about walking in the Anthropocene, 'along pipeline routes and proposed pipeline routes, and the strange and threatening paths they cut through our environments' notably the Alberta tar sands, described as the construction of a vast desert in the middle of the boreal forest. He speaks about the power of 'poetic documentary … mostly I find poetry is best at digging into motivations, into the feelings driving us to act and resist'. *Al-Iraqiya* tells us about an American poetry magazine's latest issue focused on climate change, mentioning Craig Santos Perez's poem 'Halloween In the Anthropocene' (another example of the way in which climate change is used as a metonym for the Anthropocene).

Cli-fi (climate fiction)

Although it is stretching the bounds of 'mass media' again, an article in *Conversation Africa* 'Extinction Or Survival: How Storytellers Explore the Ethics of Colonizing Other Planets' (2017) by Siobhan Lyons of Macquarie University in Australia, provides a splendid introduction to Anthropocene-related cli-fi. She begins: 'Interplanetary colonization was once the stuff of science fiction but now there are plans to colonize Mars. How have film makers and writers dealt with our rapacious Anthropocene age?' Lyons cites Kim Stanley Robinson's influential *Mars Trilogy*, Andy Weir's novel (and subsequent film) 'The Martian', and the film 'Interstellar' (see below), showing how these fictions raise questions about human survival, making an interesting comparison between the 1951 version of the film 'The Day the Earth Stood Still' and the 2008 version, the earlier focused on the nuclear threat, the later remake in 2008 on environmental catastrophe. Lyons argues that the book and film of Shute's *On the Beach* and Ballard's *The Drowned World* (and many others) pose questions about whether humanity deserves to survive, remarking that Woody Allen's 'Ozymandias Melancholia … the artist's job is not to succumb to despair, but to find an antidote for the emptiness of existence' sounds apt for the Anthropocene. Lyons concludes: 'But when the prospect of human survival intrudes upon the natural environment of other planets, which would be best left alone, the idea of colonizing other planets becomes unethical'.[14] *Seattle Post-Intelligencer* (2016), in a review of *Barkskins* by Annie Proulx, proclaims: 'And if we're lucky enough to survive the Anthropocene we've seemingly wrought, then *Barkskins* will surely survive as the crowning achievement of Proulx's distinguished career, but also as perhaps the greatest environmental novel ever written'. In *Times of Israel* 'Literary Critics Probing the Rise of Cli-Fi Novels in the 21st Century. Cli-Fi, Sci-Fi, We All Cry, The End is Nigh: What Cli-Fi Novels Say About the Anthropocene' (2017), the writer suggests that Adam Trexler (2015) led the way with his book *Anthropocene Fictions*, a sentiment most scholars share.

Film

There are two distinct types of Anthropocene films. The first comprises documentary films made explicitly to explain the Anthropocene to various publics, for example, the short but much referenced 'Welcome to the Anthropocene' for Rio +20, the longer 'Welcome to the Anthropocene' by Steve Bradshaw, the Australian Broadcasting Corporation documentary 'The Anthropocene: The age of mankind'[15] and Burtynsky's 'Anthropocene' discussed above. The best-known example of this genre is probably Louie Psihoyos' award-winning Discovery Channel documentary 'Racing Extinction', reviewed in the Argentine *La Nación,* the Bengali magazine *Outlook, La Tercera* in Chile, Pakistan's *News International*, Portugal's *Grande Reportage, Seattle Times*, and other media. The detailed evaluation in the *Cornell Chronicle* (the university paper) concludes that the film 'offers a series of recommendations about what we as individuals can do to try to slow down the tidal wave of extinction, whether it's skipping meat for a day or switching to a greener source of energy. Science is important but it's not the answer. The answer lies in human behaviour'.[16] Some feature films make little or no explicit reference to the Anthropocene, but their reception has been framed as 'Anthropocene' events.[17] The impact of Roland Emmerich's legendary 'Day after Tomorrow' has been analyzed in detail, the catastrophe is a new ice age brought about by global warming (Reusswig and Leiserowitz 2005, Hart and Leiserowitz 2009).[18] More recently, Christopher Nolan's 2014 film, 'Interstellar' raises issues of an increasingly uninhabitable planet and the search for a more liveable one (see Dickens 2015, Powell 2015, Andersen and Nielson 2018).[19] If it ever gets made, the film of Robinson's 'Mars Trilogy' will surely join this list.

Multimedia

The Lebanon Daily Star, 'The Aesthetics of extinction' (2016), reports an exhibition 'Let's Talk About the Weather' at the Sursock Museum in Beirut, including works by 17 artists from Lebanon and beyond, supplemented by a public programme of talks, workshops, films, and guided tours of toxic sites in Beirut, plus 'Elements for the World', a series of five English- and Arabic-language publications on Stone, Water, Wood, Fire, and Sky edited by Ashkan Sepahvand. Noting that most of the work linking the Anthropocene to contemporary art comes from the north, the curators, Nataša Petrešin-Bachelez and Nora Razian, see this show as speaking from the global south. 'The Touchstone For These Labors is the Anthropocene'.[20] The daily *Seattle Times* (a socially conservative newspaper) has six Anthropocene-related articles, five of which are cultural reviews, with little substantive discussion of the Anthropocene. An exception is 'Bloom and Doom: Pokémon Go-Like Art Project Explores Climate Change' (2016), reviewing 'Gardens of the Anthropocene' by Tamiko Thiel who, with help from University of Washington scientists, creates augmented-reality art, linking to articles on 'The Glaciers are Melting. Sea Levels are Rising' and 'Seattle's Summer Of 2016 is Breaking Heat-Wave Records'. The journalist remarks that this 'makes it a grimly perfect time' to visit the exhibition, also listed in the *Seattle Post-Intelligencer*. The Danish *Kristeligt Dagblad* prints a very long article on a very broad event, 'Nature and the Antropocæne man' (2017), part of the multi-media ARoS museum Triennale, a four-km long art zone along the Gulf of Aarhus, illustrating the human relationship to nature. *The Star Tribune* of Minnesota prints what is probably the weirdest title we found, 'What If You Could Understand Climate Models Better By Eating Them?' (2016), on the work of Jonathon Keats, experimental philosopher and artist, who proposes 'squishing all the continents together' into 'anthropocenic sorbets'. He calls the process 'data gastronification'. *Extra el Clarin* (2017) adds the ideas of Dr. Joachim Krausse (Humboldt University, Berlin) on anthropocene cuisine. The Scottish magazine *Skinny* reports the collaboration between Universidad Nacional de San Argentina and C-EENRG at University of Cambridge, a multimedia experience of Earth's rotation in real time, showing 'the new geological epoch that we call Anthropocene fleshing out the unprecedented humankind footprint on our planet' (part of the Edinburgh Fringe festival). The *Skinny* also reports the Anthropocene-themed exhibition 'Grike' at the Fringe. In an unusual case of contrarianism, *The Times-Picayune* (New Orleans) reports 'Ecozoic exhibition explores our connection to nature through plants, seeds, soil' (2016), explaining that the curator of the exhibition, Bobbe Besold, makes a case for the term Ecozoic Era as opposed to Anthropocene epoch. *The Houston Chronicle*, 'These Bizarre And Beautiful Persian Rug-Style Images are Made From Google Maps' (2013), explains how the artist David Thomas Smith is putting on an exhibition called 'Anthropocene'. *Washington City Paper* reports that local artist Rachel Schmidt is calling her exhibition 'Daydreams in the Anthropocene'. The government-owned paper *Rossiyskaya Gazeta* reports, 'Intermuseum-2019 Festival Opens in Manege' (2019), a major theatre, multi-site, multi-genre celebration, including 'Welcome to the Anthropocene!' in the Perm Museum of Local Lore. *The Herald* (2015) 'Ten Things To Do In Dublin This Weekend' picks the 'Riddle Of the Burial Grounds', a free, international exhibition motivated by one of the major problems facing our planet – the markings and warnings around nuclear burial sites. The journalist comments: 'Did you know that mankind is entering the 'Anthropocene' era? Nope, me neither. I guess we'll just have to allow the visual arts to educate us further, and Riddle of the Burial Grounds has it covered via the mediums of sculpture, film, photography, documentary, science-fiction, and imagined futures'. And finally, s*airblog.files* (2017) gives us the impressions of the director of a Japanese art and science interdisciplinarity initiative, who reports that what appeared most different upon arriving in Europe was the difficulty of finding an art exhibit or academic event where the 'Anthropocene' (人新世) did not figure, a fitting conclusion to this cornucopia of artists of all sorts striving to express the Anthropocene aesthetically.

Notes

1 Sofia Avila, Medine Duvarci, and Viviane Reigel also contributed to this chapter. More arts-related Anthropocene content appears in several of the regional chapters in Part Two.

2 It is not entirely clear if the Anthropocene or global warming etc. (metonym) is the hyper-object. See Boulton (2016) for what might be called a 'hyper-critique' of Morton.

3 Tavares (2014) indicates the interdisciplinary nature of his work, testifying to the interest of HKW in crossing science–arts–humanities borders. See also Anderson (2015) and Sklair (2020).

4 For another enthusiastic review and more spectacular photographs, see the *British Journal of Photography,* https://www.bjp-online.com/2018/10/edward-burtynsky-the-anthropocene-project/. For a perceptive anthropological perspective on the exhibition, see Evans and Ryan (2019).

5 It has been argued that the use of drones and satellites evident in Burtynsky's more recent work, while adding to the dramatic effect, could compromise the artistic process (Khatchadourian 2016).

6 I have not seen it myself, but my nephew Lewis Tritchler (an engineer by training) sent me these comments: 'The film showing was at a full house at the Centre for Contemporary Arts (CCA) in Glasgow. The intention of the film was to focus on the 'massive re-engineering of the surface of our planet', and it achieved this to great effect. The irony wasn't lost in the building of sea wall defences along China's shore against rising sea levels, using massive concrete barriers to protect oil production. New terms mentioned in the film that I hadn't heard of before included "technofossils", "technosphere", and "ocean acidification"'.

7 For an up-to-date digest of press coverage of Burtynsky's work, see https://www.edwardburtynsky.com/news and for an interview and film trailer, https://www.youtube.com/watch?v=594iE6hHvQA, where he deals with the tricky question of the 'good' Anthropocene. I am grateful to Alanna Joanne at the Burtynsky studio for her clarifications.

8 I saw this exhibition at the Maxxi in Rome; despite its many spectacular images the Anthropocene theme was muted, as is probably the case for other exhibitions.

9 The Anthropocene-related exhibition in Barcelona 'After the End Of the World', was reviewed in *La Vanguardia* (in Catalan), 'The CCCB invites you to experience climate change and imagine the future world' (2017).

10 An essay by the curator (https://www.guariglia.com/earthworks-mapping-anthropocene) explains the unique qualities of Guariglia's work. Like many artists, he has been inspired by Greta Thunberg, see https://news.artnet.com/art-world/greta-thunberg-climate-art-1645336

11 On photography after the Anthropocene, see Kember (2017)

12 Wodak (2018) provides an acute analysis of climate change and Anthropocene narratives in popular music.

13 A picture of this adorns the cover of Heise et al. (2017).
14 There is a flourishing scholarly literature on cli-fi, notably Johns-Putra (2010 and 2016), and Milner (2018: part 3).
15 First broadcast on ABC in 2016, a longer version from VPRO has been available since 2017 at https://www.youtube.com/watch?v=AW138ZTKioM.
16 For a penetrating critique of 'Racing Extinction' (and Naomi Klein's documentary 'This Changes Everything') as 'eco-opportunist' (see Truscello 2018).
17 Salmose (2018) labels this 'Hollywood's apocalyptic sublime', Fay (2018) adds historical depth.
18 See https://web.archive.org/web/20041011172259/http:/www.pik-potsdam.de/~stefan/tdat_review.html#expand, a review of the film by the distinguished climatologist Stefan Rahmstorf; he concludes: 'I think it would be a mistake and not do the film justice if scientists simply dismiss it as nonsense'.
19 On 'Interstellar' and other sci-fi films, see Michael Svoboda's https://www.yaleclimateconnections.org/dl/YCC_2014_Svoboda_TheCompleteCli-FiSeries.pdf.
20 On an Anthropocene-themed summer school at the museum, see https://sursock.museum/search/node/anthropocene

References

Andersen, G. & E.B. Nielson (2018) Biopolitics in the Anthropocene: On the invention of future biopolitics in Snowpiercer, Elysium, and Interstellar. *Journal of Popular Culture* 51(3): 613–634.

Anderson, K. (2015) Ethics, ecology, and the future: Art and design face the Anthropocene. *Leonardo* 48(4): 338–347.

Boulton, E. (2016) Climate change as a 'hyperobject': A critical review of Timothy Morton's reframing narrative. *WIREs Climate Change* 7: 772–785.

Dickens, P. (2015) Leaving the world instead of saving it. *Capitalism Nature Socialism* 26(4): 249–252.

Evans, M. & N. Ryan. (2019) (De)compositions: A review of 'Anthropocene'. Visual and new media review. *Fieldsights*. Society for Cultural Anthropology (29 April).

Fay, J. (2018) *Inhospitable World: Cinema in the Time of the Anthropocene*. New York: Oxford University Press.

Hart, P.S. & A.A. Leiserowitz (2009) Finding the teachable moment: Analysis of information-seeking behavior on global warming related websites during the release of 'The Day After Tomorrow'. *Environmental Communication* 3(3): 355–366.

Heise, U.K. et al. eds. (2017) *The Routledge Companion to the Environmental Humanities*. London: Routledge.

Jagodzinski, J. ed. (2018) *Interrogating the Anthropocene: Ecology, Aesthetics, Pedagogy, and the Future in Question*. Cham: Palgrave Macmillan/Springer.

Johns-Putra, A. (2010) Ecocriticism, genre, and climate change: Reading the utopian vision of Kim Stanley Robinson's science in the capital trilogy. *English Studies* 91(7): 744–760.

Johns-Putra, A. (2016) Climate change in literature and literary studies: From cli-fi, climate change theater and ecopoetry to ecocriticism and climate change criticism. *WIREs Climate Change* 7: 266–282.

Jørgensen, F.A. & D. Jørgensen (2016) The Anthropocene as a history of technology: Welcome to the Anthropocene: The earth in our hands, Deutsches museum, Munich. *Technology and Culture* 57(1): 231–237.

Kember, S. (2017) After the Anthropocene: The photographic for earthly survival? *Digital Creativity* 28(4): 348–353.

Keogh, L. & N. Möllers (2014) Pushing the boundaries: Curating the Anthropocene at the Deutsches museum. In Cameron, F. & B. Neilson eds. *Climate Change, Museum Futures*. London: Routledge, chap. 5.

Khatchadourian, R. (2016) Edward Burtynsky's quest to photograph a changing world. *New Yorker* (19–26 December).

Milner, A. (2018) *Again, Dangerous Visions: Essays in Cultural Materialism*. Leiden: Brill.

Morton, Timothy (2013) *Hyperobjects: Philosophy and Ecology After the End of the World*. Minneapolis, MN: University of Minnesota Press.

Powell, C.L. (2015) *Cli-Fi Cinema: An Epideictic Rhetoric of Blame*. Master's thesis, University of Nevada, Las Vegas.

Reusswig, F. & A.A. Leiserowitz (2005) The international impact of 'The Day after Tomorrow'. *Environment: Science and Policy for Sustainable Development* 47(3): 41–44.

Robin, L. et al. (2014) Three galleries of the Anthropocene. *Anthropocene Review* 1(3): 207–224.

Robin, L. & C. Muir (2015) Slamming the Anthropocene: Performing climate change in museums. *reCollections (Journal of the National Museum of Australia)* 10(1).

Salmose, N. (2018) The apocalyptic sublime: Anthropocene representation and environmental agency in hollywood action-adventure cli-fi films. *The Journal of Popular Culture* 51(6): 1415–1433.

Sklair, L. (2020) Globalization and the challenge of the Anthropocene. In Rossi, I. ed. *New Frontiers of Globalization Research*. New York: Springer, chap. 4.

Swanson, H.A. et al. (2015) Less than one more than many: Anthropocene as science fiction and scholarship-in-the-making. *Environment and Society: Advances in Research* 6: 149–166.

Tavares, P. (2014) Re-legislating the soil: Enclosure and exception at the amazon frontiers. *Architecture and Culture* 2(1): 94–115.

Trexler, A. (2015) *Anthropocene Fictions: The Novel in a Time of Climate Change*. Charlottesville, VA: University of Virginia Press.

Truscello, M. (2018) Catastrophism and its critics: On the genre of environmentalist documentary film. In Jagodzinski, J. ed. *Interrogating the Anthropocene: Ecology, Aesthetics, Pedagogy, and the Future in Question*. Cham: Palgrave Macmillan/ Springer, chap. 11.

Wodak, J. (2018) Shifting baselines: Conveying climate change in popular music. *Environmental Communication* 12(1): 58–70.

14 Conclusion

We need to talk about the Anthropocene

Leslie Sklair

Introduction

The main conclusion of this book is that, consciously or unconsciously, the mass media and many of the scientists and commentators they report have generally neutralized the risks of the Anthropocene, either by failing to mention them at all or by turning them into opportunities for human ingenuity and profit. In the 1994 edition of the *Oxford Paperback Dictionary* 'neutralize' was defined as follows: 'to render ineffective by an opposite force or effect'. So, despite some reporting of dire warnings, notably a few stories about the views of scientists, like the late Frank Fenner and others, who have written specifically about the possibility of human extinction,[1] and environmental humanities scholars like Clive Hamilton and Roy Scranton, the net effect has been 'reassurance narratives' encapsulated in the optimistic 'good' Anthropocene frame, accepted by governments and corporations. My intention in this concluding chapter is (paradoxically) to provide a brief introduction to the scholarly literature on various contentious aspects of the Anthropocene (and Anthropo-scene), some of which are adequately reported in the media but most of which are not.

Hope, despair, neutralizing

As discussed in previous chapters, journalists and scientists alike have the unenviable task of deciding whether to send messages of hope or despair, or both, and so it is not so surprising that reassurance narratives of various kinds dominate media coverage. There is a considerable body of literature on the relative effectiveness of messages of hope as compared to messages of despair on mobilizing people to take some sort of 'climate action' (Rayner and Minns 2015). Opinions vary all the way from the tabloid 'Hell doesn't sell' to the academic research-based 'Fear Won't Do It: Promoting Positive Engagement With Climate Change Through Visual and Iconic Representations' (O'Neill & Nicholson-Cole 2009). In 'Sixty Years of Fear Appeal Research', Ruiter et al. (2014) explain that the argument has been going on for decades. Cohen (2017) and Becker (2017) are just two in the storm of criticism that greeted an alarming article 'Uninhabitable Earth' in *The*

New York Magazine (10 July 2017), and the book that followed (Wallace-Wells 2019), though both also attracted many positive reviews.

Over-optimistic tendencies have been observed in the social sciences, for example what Hornborg (2017a) has described as 'dithering while the planet burns' in his withering review of the contributions of some influential anthropologists to Anthropocene studies. In their critical analysis of how sociologists deal with the 'climate crisis' (metonym for the Anthropocene), Leahy et al. (2010) write: 'Much social science on this topic has a kind of "boy scout" willingness to tell us that it can all be fixed. Giddens and Beck [who continue to be key references for the sociology of climate change] declare that the business class can change its spots'. There are, of course, exceptions, for example, Nigel Clark (2014) in sociology and Deborah Danowski and Eduardo Viveiros de Castro (2016) in philosophy and anthropology, but the general trend, as in the media, has been to avoid and/or neutralize the risks (for example, Grundmann and Stehr 2010, and Lidskog and Waterton 2016).

Theories of Anthropocene inertia

There are several explanations for apparent widespread inertia on ecological threats, all relevant to scientists, journalists, and everyone else. We can begin with the poignant idea of solastalgia, 'the pain or sickness caused by the loss or lack of solace and the sense of isolation connected to the present state of one's home and territory' (Albrecht 2005: 45) and the more general idea of planetary solastalgia: 'I contend that the experience of solastalgia is now possible for people who strongly empathise with the idea that the Earth is their home and that witnessing events destroying endemic place identity (cultural and biological diversity) at any place on Earth are personally distressing to them' (ibid.: 46).[2] Along similar lines, Brulle and Norgaard (2019) write of cultural trauma and social inertia in the face of anthropogenic climate change: 'The risk of cultural trauma is met with resistance and attempts to restore and maintain the status quo. These actions to avoid cultural trauma result in social inertia on climate change at the individual, institutional, and societal levels' (ibid., 887–88). They go on to say that the science of climate change is well known and understood, yet individuals continue their daily practices unchanged – such as flying frequently, buying large energy-wasteful houses and cars, and maintaining high consumption levels. But, as the research in our book has shown, it is by no means certain that the science of climate change is well known and understood, especially by most people around the world who get their information from mass and local media rather than scientific journals (with the possible exception of a few quality publications, for which see Table 2.2 in chapter 2 of this book).[3] Nevertheless, cultural trauma is, in some cases, likely to be part of the explanation of inertia, though the linguistic device of climate change as a metonym for the Anthropocene (substituting the part for the whole) needs to be raised. It may be the case that many people believe that, if we can 'fix the climate', the problem will be solved, while informed discussion of the Anthropocene tells us that it is not just climate change that poses risks, but degradation of all our life-supporting ecosystems, a point made (though not often) in media coverage of the Anthropocene. The International Psychoanalytical Association is on record (pre-Coronavirus epidemic) for stating that climate change is 'the biggest global health threat of the 21st century' and mental health professionals are increasingly referring to 'ecoanxiety' (see Castelloe 2018, and the useful links she provides).[4]

With the discourse-challenging term 'slow violence' and his creative elaboration of the idea of the 'environmentalism of the poor'[5] Rob Nixon (2013) made a notable intervention in debates over the difficulty of understanding both climate change and the Anthropocene. He argues: 'Climate change, the thawing cryosphere, toxic drift, biomagnification, deforestation, the radioactive aftermaths of wars, acidifying oceans, and a host of other slowly unfolding environmental catastrophes present formidable representational obstacles that can hinder our efforts to mobilize and act decisively' (ibid., 2). The plight of those people lacking resources who are the principal casualties of slow violence does appear in many Anthropocene-related media stories but the more shaming concept of 'slow violence' appears only twice in all our results. While Nixon's book was widely reviewed in the scholarly journals and won many prizes, this is obviously something the media are not comfortable about reporting. Another important feature is his engagement with the role of what he terms 'environmental writer-activists'.[6] Jennifer Jacquet offers us the 'Anthropocebo Effect'. This helps explain the 'psychological condition that exacerbates human-induced damage – a certain pessimism about humanity that leads us to accept humans as a geologic force and destruction as inevitable' (Jacquet 2013: 898).

All of these ideas are useful in understanding the complexities of the problem, as in statements by the distinguished conservation biologist Paul Ehrlich and the renowned science journalist Elizabeth Kolbert. Ehrlich warns: 'In pushing other species to extinction, humanity is busy sawing off the limb on which it perches' (quoted in Kolbert 2014: 268).[7] Kolbert muses: 'It seems impossible to imagine that a technologically advanced society could choose, in essence, to destroy itself, but that is what we are now in the process of doing' (2006: 189). Though disturbing, these messages seem to offer neither hope nor despair. Whereas there are many other alarming reports in the media about the risks of the Anthropocene, there are very few that present the issue as directly as these two statements, and most of those that hint at such conclusions temper their stories with the hope that human ingenuity will solve the problems – the 'boy scout' willingness to tell us how it can all be fixed, referred to above, neutralizing the risk. As documented in chapter 2 (note 8) even some prominent Earth System scientists, notably Paul Crutzen, appear to have made the journey from existential risk to geoengineering.[8] This is understandable as a strategy for responsible scientists and journalists striving to communicate risks, if we accept the social science research on the limited effectiveness of alarming messages. The problem, however, is that 'positive engagements' for the better-off, like driving and flying a bit less, eating less meat and dairy, searching out renewable sources of energy to sustain our life styles, although noble endeavours, in themselves will not make sufficient difference to avoid the potentially most catastrophic outcomes of the Anthropocene.

If all those who had the choice stopped driving and flying and eating meat and began to use genuinely emissions-neutral energy (if this is possible) only for the essentials, that might make a difference, but it would also mean the end of capitalist civilization as we know it. The media almost never spell out the radical social and economic changes necessary to avoid the risks, and almost always imply that 'renewables' are carbon neutral, which is far from the truth (see Chapter 2, note 9 and Chapter 9, note 2 of this book). Governments of all types generally raise most of their taxes, directly or indirectly, on carbon-intensive activities. It is difficult to envisage carbon-neutral states or carbon-neutral capitalism of whatever variety. Though the timescales of potential civilizational collapse are not precisely known, the potential risks of it happening are clear enough.

Where do all of these ideas, about how we arrived at this juncture of human and planetary history and what to do about it, leave us? They all sound plausible but incomplete. On a more grounded level, a research project on the floodplain along the Mississippi River in the American Midwest (Casagrande et al. 2017) introduces the concept of ecomyopia. 'We define ecomyopia as the tendency for societies to ignore, not recognize, or fail to act on new ecological information that contradicts political arrangements, social norms, or world views … The failure to meaningfully address climate change is a spectacular example of ecomyopia' (ibid.: 23). But this is not simply careless short-sightedness: 'In the case of flooding in the American Midwest, people prefer to avoid extremely difficult decisions like whether to relocate homes or villages out of floodplains and will accept large-scale infrastructural solutions promoted by special interests, even though they prefer solutions more in balance with nature. It comes as no surprise that the amount of federal funding allocated for wetland restoration and the relocation of homes throughout the US is insignificant compared to that allocated for large-scale infrastructure solutions' (ibid.: 25). The idea of ecomyopia is useful in understanding how the media (and other communicators) present climate change and the Anthropocene. It is not only along the Mississippi River that people fail to see the risks (both literally and symbolically).[9] This is sardonically expressed in a cartoon by Felice Wynham, one of the authors of the ecomyopia article (see Figure 14.1).

To appreciate this more fully, I suggest that we consider if and how these risks are communicated to the young.

Educating the young in and about the Anthropocene

While the media rarely engage, scholarly articles on education about and in the Anthropocene are more numerous (for example, Gilbert 2016). However, the political economy dimension is rarely addressed, leading to an almost complete neglect of issues arising from the fact that the Anthropocene exists in a world riven by extreme inequalities that is generally dominated by the norms of capitalism within the confines of hierarchical, competitive so-called nation-states. Nevertheless, many good ideas emerge from this research. For example, Ariyadasa (2016) writes directly on 'Empowering Children for Governing in the

Figure 14.1 'Can I get mine with long duree?' (2015) © Felice Wynham, https://kuchka.org/ecomyopia.

Anthropocene' in her study of children's homes in Sri Lanka. This is an optimistic account of teaching institutionalized children (usually completely marginalized) 'to be the future governors [more hierarchy] of the Anthropocene' (ibid.: 698). Focusing on small-scale interventions, the children are taught about how to become organic dairy and worm farmers, beekeepers, plant nursery marketers, compost and bio-gas producers, herbal medicine practitioners, environmentalists, and agricultural inspectors. Volunteering in the local community increases positive connections with the world outside the homes. The methodologically rigorous will understandably demand follow-up studies, but the research as reported is inspiring. The follow-ups in Kopnina et al. (2018) give us a reality check. Helen Kopnina and her colleagues are part of a global research project to promote the idea of childhoodnature [sic] defined in the following terms: 'children are nature … The challenge of unlearning anthropocentrism, we submit, thus lies at the crux of any project to (re)define and integrate notions of childhoodnature' (ibid: 2). So far this is all very philosophical (as much of the environmental education literature is). The research embraced nature-based schools in the Netherlands and Canada to illustrate the tenacity of what is aptly labelled anthropocentric 'common sense' and sketches out 'the beginnings of pedagogy of childhoodnatures guided by notions of rewilding and ecological humility' (ibid.).[10] The results of ethnographic studies of the children in these schools is not encouraging; students almost always resort to conventional ideas of nature in the context of 'managed environmentalism'.[11] The researchers are forced to conclude: 'For us, hyperactive pessimism is a necessary orientation to be constantly alert to these moves to innocence and to assume that any enacted habit or uncritically offered belief was likely rooted in the very culture we were aiming to transform' (ibid.:14). Teaching children to live in the Anthropocene will be a long and arduous process.[12] Laird (2017: 2) quotes Sandra Steingraber: 'Ultimately the environmental crisis is a parenting crisis. It undermines my ability to carry out two fundamental duties: to protect my children from harm and to plan for their future', inspired by Donna Haraway's 'make kin not babies' (a plea for both biological and social parenting). Laird's reflection is entitled 'Learning to Live in the Anthropocene' which may be a riposte to Roy Scranton's 'learning to die', not a popular theme in environmental education for children.

Moving up the age range, two articles in the *American Biology Teacher* address the issue of teaching about the Anthropocene for upper-level students. Wagler (2011) grasps the nettle in 'The Anthropocene Mass Extinction'. He is careful to explain that curriculum topics address a variety of anthropogenic drivers of ecosystem degradation, not simply climate change. Cohen (2016) sets out a scheme for students to do their own research on extinction and recovery. Evidence for the necessity for radical shifts in science education is provided by poll data (Leiserowitz et al. 2014), showing that 67% of Americans believed that future generations will be harmed 'a great deal' or 'a moderate amount' by 'global warming' (the phrase chosen in the survey), yet only a minority believed that they personally (38%), their family (43%) or community members (45%) will be harmed. Ecomyopia in practice?

A relatively new approach to education for the Anthropocene (though not so new for environmental education per se) is the idea that indigenous communities may have much to offer. Bardsley and Wiseman (2016) research this in the context of remote Indigenous communities of Central Australia, and Whyte surveys the literature from an American Indian First People perspective (discussed in Chapter 12 of this book). 'Indigenous Artists against the Anthropocene' (Horton 2017: 59) reinforces the message powerfully with a quote from Pablo Solón Romero, a former Bolivian diplomat:

> collectively grappling with transborder monsters, such as climate change, cannot proceed without a thorough political accounting of the entanglement of elite minorities, disenfranchized majorities, and other-than-human persons. Native intellectuals and leaders have laid the foundations for such a project by modelling worlds in which matter is not simply endowed with sensuous power but is granted a role in mediating diverse human jurisdictions. 'To speak about Mother Earth's rights challenges the entire legal system on which this capitalist system is based'.

Who should be doing what? Corporations, governments, consumers

Though our media results rarely include the term 'Capitalocene', the power of fossil fuel industries is regularly reported. As Malm (2016) has demonstrated, there is a strong case for locating the causes of the Anthropocene (though he rejects the term) in the transition from water power to coal-fired steam power in the Industrial Revolution in Europe. The great acceleration in a sizable number of anthropogenic indicators from the 1950s onwards further implicates capitalism in the ecological crisis, whatever we call it. Whereas much Anthropocene media coverage implies that hyper-consumption is a large part of the problem, the idea of degrowth occurs only three times in our results. The beautifully simple growth/degrowth graphic created by Barbara Castro (see Figure 14.2) sets out the issues clearly. Despite the necessity of degrowth as part of the response to the unfolding human-driven ecological crisis, the myth of 'sustainable development' prevails and few scholars and even fewer policy makers seem willing to grapple with this difficult reality.

Several scholars have begun to identify the responsibility of specific industries, corporations, and governments for their destructive impacts on specific ecosystems. For example, Pullman (2012) deconstructs the 'Cozy Ties' that bind Canadian conservatives and corporations promoting the tar sands expansion in Alberta – an enterprise that will damage the environment to bring profits to the foreign corporations invested in the project. Though it is difficult to measure ecological footprints, it is a valuable endeavour.[13] Österblom et al. (2015) introduce the inventive concept of 'keystone actors' to explain the degradation of marine ecosystems. On the analogy of keystone species, they present data from 2012 to show how 13 corporations controlled up to 16% of the global marine catch, and Licker et al. (2019) do the same for ocean acidification. These companies, they argue, act as keystone actors of the Anthropocene, an increasingly important feature of the human-dominated planet.[14] Ekwurzel et al. (2017) trace the historic rise in global warming to the emissions of major carbon producers, demonstrating that around two-thirds of greenhouse gas emissions come from 83 producers of coal, oil, and natural gas, and seven cement manufacturers (private and

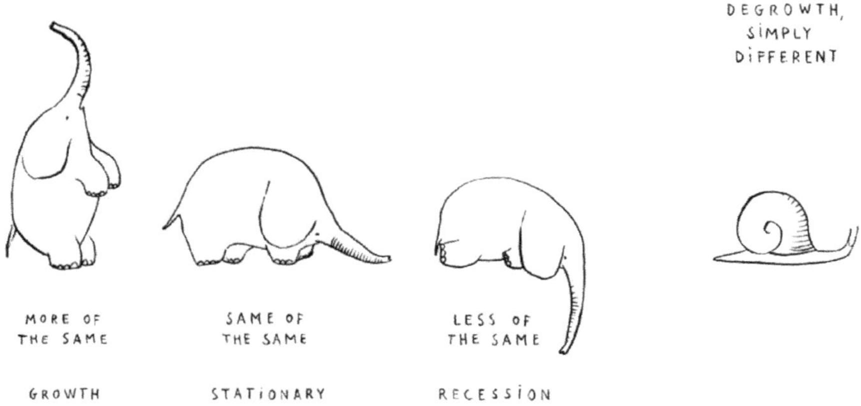

Figure 14.2 'Degrowth' (2015) © Barbara Castro (Barbaracastrourio.com).

state-owned). The media rarely name corporations, which suggests that most consumers are not well-informed on the links between corporate profits and ecological damage.[15]

Even from this small, but probably not unrepresentative sample of the literature, it must be concluded that there are few clear answers to the question 'who should be doing what?' In the case of those most-commonly identified as the main culprits, there is growing evidence that many corporations, prompted by increasing difficulties in raising finance, are moving out of fossil fuels, and transitioning to renewables. However, in lifecycle terms, no renewables are carbon neutral and there is always the risk of the Jevons effect: we have electric cars, so we can drive as much as we like, and we have solar/wind/biomass energy, so we can use as much energy as we like, and so on, issues on which the media are mostly silent. Governments speak fine words and pass laws to deal with the 'climate crisis', but, in most countries, government revenues are dependent on high-emission industries. Many worthy initiatives (reduce single-use plastic products, recycle, use more efficient light bulbs, and so on) are often considered to be 'too little, too late'. British green guru George Monbiot made himself very unpopular with the global environmental establishment for campaigning in favour of nuclear power as a strategy to reduce emissions. Complexity blossoms and wilts in the Anthropocene.

If it looks unlikely that corporations or governments will take truly decisive action to deal with the potential risks of the Anthropocene over the foreseeable future, what about consumers? After decades of relentless indoctrination (I have labelled this the 'culture-ideology of consumerism'), it is not surprising that many consumers at all levels in all societies are reluctant to stop consuming as they like in the face of exhortations to 'save the planet'. The beginning of the 21st century may prove to be a particularly inauspicious time for humanity to find itself in the Anthropocene.[16] In his study of what appears to be an international surge of right-wing populism, Lockwood (2018) provides evidence that the leaders and supporters of these movements tend to be on the climate denial spectrum. Calls to curb consumption of goods and resources are likely to be ignored by such people. However, there are probably many millions of people worldwide who do take the situation seriously and wonder about what needs to be done by whom. This inevitably leads to the question of blame in the Anthropocene. In 'The Social Life of Blame in the Anthropocene', the anthropologist Peter Rudiak-Gould (2015: 48) confronts this dilemma directly. He argues: 'Treating global climate change as a metonym for this predicament [a rare acknowledgement of a common practice], I show how life in the Anthropocene reconfigures blame in four ways: it invites ubiquitous blame, ubiquitous blamelessness, selective blame, and partial blame'.[17] Taking Rudiak-Gould's argument to its logical conclusion, the rationale for replacing the Anthropocene with the Capitalocene (or any other – scene) dissolves for two reasons. The first is clearly expressed by Donna Haraway, as documented above (Chapter 12, note 7 of this book). In an unlikely pairing, in a critique of the politicization of Green thought, James Lovelock argues: 'It stops us from realizing that it is not them, the multinational companies, or the state industries of Russia and China, that are wholly to blame for our fast-degrading world. Our much-too-vociferous advocates, the consumer lobbies, and we, the consumers, are equally responsible for the gaseous greenhouse and the extinction of wildlife' (Lovelock 2016: xiv).[18] A third reason to reject the name 'Capitalocene' is that the effects of the Anthropocene will probably last for thousands, if not millions of years, few predict the future of capitalism on this sort of timescale. Given the complexities of measuring ecological footprints, and the pressing need for more ecological justice, the understandable but blinkered thinking that lies behind the Capitalocene and its variants is not helpful. The task, as set out in this book and embarked upon by many others, is to create a more critical theory of the Anthropocene, not to abandon the concept, properly applied.[19]

Many will no doubt disapprove that this book concludes with the key anthropocentric question posed by the Anthropocene, namely the potential existential threat to the continued existence of the human species on our home planet. As documented throughout, mass media responses to this challenge are overwhelmingly in terms of neutralizing or reassurance narratives. Despite alarms raised in the media by Earth scientists and other concerned parties, and on the streets by Extinction Rebellion and other environmental activists, including increasing numbers of children, the dominant message is 'business as usual' (with renewables, more recycling, and geoengineering). Given this widespread inability to think beyond the growth-obsessed status quo, embodied in consumer capitalism and the hierarchic state, it is not surprising that governments and corporations have made so little genuine progress in dealing with the potential existential risks of the Anthropocene and why most publics all over the world have failed to grasp them. It is ironic that car insurance is mandatory in most countries and that many millions of people insure themselves against unlikely events, but governments everywhere seem unwilling to spend the money to mitigate the risks of potential existential threats to human survival on our home planet (see Herbstein 2015 and Keys et al. 2019). As evidence mounts on the likely connections between the Anthropocene and the Coronavirus pandemic (see Morand and Walther 2020) the strategies of governments seem more and more chaotic.

An alarming precedent

Unsurprisingly, comparisons have been made between climate change denial and Holocaust denial (and denial of other genocides). Jacques argues cogently: '*Climate change and the Holocaust are not equivalent,* but that does not mean there is no climate denial' (2012: 10, italics in original). Whereas it would be mistaken to identify the slogan 'Welcome to the Anthropocene' always with climate change or Anthropocene denial, I would argue that, even when used ironically, it can create an atmosphere of scepticism about the severity of the situation. As documented at various points throughout this book, 'Welcome to the Anthropocene' gives the impression that there is a middle way between the Earth System science Anthropocene and Anthropocene scepticism and that the idea of the 'good' welcoming Anthropocene provides it. It is possible that we may be having to think about the survival of humanity sooner rather than later; indeed some communities around the world are already facing actual existential risks. So, for example, imagine, in terms of the responses of citizens in Germany to potential threats from the Nazis after Hitler came to power, and compare: 'Welcome to the Anthropocene' with 'Welcome to the Holocaust'; or the prizewinning book 'Adventures in the Anthropocene' with 'Adventures in the Holocaust'; or the art exhibition and conference entitled 'Postcards from the Anthropocene' with 'Postcards from the Holocaust'; or 'Romancing the Anthropocene' with 'Romancing the Holocaust'.[20] These comparisons might seem ridiculous (even offensive) now, but if the worst comes to the worst … .

Notes

1 Ceballos et al. argue: 'All signs point to ever more powerful assaults on biodiversity in the next two decades, painting a dismal picture of the future of life, including human life' (2017: 7).
2 Both Albrecht and Leahy chose the distressed coal-mining region of Hunter Valley in Australia as their research site. *Le Devoir* (Quebec) published a profile of Albrecht, 'Do you suffer from "solastalgia"?' in 2015.
3 Instructive in this context is the research of Wynes and Nicholas (2017: Figure 1) calculating the CO_2 emissions prevented in rich countries by simple, if problematic, changes in behaviour. The top seven are, in descending order of impact: have one fewer child, live car free, avoid one transatlantic flight, buy green energy, a more efficient car, switch from electric to being car free, switch to a plant-based diet. The first is by far the most effective, by a factor of more than 70:1 over the last.
4 Robbins and Moore (2012) anticipate the idea. See also: https://edgeeffects.net/who-gets-to-have-ecoanxiety. There is little doubt that the Coronavirus crisis will displace the 'climate' crisis for the foreseeable future: we must not fail to learn from the connections between them Kothari et al. 2020 and updates on the websites of *Carbon Brief* and *Desmog*.
5 Dauvergne (2017) explores the 'Environmentalism of the Rich'.
6 See Farrell et al. (2019) and global media coverage of Extinction Rebellion all over the Internet in 201819.
7 There are various versions of this quote. Ehrlich is best known for his long-standing warnings about pressures of population, a theme that features regularly in media coverage of the Anthropocene.
8 On 'scientific reticence' see Wallace-Wells (2019: 155ff.).
9 It is also true that many people all around the world have little choice in the matter. See also Goleman at https://www.danielgoleman.info/anthropocene-thinking/
10 This theoretical initiative is inspired by Donna Haraway's engagement with the 'troubles' that the debate around the Anthropocene has highlighted. See also Leichenko and O'Brien (2020).
11 See, again, Wynes & Nicholas (2017) on how climate change is presented in Canadian school textbooks, https://www.teenvogue.com/story/how-climate-change-taught-schools and https://nexusmedianews.com/climate-change-is-conspicuously-absent-from-college-textbooks-94833d325823. Widener et al. (2016) outlines a different and more promising approach.
12 For some rays of hope, see Malone (2013) and Salaün (2019).
13 A technical discussion comparing ten different approaches to the idea concludes: 'The present paper will hopefully contribute useful elements to help the reader form her/

his own opinion on the merits of and the need for an ecological footprint' (Galli and Giampietro 2016: 231–2).

14 See also the work of the eminent fisheries scientist Daniel Pauly (2010). Wodak (2018: 66ff.) amplifies Pauly's path-breaking concept of 'shifting baselines' in the context of anthropogenic change.

15 One exception is a report in *The Guardian* 'Revealed: the 20 firms behind a third of all carbon emissions' (October 9, 2019) on the research of Richard Heede. See also Heede (2014).

16 The Coronavirus crisis that began in 2020 certainly led to a reduction of greenhouse gas emissions; it remains to be seen if this becomes permanent.

17 For an interesting empirical study on climate change blame in the Swedish media, the land of Ingmar Bergman, see Olausson (2011).

18 This recalls a wise saying attributed to Gandhi: 'if you want to change the world, start with yourself'. Greta Thunberg's decision in 2019 to travel to and from New York in a more-or-less zero-emission boat is inspirational. The aviation industry is predicting both increased demand and zero-carbon flying.

19 Readers at this point may justifiably ask, 'what would you do about it?' For my answer, see Sklair (2019b).

20 These are all adapted from titles of Anthropocene events and publications.

References

Albrecht, G. (2005) 'Solastalgia': A new concept in health and identity. Philosophy Activism Nature 3: 41–55.

Ariyadasa, A. (2016) Educating and empowering children for governing in the Anthropocene: A case study of children's homes in Sri Lanka. *Systems Research and Behavioral Science* 33(5): 691–700.

Bardsley, D. & N. Wiseman (2016) Socio-ecological lessons for the Anthropocene: Learning from the remote indigenous communities of central Australia. *Anthropocene* 14: 58–70.

Becker, R. (2017) Why scare tactics won't stop climate change: Doomsday scenarios don't inspire action. *The Verge* (11 July).

Brulle, R.J. & K.M. Norgaard (2019) Avoiding cultural trauma: Climate change and social inertia. *Environmental Politics* 28(5): 886–908.

Casagrande, D.G. et al. (2017) Ecomyopia in the Anthropocene. *Anthropology Today* 33(1): 23–25.

Castelloe, M. (2018) Coming to terms with ecoanxiety: Growing an awareness of climate change. *Psychology Today* (9 January).

Ceballos, G.P. et al. (2017) Biological annihilation via the ongoing sixth mass extinction signaled by vertebrate population losses and declines. *PNAS* 114(30): E6089–E6096.

Clark, N. (2014) Geo-politics and the disaster of the Anthropocene. *The Sociological Review* 62(S1): 19–37.

Cohen, D.A. (2017) New York mag's climate disaster porn gets it painfully wrong. *Jacobin* (July).

Cohen, J. (2016) Biodiversity education & the Anthropocene: An indicator of extinction or recovery. *The American Biology Teacher* 78(4): 293–299.

Danowski, D. & E.V. de Castro (2016*) Is there any world to come?* Cambridge: Polity Press.

Dauvergne, P. (2017) *Environmentalism of the rich.* Cambridge: MIT Press.

Ekwurzel, B. et al. (2017) The rise in global atmospheric CO_2, surface temperature and sea level from emissions traced to major carbon producers. *Climatic Change* 144(4): 579–590.

Farrell, C. et al. eds. (2019) *This is not a Drill: An Extinction Rebellion Handbook.* London: Penguin.

Galli, A. & M. Giampietro (2016) Questioning the ecological footprint. *Ecological Indicators* 69: 224–232.

Gilbert, J. (2016) Transforming science education for the Anthropocene—is it possible? *Research in Science Education* 46: 187–201. doi:10.1007/s11165-015-9498-2

Godfrey, P. & D. Torres eds. (2016) *Systemic Crises of Global Climate Change: Intersections of Race, Class and Gender*. London: Routledge.

Grundmann, R. & N. Stehr (2010) Climate change: What role for sociology? A response to Constance Lever-Tracy. *Current Sociology* 58(6): 897–910.

Heede, R. (2014) Tracing anthropogenic carbon dioxide and methane emissions to fossil fuel and cement producers, 1854–2010. *Climatic Change* 122: 229–241.

Herbstein, T. (2015) *Insurance and the Anthropocene: Like a Frog in Hot Water*. Ph.D. thesis, Law Faculty, University of Cape Town.

Hornborg, A. (2017a) Dithering while the planet burns: Anthropologists' approaches to the Anthropocene. *Reviews in Anthropology* 46(2–3): 61–77.

Horton, J.L. (2017) Indigenous artists against the Anthropocene. *Art Journal* 76(2): 48–69.

Jacques, P.J. (2012) A general theory of climate denial. *Global Environmental Politics* 12(2): 9–17.

Jacquet, J. (2013) The anthropocebo effect. *Conservation Biology* 27(5) 898–901.

Keys, P.W. et al. (2019) Anthropocene risk. *Nature Sustainability* 2: 667–673.

Kolbert, E. (2006) *Field Notes from a Catastrophe: Man, Nature, and Climate Change*. London: Bloomsbury.

Kolbert, E. (2014) *The Sixth Extinction: An Unnatural History*. London: Bloomsbury.

Kopnina, H. et al. (2018) Moving beyond innocence: Educating children in a post-nature world. In Cutter-Mackenzie, A. et al. eds. *Research Handbook on Childhoodnature*. New York: Springer. https://link.springer.com/referenceworkentry/10.1007%2F978-3-319-51949-4_40-1

Kothari, A. et al. (2020) Coronavirus and the crisis of the Anthropocene. *The Ecologist* (27 March).

Laird, S. (2017) Learning to live in the Anthropocene: Our children and ourselves. *Studies in Philosophy and Education* 16: 1–18.

Leahy, T.V. et al. (2010) Stumbling towards collapse: Coming to terms with the climate crisis. *Environmental Politics* 19(6): 851–868.

Leichenko, R. & K. O'Brien (2020) Teaching climate change in the Anthropocene: An integrative approach. *Anthropocene* 30: 1–4.

Leiserowitz, A.A. et al. (2014) *Climate Change in the American Mind.* (Yale University and George Mason University). New Haven, CT: Yale Project on Climate Change Communication.

Licker, R. et al. (2019) Attributing ocean acidification to major carbon producers. *Environmental Research Letters* 14: 124060.

Lidskog, R. & C. Waterton (2016) Anthropocene – a cautious welcome from environmental sociology? *Environmental Sociology* 2(4): 395–406.

Lockwood, M. (2018) Right-wing populism and the climate change agenda: Exploring the linkages. *Environmental Politics* 27(4): 712–732.

Lovelock, J. (2016) *Gaia: A New Look at Life on Earth*. Oxford: Oxford University Press.

Malm, A. (2016) *Fossil Capitalism: The Rise of Steam Power and the Roots of Global Warming*. London: Verso.

Malone, K. (2013) 'The future lies in our hands': Children as researchers and environmental change agents in designing a child-friendly neighbourhood. *Local Environment* 18(3): 372–395.

Morand, S. & B.A. Walther (2020) The accelerated infectious disease risk in the Anthropocene: more outbreaks and wider global spread. *bioRxiv*. https://doi.org/10.1101/2020.04.20.049866 (preprint)

Nixon, R. (2013) *Slow Violence and the Environmentalism of the Poor*. Cambridge: Harvard University Press.

Olausson, U. (2011) 'We're the Ones to Blame': Citizens' representations of climate change and the role of the media. *Environmental Communication* 5(3): 281–299.

O'Neill, S.J. & S. Nicholson-Cole (2009) 'Fear won't do it' promoting positive engagement with climate change through visual and iconic representations. *Science Communication* 30(3): 355–379.

Österblom, H. et al. (2015) Transnational corporations as 'Keystone Actors' in marine ecosystems. *PLoS ONE* 10(5): 0127533. doi:10.1371/journal.pone

Pauly, D. (2010) *5 Easy Pieces. The Impact of Fisheries on Marine Ecosystems*. Washington and London: Island Press.

Pullman, E. (2012) *Cozy Ties*. https://rabble.ca/blogs/bloggers/emmapullman/2012/01/cozy-ties-astroturf-ethical-oil-and-conservative-alliance-promote

Rayner, T. & M.A. Minns (2015) *The Challenge of Communicating Unwelcome Climate Messages*. Norwich: Tyndall Centre for Climate Change Research.

Robbins, P. & S.A. Moore (2012) Ecological anxiety disorder: Diag*nosing the politics of the Anthropocene*. Cultural Geographies 20(1): 3–19.

Rudiak-Gould, P. (2015) The social life of blame in the Anthropocene. *Environment and Society: Advances in Research* 10: 48–65. https://doi.org/10.3167/ares.2015.060104

Ruiter, R.A. et al. (2014) Sixty years of fear appeal research: Current state of the evidence. *International Journal of Psychology* 49(2): 63–70.

Salaün, R. (2019) The children who accuse us. *Green European Journal* (13 November). https://www.greeneuropeanjournal.eu/the-children-who-accuse-us/

Sklair, L. (2019b) World revolution or socialism, community by community, in the Anthropocene? *Globalizations* 16(7): 1012–1019.

Wagler, R. (2011) The anthropocene mass extinction: An emerging curriculum theme for science educators. *The American Biology Teacher* 73(2): 78–83.

Wallace-Wells, D. (2019) *Uninhabitable Earth*. London: Penguin.

Widener, P. et al. (2016) Climate action and literacy through creativity and conversations. In Godfrey, P. & D. Torres eds. *Systemic Crises of Global Climate Change: Intersections of Race, Class and Gender*. London: Routledge, chap. 18.

Wodak, J. (2018) Shifting baselines: Conveying climate change in popular music. *Environmental Communication* 12(1): 58–70.

Wynes, S. & K.A. Nicholas (2017) The climate mitigation gap: Education and government recommendations miss the most effective individual actions. *Environmental Research Letters* 12: 074024.

Appendix 1

Countries in Regions

AFRICA (37) Algeria; Angola; Benin; Botswana; Burundi; Burkina Faso; Cameroon; Central African Republic; Congo (Brazil); Congo (DRC); Ethiopia; Gambia; Ghana; Guinea; Ivory Coast; Kenya; Mali; Lesotho; Liberia; Libya; Madagascar; Malawi; Mauritius; Morocco; Namibia; Niger; Nigeria; Rwanda; Senegal; South Africa; Sudan; Tanzania; Tunisia; Uganda; Zambia; Zimbabwe; Pan-African Media

ASIA SOUTH (14) Bangladesh; Bhutan; Brunei; India; Laos; Malaysia; Maldives; Nepal; Pakistan; Philippines; Singapore; Sri Lanka; Thailand; Vietnam

ASIA NORTH (6) Afghanistan; China; Japan; Korea; Mongolia; Taiwan

NORTH AMERICA (2) Canada; USA

LATIN AMERICA/CARIBBEAN (25) Argentina; Bolivia; Brazil; Chile; Colombia; Costa Rica; Ecuador; Mexico; Nicaragua; Panama; Paraguay; Peru; Uruguay; Venezuela; Bahamas; Bermuda; Cuba; Haiti; Honduras; Dominican Republic; Trinidad and Tobago; Guyana; Grenada; Jamaica; Puerto Rico

CENTRAL/EASTERN EUROPE (12) Croatia; Czech Republic; Estonia; Hungary; Latvia; Lithuania; Poland; Romania; Russia; Serbia; Slovakia; Ukraine

WESTERN EUROPE (22) Austria; Belgium; Cyprus;
Denmark; Finland; France; Germany; Gibraltar; Greece; The Netherlands; Iceland; Republic of Ireland; Italy; Luxembourg; Malta; Monaco; Norway; Portugal; Spain; Sweden; Switzerland; United Kingdom.

MIDDLE EAST (15) Bahrain; Dubai; Egypt; Iran; Iraq; Israel; Jordan; Lebanon; Palestine; Qatar; Saudi Arabia; Syria; Turkey; Yemen; Pan-Arab Media

Index

(media sources *in italics*)

9 Media (New Zealand) 202
20 Minutos (Spain) 175
21st Century Business Herald (China) 101
24 Heures (Switzerland) 172

Abend Zeitung (Germany) 235
Ackerman, D. 58, 99–100, 120
adaptation 51, 97, 184, 202, 212n10, 226
Addis Fortune 41
AFPBB News (Japan) 110
Africa Science News Service 49
Aftenposten (Norway) 180, 218
Aftonbladet (Sweden) 179, 221
Agarwal & Narain 24–25, 52n17
age of man 71, 84, 104, 119, 127, 131, 136, 146, 179, 192, 206, 220
agriculture 51, 72, 121, 180, 224, 235
Ahora (Cuba) 88
Akhbar Libya 44
Albrecht, G. 253, 261n2
Albuquerque Journal 66
Amazon 11, 92
Ambito Financiero (Argentina) 93, 238
American Scientist 66
Anchorage Daily News 70
anthrocene 62, 64, 242
anthromes 62, 72
Anthropocene: African 51; American 81n28; Anthropocebo effect 254; Anthropo-scene 13–14, 16n31, 170, 178; Appalachia 75–76; Arab 187ff; Asian 118ff, 221–224; Australian 202–207; blame 6, 31n7, 58, 81n24; Caribbean 83, 88–91; catastrophe 7, 69, 194, 238, 247; children 44, 56, 107, 134, 255–257; Chinese 97, 102, 146; colonialism 13, 24–25, 30, 43, 48, 222–227; conservation 39, 43, 52n11, 59, 112, 118–119, 175, 206; 257–260; creative arts 238–249; dance 212, 238, 243; decolonizing 11, 42, 226–227; denial 28–29, 31n16; education 23, 45, 123, 129, 130, 134, 257; *see also* children; environmental humanities 12–14, 28, 153, 197; 217ff; Eurocentrism 58, 155, 224; existential risk 5, 254; extinction 7–8, 31n15, 57, 69, 70–72, 99–100, 138n7, 206, 244; feedback loops 4, 15n1; feminist 151, 157n16; film 131, 137, 160, 236–237, 247; fossil fuel 6, 42, 71, 160, 185n19, 200, 212n5, 259; gender 115n25, 164; geoengineering 31n8, 60, 77, 86, 254; 'good' 26–28; health 74, 100–101, 114n6, 132, 189, 254; history 13, 27, 45, 50–51, 58, 70, 79, 102, 168–169, 173, 203, 221–223; Holocaust 260–261; imperialism 30, 121; Indian 221–223; indigenous communities 11, 42, 84, 226–227, 257; international law 48, 212n8; Latin American 84–85; localizing 13, 72–75; Manthropocene 157n16; metaphors 24, 48, 156, 243, 246; music 242–243; narratives/frames 12–13, 24–26; New Zealand 207; opera 243; poetry 245–246; politics 7, 9, 38, 110, 136, 151, 172, 184n9, 199; pop culture 16n25; population 46, 56, 93, 197; provincializing 13, 25, 48, 72–75; racial 52n15; religion 179, 227–228; safe operating space 4–5, 124; sculpture 243–244; security 80n8, 112, 206, 212n10; social sciences 181, 218–221; soil 31n3, 100, 114n3, 180, 189; start

dates 71, 177, 179; style 6; tipping points 4, 124; urban 42–43, 52n10, 111, 120, 126; water 40, 52n5, 78, 87, 110, 120, 190
Antilla, L. 8, 16n24
Antinews (Greece) 177
A2 (Czech Republic) 144–145
Apple Daily (Taiwan) 105
Al Arabiya 188, 194
Arab News 195, 200
Al Araby (TV) 193, 200
Arctic 81n25, 178, 210
Asahi Shimbun (Japan) 110
Asian Age 120, 223
astrobiology 134, 138n12
Atlantic 15n17
Austin Daily Texan 69
Australian 202–207, 210, 212n9
Australian Financial Review 203, 205
Avgi (Greece) 177
AWG 5, 22, 28, 89, 99, 110, 164, 224, 233, 237
Ayn Libya 44

balance as bias 14, 85
Baltimore Sun 71
BBC 52n13, 62, 85, 88, 110, 123, 150, 166, 181, 226
Beck, U. 15n5, 253
Beeld (South Africa) 40
Belt and Road 114n10
Bennett, E. 56, 64
Berlingske Tidende (Denmark) 179
Bild (Germany) 169, 233
biodiversity 39, 66, 68–69, 118–119, 189, 261n1
Boff, L. 227
Borneo Bulletin 133, 228, 236
Borneo Post 133
Boykoff, M. 14, 16, 31n16, 78
Breakthrough Institute 111, 115n22, 183
Burtynsky, E. 60, 233, 236–237
business as usual (BAU) 26, 86, 125, 173, 183, 199, 260
Business Mirror (Philippines) 131
Business Times (Singapore) 137

Calgary Herald 60
Canberra Times 24, 204, 206, 243, 244
Cantrik (Indonesia) 208
Cape Town Congress 22, 28, 40, 59, 89, 104, 112, 121, 126, 127, 147–150, 155, 161, 172–174, 181, 233
Capital Ethiopia 41
capitalism 6, 55, 91, 98, 142, 151, 160, 163, 182, 199, 221, 226, 229, 235, 260
Capitalocene 6, 15n8, 64, 146, 178, 219–221, 230n7, 260
Carrington, D. 164, 183
Castree, N. 13, 184n9
Castro, B. 258
Castro, E.V. 87, 221, 253
Ceballos, G.P. 138n7, 261n1
celebrities 182, 212, 218
Chakrabarty, D. 13, 221–222, 224, 230n14
Charleston Post & Courier 69–70
Chattanooga Times-Free Press (Tennessee) 73
Chernobyl 143, 155
Chicago Tribune 77, 78
China Daily 98, 104
China Times (Taiwan) 105
China Youth Daily 99
China Youth News 98
Chosun Ilbo (South Korea) 105
Christian Science Monitor 70
City Press (South Africa) 40
Clarín (Argentina) 93, 234
Cleveland Plain Dealer 77
cli-fi 12, 246
climate change: denial 28–29, 80n3, 175, 199, 228, 260; indigenous conceptions 11, 42, 212n14, 226, 257; litigation 52n18, 95n6; as metonym for Anthropocene 3, 24, 48, 220, 253, 259
climate lobby: 28, 199, 212n5
Columbian exchange 71, 100, 154, 162, 176, 177, 179, 181, 208, 224–249, 230n15; *see also* Lewis & Maslin
Comercio (Peru) 85, 87
Communism 55–56, 92
consumerism 6, 144, 183, 227, 259
Conversation 49, 166, 246
Conversation Africa 37, 49, 50, 51n2
Corneliussen, S. 9, 29
Coronavirus 52n8, 81n23, 114n6, 185n19, 213n16, 254, 260, 261n4, 262n16
Corriere Della Sera (Italy) 81n19, 173, 174
Counterpunch (USA) 50, 124
Courier-Mail (Brisbane) 203
Croatian News Agency 149
Cronon, W. 220
Crutzen, P. 10, 22, 27, 45, 77, 92, 162, 203, 238
Cumhuriyet (Turkey) 198

Dagens Næringsliv (Norway) 180
Dagens Nyheter (Sweden) 179, 221
Dagsavisen (Norway) 180
Daily Express (UK) 166
Daily Mail (UK) 24, 166, 197, 237
Daily Maverick (South Africa) 52n9
Daily Mirror (Sri Lanka) 122
Daily Nation (Kenya) 43
Daily News (New York) 65
Daily News (Sri Lanka) 122
Daily News Egypt 193–194
Daily Talk (Liberia) 45
Danowski, B. 87, 227, 253
deep time 113, 130, 136, 138n16, 211
degrowth 151, 172, 174, 184n14, 258
Democrat-Gazette (Arkansas) 73
Deseret News (Salt Lake City) 70
Deutsches Museum (DM) 169, 233–236
De Volkskrant (Holland) 183
Diário de Notícias (Portugal) 176, 224
Diario las Américas (Miami) 70
Diario Libre (Dominican Republic) 89, 91
Die Presse (Austria) 171
Discover 24, 66, 220
Diversity and Distributions (Russia) 154
Dong AIlbo (South Korea) 106
DV (Iceland) 243
D/W (Indonesia) 207

Earth System science 4–5, 13, 27, 50–51, 224
ecoanxiety 254, 261n4
ecocide 152–153, 158n17, 161–162, 184n4
ecomyopia 255–256
Economic Daily (China) 101
Economist (UK) 70, 84, 164
Eco Peace Middle East 190
Edmonton Journal (Canada) 60, 245
Ekonom (Czech Republic) 146
Eleftheria (Greece) 177
Eleftheros Typos (Greece) 177
Elite Reference (China) 99
Ellis, E. 27, 62, 80n12, 119, 182, 185n25
emissions 24–25, 30, 31nn3, 4, 9, 48, 79, 80n4, 91, 114n10, 175, 203, 208, 258–259, 261n3, 262n15
environmental movements: Africa 52n4; Caribbean 90, 91; Central/East Europe 141–143, 156nn3, 5; Middle East 200n3; North America 80nn8, 15; North Asia 114n13, 115n25; Oceania 212n14; South Asia 119, 127; Western Europe 185n16

Evening Standard (London) 166
Expansión (Spain) 175
Express.hr (Croatia) 150
Extinction Rebellion 167, 260, 261n6
Extra El Clarin (Argentina) 93, 236, 248

Femina (Indonesia) 208
Fenner, F. 98, 126, 173, 252
Figaro (France) 161
Financial Post (Canada) 60
Financial Times (UK) 162, 163
Finney, S. 29, 31n17, 166
First Financial Daily (China) 101
Flannery, T. 31n8, 204–205, 212n7
Focus (Germany) 24, 170
food 180, 195, 223
forests 118–119, 181
Fort McMurray Today (Canada) 61, 73, 80n9, 244–245
Fox News 59, 71
Frankfurter Allgemeine Zeitung 24, 146, 149, 167–168, 235

Gaia 4, 89, 162, 169; *see also* Lovelock
Gazeta po Ukrainski 153
Gazeta.ru (Russia) 153–155, 157n18
Gazeta Wyborcza (Poland) 151, 157n13
Georgia Straight (Canada) 56
Ghosh, A. 222–223, 230nn10, 12, 14
Gladding, J. 90
Global Times (China) 97, 99, 101
global warming 3, 14, 59–60, 74–75, 168, 191, 257
Globe and Mail (Toronto) 57, 58, 60
Grande Reportagem (Portugal) 176
Granma (Cuba) 85, 88
Grapevine (Iceland) 243
Great Acceleration 4, 62, 105, 131, 191–192, 225
Greenock Telegraph (Scotland) 239
growth 6, 40, 42, 46, 47, 83, 107, 130–131, 227
Guangming Daily (China) 100, 102
Guardian (UK) 9–11, 16n23, 24, 29, 47, 99, 125, 164–165, 199, 200, 203, 229, 236, 262n15
Guariglia, J.B. 72, 240, 242, 249n10
Gulf Daily News 189, 197
Gulf Times 197

Haaretz (Israel) 189–190, 198, 199
Habarileo (Tanzania) 48
H-Alter.hr (Croatia) 149, 150

Hamilton, C. 29, 50, 125, 162, 198, 204, 212n3, 228
Hangeorae (South Korea) 106
Hangzhou Daily (China) 101, 239
Harare News 46
Haraway, D. 220–221, 230n7, 257, 259–260, 261n10
Haus der Kulturen der Welt (HKW) 169, 178, 233–236
Helsingin Sanomat (Finland) 180
Herald (Ireland) 248
Herald Sun (Melbourne) 205
High Country News (USA) 66
Himalayan Times 134
Hobart Mercury 204, 206
Holocene 5, 44, 62, 70, 85, 104, 110, 128, 148, 166–168, 183, 225
Hong Kong Economic Times 102, 236
Hornborg, A. 179, 221, 230n8, 253
humanity: conceptions of 6, 24, 39, 41, 44, 55, 62, 80, 86, 134, 154, 195, 199, 229, 246, 254; posthumanism 220–221; survival 4, 39, 52n13, 125, 221, 246, 261; *see also* extinction

Igihe (Rwanda) 43
Illawarra Mercury (Australia) 210
Independent (UK) 24, 127, 128, 132, 150, 163, 191
India Today 121, 222
industrial revolution 49, 55–56, 60, 66, 102, 163, 176, 225
inequality 49, 111, 200n2
Information (Denmark) 178
inky.cz (Czech Republic) 144
Interstellar (film) 135, 246, 247, 250n19
IPCC 4, 6–7, 15nn9, 12, 31n5, 150, 160, 167
Iran Daily 193
Iraq Independent News Agency 194
Irish Examiner 182
Irish Independent 182
Irish Times 182, 228
Izvestia (Russia) 153, 154

Jacobson, M. 67, 80n17, 184n2
Jakarta Globe 208
Jakarta Post 210
Jamaica Observer 89
Japan Times 111–113
Al Jazeera 188, 191, 192, 194–196
Jerusalem Post 191
Joongang Daily (South Korea) 107, 108
Joongang Ilbo (South Korea) 106
Jordon Times 195–197, 201n10
Journal de Montréal 56
journalists 5–7, 13, 16n29, 65–67, 74, 97, 114n7, 123, 138n8, 174, 206, 232, 252–254
Jurnal Balairung (Indonesia) 208, 209
justice: ecological 6, 30, 74, 81n24, 233, 235, 260; social 47, 122
Juventud Rebelde (Cuba) 88

Kansan uutiset (Finland) 181
Kassioun (Syria) 198–199
Kathimerini (Greece) 177
Kathmandu Post 135
Kathmandu Tribune 134
Kelowna Capital News (Canada) 60
Khaleej Times (Dubai) 195–197
Klassekampen (Norway) 180
Kolbert, E. 28, 44, 61, 64, 65, 99, 132, 199, 254
Korea Herald 106
Korea Times 106
Kristeligt Dagblad (Denmark) 179, 218, 248
Krytyka Polityczna (Poland) 151
Kuensel (Bhutan) 118
Kunkumam (Sri Lanka) 121

La Caille, N. 161
L'Actualité (Canada) 56
La Depeche de Tahiti 210, 211
La Libre Belgique 159
La Montagne (France) 162
La Prensa (Argentina) 92, 238
La Prensa (Honduras) 89
La Presse (Canada) 56
La Repubblica (Italy) 173, 237–238
La Stampa (Italy) 159, 173
Latour, B. 7, 169, 217–218, 229
Lebanon Daily Star
L'Echo (Belgium) 159
Le Jeudi (Luxembourg) 183
Le Matin (Switzerland) 172
Le Mauricien 38, 39
Le Monde 24, 55, 160, 161, 177
Lenta (Russia) 153, 225
Les Affaires (Canada) 79–80
Le Soir (Belgium) 236, 244
L'Espresso (Italy) 159
Le Temps (Switzerland) 159
Lewis & Maslin 5, 162, 224–226, 230n15, 16

Lianhe Zaobao (Singapore) 135, 136
Libération (France) 159, 161, 184n4, 221
Liberation (Morocco) 43
L'Obs (France) 159, 162
Lorimer, J. 13
Los Angeles Times 14, 70, 74, 78
Lovelock, J. 163, 182, 259–260
Lynas, M. 31n10, 166, 184n9

Macfarlane, R. 164, 229, 235
Maclean's (Canada) 59, 60, 236, 240
Mahalakshmi, S. 122
Mainichi Shinbun (Japan) 108
Malay Mail 132
manga 113, 115n24
Manila Times 131, 228
Martínez-Alier, J. 227
Matangi Tonga 210, 211
Mbembe, A. 227
media: cycle 6, 28, 30; discourses 38, 51, 141; local 10, 39, 69, 91, 232, 253; quality 160 ff, 253; regional 62, 75–76, 77, 87, 88, 119, 130, 149, 162, 175, 209; social 16n19, 23; tabloid 58, 65, 97, 106, 131, 154, 159, 166, 169, 180, 205, 252
Melbourne Age 203
Metro (various) 56, 58, 59, 159, 167
Miami Rail 87
minerals 44, 75, 111, 121, 137, 176–177
Minneapolis Post 70, 225
mitigation 48, 51, 97, 202
Moore, J. 15n8, 220
Moore, M. 185n19
Morgenbladet (Norway) 24, 180
Morgunblaðið (Iceland) 181
Morrison, K.D. 88, 224
Morton, Timothy 218–219, 232
Morton, Tom 16n29
Moskovskiy Komsomoletz (Russia) 154, 230n16
Mundo (Spain) 159, 174–175
Museo Subacuático de Arte 84, 244
Museum of Tomorrow (Rio) 84, 92–93
Mwananchi (Tanzania) 48

Nación (Argentina) 85, 219, 247
Nang Fang Du Shi Bao (China) 99
National (Dubai) 189, 193, 195, 197
National Geographic (Indonesia) 207
National Geographic (Serbia) 147
National Geographic (USA) 62, 84, 99, 220
National Geographic Croatia 150
National Geographic Japan 110, 111
National Post (Canada) 59
Nationen (Norway) 180
nature 13, 38–39, 40, 42, 56, 59, 71, 74, 83–84, 99–100, 112, 133, 150, 167, 171, 182, 199, 220, 223, 240, 256
Nepali Times 134, 222
Neue Zürcher Zeitung (Switzerland) 171
New African Magazine (Zimbabwe) 46
New Dawn (Liberia) 45, 46, 228
new normal 40, 52n8, 213
NewsCorp (Australia) 202, 206, 211
Newsday 69
New Statesman (UK) 24, 159, 180
New Straits Times (Malaysia) 133, 138n11
New Today (Grenada) 87
New York Post 65
New York Times 9, 23, 24, 29, 61–65, 72–73, 78, 80n10, 107, 124, 164, 187–188, 206
New Zealand Herald 207
Nezavisimaya Gazeta (Russia) 153
Nhân Dân (Vietnam) 137
Nikkei Asian Review (Japan) 110
Nikkei Science 111
Noseweek (South Africa) 39
Nouveliste (Haiti) 91

oceans: acidification 13, 15n4, 65, 124, 132, 193, 249n6, 258; coral reefs 13, 75, 130; fishing 209, 262n14; pollution 15n4, 57, 73, 74, 114n3, 147, 174, 193, 197, 210, 212n10; rising sea levels 72, 73, 115n23, 138n2, 205, 206, 209, 211, 212n9, 230n17, 249n6; 'sinking' islands 205–206
O Clarim (Macao) 104, 228
Oreskes & Conway 7, 15n11
Orlando Sentinel 69
Oxford Mail (UK) 167

Pais (Spain) 159, 174, 218
País Brasil 227
Palm Beach Daily News 73
Pan-African media 48–49
Paris agreement on Climate Change 69, 90, 120, 137, 193
Patria (Colombia) 86
Pejac 239–241
People's Daily (China) 97, 98, 104
Periodico de Catalunya 176
Philippine Daily Inquirer 129
Philippine Star 130
Physics Today (USA) 9, 29

Pina (Pacific Islands News Association) 209
Pittsburgh Post-Gazette 75, 78
planetary boundaries 15n4, 48, 58, 112, 137, 189
Plantationocene 15
pollution: air 74, 114n3, 118; chemical 15n4, 73, 74; corporations 88, 151–152; plastics 75–76, 85, 99, 101–102, 112, 119, 151, 167, 168, 173, 197, 210; *see also* oceans
Pope Francis 45, 104, 129, 132, 134, 196, 201n10, 222, 228
Project Syndicate 108, 110, 135, 136, 172, 194, 195, 197
Public Source (Pittsburgh) 74
Público (Portugal) 176, 224

Al Rakoba (Sudan) 44
Rapport Netwerk 24 (South Africa) 40
Reforma (Mexico) 86, 92, 94, 236
renewable energy 31n9, 66–67, 114n12, 174, 197, 259
resilience 31n12, 101, 136, 202–203, 207
Respekt (Czech Republic) 146
Reuters 15n16, 128, 137, 138n10
Revista Época (Brazil) 85, 87, 93, 236
Revkin, A. 61–64, 80n10, 203, 205, 242
Robin, L. 211, 232, 236
Rockström, J. 107, 124, 129, 147, 196; *see also* planetary boundaries
Rogers, A. 133
Rossiyskaya Gazeta (Russia) 155, 248
RTE (Ireland) 183
Ruddiman, W. 59, 71–72, 110, 153, 168, 171
Rustbelt 77–78
RÚV (Iceland) 181

Sabado (Portugal) 245
Sachs, J. 44, 52n13, 86, 107, 177, 194–195
Samoa Observer 209
San Antonio Current 67
Sankei News (Japan) 110
Sanlian Life Week (China) 99
Santa Barbara Independent 69
Scranton, R. 64, 127, 136, 257
Seimetz, R. 50
Shapin, S. 12
Shenzhen 99, 114n4
Shenzhen Du Shi Bao 99
Shenzhen Shang Bao 99
Shenzhen Te Qu Bao 99, 100
Sixth Extinction 28, 55–57, 61, 69, 99, 199, 226; *see also* Kolbert
Skinny (Scotland) 248
Slate 24, 70, 220, 235
Slobodna Dalmacija (Croatia) 149
SME (Slovakia) 146, 147
Smithsonian 131, 134, 236
solastalgia 253
Solomon Star 209
Sörlin, S. 179, 221
South China Morning Post (Hong Kong) 102
Standard (Gambia) 42
Star (Kenya) 42
Steffen, W. 4, 58, 99, 124, 192–193, 202–204, 208, 212n3
Stengers, I. 7
Stern (Germany) 169
Stoermer, E. 27, 45, 62, 147
Straits Times (Singapore) 135, 138n11, 233
Stuff (New Zealand) 207, 212n11
Sudan Vision 44
Süddeutsche Zeitung 23, 24, 168, 169, 235, 243, 244, 246
Sun (UK) 166
Sun Daily (Malaysia) 132
Surging News (China) 100, 101, 225
sustainable development 27, 46, 55, 98, 107–110, 114n5, 125, 127, 172, 193, 195, 196, 201n7, 258
Svampa, M. 91, 219–220
Sydney Morning Herald 24, 210, 235

Taiwan Times 105
Tampa Bay Times 71, 72, 74
technofossils 70, 81n19, 173, 225
Tehran Times 192, 197
Telegraph (UK) 24, 159, 163, 228, 245
terraforming 81n20, 119
Thinakaran (Sri Lanka) 121
Thunberg, G. 138n13, 196, 249n10, 262n18
timelines: Anthropo-scene 14; Geological Anthropocene 5
Times of Malta 177, 178
tipping points 4, 124, 131
Topix Africa 50
Toronto Star 57, 58, 237
Trischler, H. 234–235

United Nations 4, 7, 15n3, 46, 80n14, 84, 95n2, 97, 106, 107, 111, 114n5, 122, 166, 195, 201n7, 209–210, 212n8; *see also* sustainable development
Universal (Mexico) 87, 93, 227, 238

Urbanization 42–43, 52n10, 126, 128
USA Today 71, 78
Utusan Melayu (Malaysia) 131

Vanguardia (Spain) 75, 249n9
Verdens Gang (Norway) 180
Vernadsky, V.I. 154, 157n19
Vesmir (Czech Republic) 146
Vince, G. 112, 164, 196
violence 50, 179, 227, 254
Visao (Portugal) 176
Vokrug Sveta (Russia) 80, 153–155
Voz del sandinismo (Nicaragua) 86, 221

Wall Street Italia 174
Washington Post 58, 59, 66, 70–72, 75, 133, 240
Water 40, 52n5, 78, 87, 106, 110, 135, 172, 190, 200
Waters, C. 136, 147, 149, 167, 177, 237
Weisman, A. 62, 100, 166
'Welcome to the Anthropocene' 29, 46, 63, 70, 73, 84, 89, 92, 106, 164, 178, 192, 228, 233–235, 237, 245, 247, 248, 260
Welt Am Sonntag 235
Wenhui Daily (China) 101
Whyte, K. 11, 226, 230n17
wildfires 80n9, 213n16, 245
Wilson, E.O. 71, 134, 164, 182
Wired 24, 185n25
Wired (Japan) 111
Wodak, J. 64, 212n9
Wood, R. 239–240
Woolston, R. 62–63
Wynham, F. 255–256

Xi Jin Bao (Beijing) 99
Xi Jinping 52n8, 101
Xinhua Daily (China) 100

Yang Chen Wan Bao (China) 98, 100
Yemen Today 195
YLE-Yleisradio Oy (Finland) 181
Ynet (Israel) 191

Zalasiewicz, J. 5, 10, 15n6, 28, 58, 71, 85, 99, 144, 169, 205, 208
Zhao Defa 102–103
Zolnai, A. 25
Zoom Tunisia 44

For Product Safety Concerns and Information please contact our EU
representative GPSR@taylorandfrancis.com
Taylor & Francis Verlag GmbH, Kaufingerstraße 24, 80331 München, Germany

www.ingramcontent.com/pod-product-compliance
Ingram Content Group UK Ltd.
Pitfield, Milton Keynes, MK11 3LW, UK
UKHW022220190926
475702UK00003B/465
* 9 7 8 0 3 6 7 3 7 5 9 7 3 *